NATURAL PROCESSES AND SYSTEMS FOR HAZARDOUS WASTE TREATMENT

SPONSORED BY
Natural Processes and Systems for Hazardous Waste Treatment Task
Committee of the Environmental Council

Environmental and Water Resources Institute (EWRI)
of the American Society of Civil Engineers

EDITED BY
Say Kee Ong
Rao Y. Surampalli
Alok Bhandari
Pascale Champagne
R. D. Tyagi
Irene Lo

Published by the American Society of Civil Engineers

Library of Congress Cataloging-in-Publication Data

Natural processes and systems for hazardous waste treatment / sponsored by Natural Processes and Systems for Hazardous Waste Treatment Task Committee of the Environmental Council [and] Environmental and Water Resources Institute (EWRI) of the American Society of Civil Engineers ; edited by Say Kee Ong ... [et al.].
    p.cm.
  Includes bibliographical references and index.
  ISBN-13: 978-0-7844-0939-8
  ISBN-10: 0-7844-0939-0
  1. Hazardous wastes--Natural attenuation. 2. Hazardous wastes--Biodegradation. 3. Hazardous wastes--Environmental aspects. I. Ong, Say Kee. II. Environmental and Water Resources Institute (U.S.). Natural Processes and Systems for Hazardous Waste Treatment Task Committee.

  TD1060.N386 2008
  628.4'2--dc22
                                     2007041025

American Society of Civil Engineers
1801 Alexander Bell Drive
Reston, Virginia, 20191-4400

www.pubs.asce.org

Any statements expressed in these materials are those of the individual authors and do not necessarily represent the views of ASCE, which takes no responsibility for any statement made herein. No reference made in this publication to any specific method, product, process, or service constitutes or implies an endorsement, recommendation, or warranty thereof by ASCE. The materials are for general information only and do not represent a standard of ASCE, nor are they intended as a reference in purchase specifications, contracts, regulations, statutes, or any other legal document. ASCE makes no representation or warranty of any kind, whether express or implied, concerning the accuracy, completeness, suitability, or utility of any information, apparatus, product, or process discussed in this publication, and assumes no liability therefore. This information should not be used without first securing competent advice with respect to its suitability for any general or specific application. Anyone utilizing this information assumes all liability arising from such use, including but not limited to infringement of any patent or patents.

ASCE and American Society of Civil Engineers—Registered in U.S. Patent and Trademark Office.

Photocopies and reprints.
You can obtain instant permission to photocopy ASCE publications by using ASCE's online permission service (http://pubs.asce.org/permissions/requests/). Requests for 100 copies or more should be submitted to the Reprints Department, Publications Division, ASCE, (address above); email: permissions@asce.org. A reprint order form can be found at http://pubs.asce.org/support/reprints/.

Preface

Man is constantly creating new chemical compounds to meet specific industrial applications and for human uses. Release of these compounds into the environment is inevitable resulting in contamination of natural resources, including soils and groundwater. The fate of these compounds and their impact on human health and the environment will remain an important global ecological concern. There is a need to understand whether these compounds will persist in the environment or will eventually break down to innocuous compounds. Many compounds degrade in the presence of various natural processes.

The ASCE's Technical Committee on Hazardous, Toxic and Radioactive Waste identified natural processes as an important area that will be of interest to practicing professionals and the book will serve as a reference and as a text for undergraduate or graduate courses. This book presents a discussion of the various natural processes for the attenuation and degradation of hazardous compounds and its application in inexpensive natural systems.

The organization of this book is based on the types of natural processes with Chapter 1 introducing the topic. Chapter 2 presents a brief discussion on sorption, sequestration and binding of organic compounds to soils and sediments. Chapter 3 explores oxidation-reduction reactions and precipitation while Chapter 4 discusses biodegradation and bioassimilation of hazardous compounds. Chapter 5 discusses photolysis and photocatalytic processes. Phyto-process and phyto-assimilation are discussed in Chapter 6. An extension of phyto-processes is the application of wetlands for the treatment of wastewater and contaminated groundwater. This is discussed in Chapter 7. The final chapter discusses the various physical processes such as diffusion and dispersion and the application of natural attenuation.

The editors acknowledge the hard work and patience of the all authors who have contributed to this book.

- SKO, RS, AB, PC, RDT and IL

Contributing Authors

Alok Bhandari, *Department of Civil Engineering, Kansas State University*

Kang Xia, *Department of Crop and Soil Sciences, University of Georgia*

Say Kee Ong, *Department of Civil, Construction and Environmental Engineering, Iowa State University*

Irene M. C. Lo, *Department of Civil Engineering, Hong Kong University of Science and Technology*

Keith C. K. Lai, *Department of Civil Engineering, Hong Kong University of Science and Technology*

Feng Mao, *Department of Civil, Construction and Environmental Engineering, Iowa State University*

Pascale Champagne, *Department of Civil Engineering, Queen's University, Kingston, Canada*

Shankha K. Banerji, *University of Missouri, Colombia*

Rao Y. Surampalli, *US Environmental Protection, Kansas City, MO*

R. D. Tyagi, *Institut National de la Recherche Scientifique, Université du Québec, Québec, Canada*

Bala Subramanian, *Institut National de la Recherche Scientifique, Université du Québec, Québec, Canada*

S. Yan, *Institut National de la Recherche Scientifique, Université du Québec, Québec, Canada*

Kshipra Misra, *Naval Materials Research Laboratory, Defense Research and Development Organization, India*

Satinder K. Brar, *Naval Materials Research Laboratory, Defense Research and Development Organization, India*

Contents

CHAPTER 1

Introduction

SAY KEE ONG

1.0 Background

There are about 87,000 chemicals listed in the inventory of the Toxic Substances Control Act and about 10% or about 8,300 of the listed chemicals are currently being manufactured and sold in significant amounts (Hogue, 2007). The acute and long term effects of some of these chemicals are not known although the physical-chemical and biological properties of these chemicals may provide clues as to their persistency and impact on humans and the environment. Of interest is the fate and transport of these chemicals when they are dispersed in the environment. Chemical are released into the environment through many routes: during the manufacture and processing of the chemicals, industrial use of these chemicals, atmospheric discharge from industrial facilities, wastewater discharges, disposal of products containing these chemicals, and use of products containing these chemicals. In the environment, there are many natural processes that can attenuate these chemicals. Some chemicals are completely attenuated while some chemicals are sufficiently persistent and do not degrade in the environment. Some chemicals are partially attenuated. The attenuated compound or daughter compounds may undergo further attenuation or become more susceptible to attenuation and degradation by processes that are unable to degrade the parent compound. In some cases, these daughter products formed may be more toxic than the parent chemicals.

2.0 Natural Processes

Figure 1 provides a general illustration of the different natural processes in the environment that may attenuate organic and inorganic chemicals. Chemicals dispersed into the media (air and water) are diluted through mixing which reduces the concentration of the chemicals in the environment but the chemicals are not destroyed in the process. Dispersion plays an important role in distributing the chemicals in air and water. Lowering the chemical concentration reduces their impact on the environment and humans and reduces acute toxicity situations. However, the total mass of the chemical in the environment remains the same unless it is degraded. Chemicals that are persistent may move up the food chain, eventually manifesting in humans and resulting in chronic toxicity. Photolysis and photocatalysis are important attenuation pathways in the environment for the degradation of organic chemicals. Chemicals dispersed in lakes and sediments are subjected to oxidation or reduction environments which may impact the chemical. Organic chemicals that are highly oxidized are more likely to degrade in a reductive environment while organic chemicals that are neutral or reduced are more likely to be oxidized in an oxidizing environment. Inorganic chemicals which are toxic in one form may be made less toxic or more toxic under different oxidizing/reducing conditions. In addition, under different redox environment and in the presence of other chemicals, inorganic chemicals may precipitate and be removed from one medium into another, making them less available. Other than precipitation, many chemicals are sorbed by the various solid media such as soils and sediments. These chemicals may be strongly sorbed to soils or sediment making them unavailable and therefore their impacts are diminished.

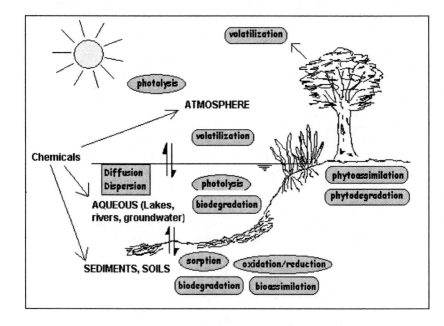

Figure 1.1 Various Natural Attenuation Processes

Degradation by various organisms is an important natural process for the attenuation of organic chemicals. Microorganisms in the environment can acclimatize to these organic chemicals by using them as carbon and energy sources. In some cases, these organic compounds may be degraded fortuitously through cometabolism. Higher forms of lives such as plants can be used to attenuate both inorganic and organic chemicals through accumulate of the chemicals in plant tissues or degraded at the roots of the plants or within the plant. One common system that has been used for the attenuation of waste is using wetlands plants.

This book describes the various natural processes providing information on the basic principles of each processes and examples of chemicals that are attenuated by these processes. In some chapters, these processes are elaborated further by illustrating the use of these processes in an engineered system. There is much to be learned from nature. We still do not sufficiently understand some of these processes or have harness their potential. We also have not fully identified or characterized the various microorganisms in the environment that can assimilate and degrade these anthropogenic chemicals. Learning from the natural processes of the environment and applying them will provide a more sustainable approach in dealing with the problem of hazardous compounds in the environment.

References

Hogue, C., (2007) The future of US chemical regulation, *C&EN*, 85(2):34-38

CHAPTER 2

Sorption, Sequestration and Binding of Contaminants to Soils

ALOK BHANDARI AND KANG XIA

2.1 Introduction

Sorption is a mass-transfer process resulting from the migration of solute molecules from a fluid phase to an adjacent solid surface or a second fluid phase accompanied by a reduction in the solute's localized entropy (Adamson, 1990). During sorption, a relatively minor component of one fluid is transferred to a second fluid phase or a solid surface without significant changes in thermodynamic properties of either phase (Weber, 2001). In dilute solutions, such as those associated with trace environmental pollutants, it is assumed that no net heat flow occurs during the establishment of sorption equilibrium between the fluid and solid phases. The chemical potential of the solute remains unchanged in each phase and the solute is assumed to obey Henry's Law.

The term "sorption" is often used to describe processes that may include absorption, adsorption, or a combination of the two. Absorption describes the inter-phase dissolution resulting from the complete mixing of solute molecules throughout the sorbent phase. Such dissolution or "partitioning" of a solute can occur between gas-gas, gas-liquid, liquid-liquid, gas-solid or liquid-solid phases. Adsorption, on the other hand, is a surface phenomenon and refers to mass-transfer of a solute from a fluid to the surface of a solid or its accumulation at the interface between two phases. Adsorption processes are further described as physisorption if the solute-sorbent interactions are primarily through weak van der Waal's forces, or chemisorption if stronger chemical bond formation occurs between the solute and sorbent surfaces. Sorption processes occurring in natural systems may include complex combinations of absorption, physisorption and chemisorption, and can exert significant control on the environmental fate of the pollutants by altering their mobility and biodegradability. The objective of this chapter is to provide the reader with a basic understanding of the sorption-desorption dynamics and equilibria of organic contaminants and heavy metals in the context of natural treatment processes for hazardous waste remediation.

2.2   Organic Contaminants

Organic contaminants at hazardous waste sites may include petroleum hydrocarbons, polynuclear aromatic hydrocarbons (PAHs), chlorinated solvents and persistent organic contaminants including polychlorinated biphenyls (PCBs), halogenated pesticides, and dioxins. These solutes can participate in a variety of physical and chemical interactions with soil and sediment components. Some of these interactions are illustrated in Figure 2.1 and include (i) surface adsorption, electrostatic interactions, hydrogen bonding, or ion-exchange of the solute with mineral surfaces, (ii) precipitation, crystallization or nonaqueous liquid formation on soil or sediment surfaces, (iii) hydrophobic interactions, sequestration, ion-exchange, proton-transfer, coordination reactions or covalent bond formation with soil or sediment organic matter, and (iv) diffusion into mineral or organic matter micropores. This section will focus on the impact of sorption, sequestration and humification processes on the fate and transport of organic solutes as related to natural processes and systems for hazardous waste treatment.

2.2.1 Sorption

Soils, sediments, aquifer material, and plant-derived detritus constitute the principal sorption domains in natural systems by providing large surface areas and secondary phases for the mass transfer of hydrophobic organic solutes from aqueous media. At the particle-scale, sorption of organic contaminants is affected by their interactions with mineral surfaces, soil organic matter, and microporosity or crystallinity associated with these domains (Figure 2.2). Various modes of interaction between organic solutes and soil components, and the impact of sorption rates and equilibria on such interactions are described in this section.

2.2.1.1 Sorption Mechanisms

In natural environments dominated by minerals with low organic contents, such as aquifer systems, the sorption of organic solutes is affected by the presence of polar hydroxyl- and oxy- groups on the surfaces. These surface moieties readily bind water molecules via hydrogen-bonding forming a clathrate structure that extends several molecule layers into the bulk aqueous medium. Sorption of organic solutes to such mineral surfaces requires the displacement of the tightly-bound water molecules. Hydrophobic organic contaminants exist in thermodynamically unfavorable surroundings in aqueous solutions and are attracted toward soil mineral surfaces by van der Waals forces. The energy released during sorption of the organic solute to the mineral surface is compensated by the energy required to displace the vicinal water molecules, and the overall sorption process is often thermodynamically neutral or even slightly endothermic. The vicinal water layer often plays an important role in the sorption of hydrophobic organic compounds on mineral surfaces by providing a "partitioning" domain for solute accumulation prior to adsorption to surface sites.

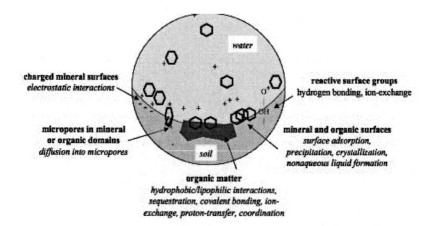

Figure 2.1 Physical-chemical interactions between organic solutes and soil components

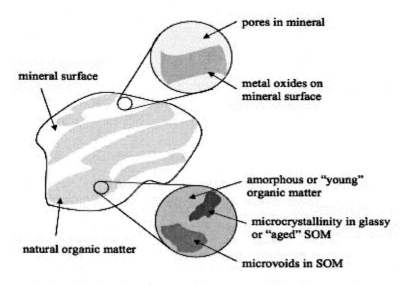

Figure 2.2 Deconstruction of a soil particle showing components of the two major domains capable of interacting with organic solutes: mineral (surface, pores and metal oxides), and organic matter (amorphous and crystalline SOM and microvoids)

The thermodynamic tendency of organic solutes to escape the entropically unfavorable aqueous environment is often the driving force for solute sorption on mineral surfaces. This propensity is inversely related to the solute's aqueous activity coefficient or solubility. The following empirical correlations can be used to estimate an organic solute's distribution coefficient (K) between water and silica or kaolin surfaces (Schwarzenbach et al., 1993):

$$\text{silica: } K = 1.01 \times 10^{-5} \left(\gamma_w \right)^{1.37} \tag{Eq. 2.1}$$

$$\text{kaolin: } K = 1.67 \times 10^{-5} \left(\gamma_w \right)^{1.37} \tag{Eq. 2.2}$$

where γ_w is the solute's activity coefficient.

Organic solutes interact with mineral surfaces in a variety of ways. Mineral oxides and non-expandable clays such as kaolin, illite and chlorite interact weakly and nonspecifically with organic chemicals resulting in very low sorption capacities for these compounds. Expandable clays such as montmorillonite and vermiculite have higher surface areas and cation exchange capacities than nonexpendable clays, and are therefore, capable of retaining large amounts of organic solutes via electrostatic and van der Waal's attractive forces.

Metal oxides and hydroxides such as those of aluminum, iron and silica can coat minerals resulting in charged surfaces. Surface charge may also arise from isomorphous substitution in the mineral's crystal lattice. Such charged mineral surfaces interact strongly with polar or ionizable organic solutes through coulombic or ligand exchange reactions. For example, Bohn et al. (1985) showed that pesticides such as 2,4-dichlorophenoxyacetic acid, 2,4,5-trichlorophenoxyacetic acid, and malathion bound strongly to charged mineral surfaces via coulombic attraction and hydrogen bonding. Polar organic contaminants can participate in dipole-dipole interactions such as hydrogen bonding with functional groups such as edge hydroxyls and oxygen on silicate surfaces. Hydrogen bonding can also occur between a polar organic solute and hydrated exchangeable cations. Burgos et al. (2002) reported that charged siloxane groups influenced adsorption of quinoline, a N-heterocyclic compound, to mineral surfaces. The sorption of polar or charged organic solutes to charged mineral surfaces can be affected by solution chemistry such as pH, and ionic strength (Schwarzenbach et al., 1993).

The presence of micrometer or nanometer size pores in minerals can affect solute mass-transfer from the aqueous phase. In such surfaces, pore diffusion followed by sorption to intrasorbent sites can exert great control on the sorption behavior of organic solutes. Ball and Roberts (1991) contended that slow sorptive uptake of tetrachloroethene and tetrachlorobenzene by a sandy aquifer material was retarded as a result of solute pore diffusion. Other researchers, however, have argued that phenanthrene sorption to mineral surfaces was rapid and intraparticle solute diffusion was insignificant (Huang et al., 1996).

Increasing organic matter content of a mineral soil results in progressive changes in the mechanisms controlling solute-sorbent interactions. Karickhoff (1984) demonstrated that soil organic matter (SOM) content exerted a dominating influence on the sorption behavior of organic solutes when it constituted > 0.1% of the sorbent mass. In contrast to mineral surfaces, the sorption of organic solutes to natural organic matter often does not require the displacement of water molecules from the sorbent surface. Mass-transfer into the organic phase is believed to occur primarily by absorption, a mechanism analogous to liquid-liquid partitioning between water and an organic solvent. Natural sorbents that absorb solutes exhibit a mechanistically similar behavior as rubbery polymers (Weber et al., 2001). Such sorbents are assumed to comprise of an organic phase that is predominantly amorphous or gel-like and relatively young in its degree of diagenetic alteration. SOM associated with geosorbents such as surface soils, peat, and humic acids is believed to be primarily amorphous at ambient temperatures and pressures (LeBoeuf and Weber, 1997). "Dissolution" of the organic solute into the gel-like SOM occurs via interpenetration of the two phases at the nanometer scale. In such situations, solute sorption occurs in a two-phase system – water and SOM – and the distribution of the organic solute between the two phases depends on its relative thermodynamic compatibility in those phases. The distribution of a hydrophobic organic solute between the two phases is represented by the ratio of the solute concentration in each phase:

$$\frac{q_e}{C_e} = K_d \qquad \text{(Eq. 2.3)}$$

where q_e is the solid phase concentration, C_e is the gas or liquid phase concentration at equilibrium, and K_d is the distribution coefficient and is analogous to a liquid-liquid partitioning coefficient such as K_{OW}, the octanol-water coefficient of an organic solute. Absorption of organic contaminants to sorbents containing amorphous SOM is rapid, reversible and noncompetitive (Huang and Weber, 1997; LeBoeuf and Weber, 2000b).

The natural organic matter associated with most geosorbents is comprised of materials that are in various stages of diagenetic alternation – from the very "young" and biopolymer-like macromolecules to the very "old" and coal-like forms. SOM comprised of diagenetically aged organic matter is described as "condensed" or "crystalline" based on the analogy of organic solute sorption behaviors observed in "glassy" polymers (Xing and Pignatello, 1997; LeBoeuf and Weber, 2000a). Glassy polymers consist of rigid structures that allow limited conformational flexibility at the molecular level. Such polymers cannot interact with the solute via phase interpenetration but contain three-dimensional microcrystalline regions or microvoids. Interactions between organic solutes and void spaces or crystalline sites in condensed SOM occur by site-specific adsorption mechanisms. Long sorption equilibrium periods observed in several natural sorbents have been attributed to the slow non-Fickian diffusion of the organic solute molecules into microvoids associated with SOM (Huang and Weber, 1998). Sorption to condensed SOM is often hysteretic and competitive (McGinley et al., 1993; Xing and Pignatello, 1997; Huang and Weber, 1997).

2.2.1.2 Sorption Rates and Equilibria

Sorption of organic pollutants on nonporous mineral surfaces is usually fast, i.e., of the order of minutes to hours (Huang and Weber, 1996). Sorption processes may be mass-transfer limited and occur slowly when solute phase distribution is controlled by the diffusion of molecules into intraparticle regions or intraorganic matter regions of soils and sediments (Weber et al., 2001). A biphasic solute phase distribution behavior is observed in most soil-water or sediment-water systems. For example, Bhandari and Lesan (2003) reported that although a large amount of atrazine sorbed to surface soil within the first hour of contact, slow sorption of the herbicide continued to occur throughout the 84-day study (Figure 2.3). An early period of rapid sorption to easily accessible near-surface sites appeared to be followed by a slow and prolonged migration of the solute into more remote sorption surfaces. Schwarzenbach et al. (1993) used a porous sphere model to describe such time-dependent sorption processes. Solute molecules were assumed to sorb on the spherical outer surface of the soil particle and diffuse radially via interconnected pore-water channels into intraparticle sorption sites. According to this model, the concentration, C, of solute (on a particle volume basis) at a distance r from the center of the particle is described by the following equation:

$$\frac{\partial C}{\partial t} = \phi D^* \left(\frac{\partial^2 C'_w}{\partial r^2} + \frac{2}{r} \frac{\partial C'_w}{\partial r} \right)$$ (Eq. 2.4)

where D^* is the effective diffusivity of the solute and C'_w is the solute concentration inside the soil particle at equilibrium with an internal sorbed phase concentration C'_s. Ball and Roberts (1991a; 1991b) used a similar model to describe the long-term sorption of chlorinated aliphatic and aromatic contaminants in porous aquifer solids. In most fluid-solid mass-transfer processes encountered in natural environments, the diffusive transport of solute molecules into a heterogeneous domain occurs concurrently with reaction (sorption) on the sorbent surfaces. Diffusion through such domains is impeded by solvent resistance, tortuosity, and sorption (Weber and DiGiano, 1996).

When sorption equilibrium has been achieved, the phase-distribution relationships (PDRs) between an organic solute in the fluid and solid phases at a given temperature can be described using sorption isotherms such as those illustrated in Figure 2.4. Absorption or "dissolution" of a solute into a three-dimensional phase, such as amorphous SOM or a gel-like polymer, is represented by Type I isotherms. In Type I PDRs, absorption equilibria are controlled by linear free energy chemical potential relationships and the assumption of dissolution allows relating the extent of absorption of a solute in a second phase to its solubility in that phase.

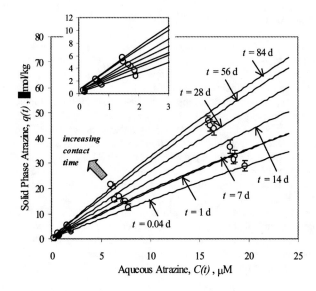

Figure 2.3. Sorption of the herbicide, atrazine, to a surface soil occurred slowly and required more than 84 days to reach equilibrium (Lesan and Bhandari, 2003)

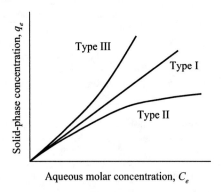

Figure 2.4 Isotherms representing solid-liquid phase distribution of solutes at equilibrium

Type I sorption behavior is described by a "partitioning-type" linear free energy relationship equating the concentrations of the solute in each of the two phases. For relatively small concentration ranges and dilute pollutant concentrations, sorption equilibria can be effectively represented by the simple linear phase distribution relationship illustrated in Eq. 2.3. When absorption of the solute into soil or sediment organic matter is the dominant mode of solute-sorbent interaction, a rough estimate of the equilibrium relationship can be obtained by normalizing K_d to the fraction of organic matter (f_{om}) or organic carbon (f_{oc}) content of the sorbent, and relating it to the molar solubility (C_s) or the octanol-water partition coefficient (K_{OW}) of the solute:

$$\left(\frac{K_d}{f_{om}}\right) = b(C_s)^a \qquad\qquad\qquad \text{(Eq. 2.5)}$$

$$\left(\frac{K_d}{f_{om}}\right) = c(K_{OW})^d \qquad\qquad\qquad \text{(Eq. 2.6)}$$

where, a, b, c and d are coefficients that depend on the class of the organic solute and the natural organic matter (Schwarzenbach et al., 1993).

Sorption of organic contaminants to Type I domains such as rubbery polymers or geosorbents containing predominantly "young" SOM, is rapid and often of the order to minutes to hours (Huang and Weber, 1996; Palomo and Bhandari, 2005). Such partitioning-type solute-sorbent interactions have been shown to be reversible and noncompetitive at dilute solute concentrations (Huang and Weber, 1996; Bhandari and Xu, 2001; Palomo and Bhandari, 2005). In a process analogous to the swelling of synthetic polymers by the action of a permeant, water is believed to expand the amorphous natural organic matter resulting in the exposure of a nonspecific sorption domain that behaves like a micelle or membrane-like bilayer (Carroll et al., 1994; Wershaw, 1981). At low solute concentrations, hydrophobic organic contaminants are absorbed into this gel-like structure rapidly, reversibly and without significant competition from similar solutes.

The Type II curve in Figure 2.4 describes a saturation-type equilibrium PDR in which solute transfer from a fluid phase to a solid surface decreases with increasing solute concentration. Type II sorption or *adsorption* occurs when solute molecules interact with specific surface sites and the surface has a finite number of such sites. Solute molecules are assumed to produce a monolayer surface coverage resulting in a concave isotherm. Type II behavior has been widely observed in the adsorption of organic solutes on mineral surfaces and rigid or glass-like, synthetic polymers (Carroll et al., 1994; Weber and Huang, 1996; Xing and Pignatello, 1997). Site-specific adsorption is rapid and often competitive in nonporous sorbents (Adamson, 1990). Sorption is generally fast if mass-transfer to the surface and into pores of the mineral or organic domains is faster compared to macroscopic solute transport in the bulk aqueous phase (Weber and DiGiano, 1996). Conversely, sorption is protracted if it is controlled by diffusion into intraparticle or intraorganic regions (Pignatello and Xing, 1996).

Type III sorption behavior results in a convex isotherm in which the tendency of the solute molecules to associate with a second phase, a surface or an interface increases with solute concentration. Type III behavior is not common in the interaction of organic solutes with geosorbents and occurs when solute partitioning or adsorption results in the modification of the sorbent surface in a manner that allows more favorable solute transfer from the fluid phase. It should be noted, however, that irrespective of the type of sorption behavior, all isotherms appear linear at low solution concentrations and over narrow concentration ranges.

A variety of models have been used to describe the phase distribution of organic solutes between gas/liquid and solid phases. Five widely used sorption models are summarized in Table 2.1 and various sorption mechanisms are graphically illustrated in Figure 2.5. The Type I isotherms discussed earlier describe absorption or partitioning-type solute transfer and are modeled according to the linear phase-distribution model. The slope of the phase-distribution relationship, K_d, represents the distribution constant of a solute between the two phases under specified conditions. K_d values are unique for each combination of solute, fluid and sorbent.

The Langmuir model, developed originally to describe the distribution of solutes between gaseous and solid phases is based on the assumption of solute adsorption on a finite number of sorption sites. The parameter q_o in the Langmuir equation is related to the solute's heat of adsorption and is a measure of the sorbent's sorption capacity. The Langmuir model assumes that (i) sorption occurs on localized sites and there is no interaction between solute molecules, (ii) sorption energy is constant and independent of surface coverage, and (iii) sorption is limited to a monolayer coverage. Sorption of organic contaminants on mineral surfaces can often be adequately described by the Langmuir model. The Brunauer-Emmett Teller (BET) model for solute-sorbent phase distribution is also based on the hypothesis that the sorbent consists of a finite number of sorption sites, but accounts for multi-layer sorption by assuming that underlying layers of the sorbate act as sorption sites for upper layers. The solute-sorbent interaction energy for the first layer is assumed to be analogous to the heat of adsorption while solute interactions with subsequent layers occur through condensation-type reactions.

Table 2.1 Commonly used sorption models for organic and inorganic solutes

Sorption Model	Equation	Coefficients
Linear Phase-Distribution Model	$$q_e = K_d C_e$$	K_d = linear phase distribution constant
Langmuir Model	$$q_e = \frac{q_0 K_L C_e}{1 + K_L C_e}$$	q_o = sorption capacity K_L = sorption energy factor
Brunauer-Emmett-Teller (BET) Model	$$q_e = \frac{q_0 K_{BET} C_e}{(C_s - C_e)\left[1 + (K_{BET} - 1)(C_e/C_s)\right]}$$	q_o = sorption capacity K_{BET} = sorption energy factor C_s = molar solubility
Freundlich Model	$$q_e = K_F C_e^n$$	K_F = specific capacity n = sorption energy and heterogeneity factor
Dual Mode Model	$$q_e = K_d C_e + \sum_1^N \frac{q_{0,i} K_{L,i} C_{e,i}}{1 + K_{L,i} C_{e,i}}$$	N = types of sorption-sites with in heterogeneous SOM

The Freundlich model is among the most widely used models to describe sorption of environmental organic pollutants to soils and sediments from dilute solutions. The curves shown in Figure 2.3 represent model fits using a Freundlich-type, non-equilibrium phase distribution relationship between atrazine and the surface soil. This model expresses the overall solute sorption pattern on heterogeneous surfaces consisting of a variety of sorption sites, each with its own sorption characteristics. Under such an assumption, the Freundlich isotherm represents the summation of several Langmuir isotherms, each resulting from the adsorption of the solute to an individual type of sorption site. The model assumes that most surfaces are energetically heterogeneous, and at low surface coverages adsorption takes place on sites possessing the highest energies. The exponent n is a joint measure of the magnitude and diversity of sorption energies associated with the sorption process, and is called the linearity, affinity or sorbent heterogeneity factor. The parameter K_F is the specific capacity and represents the sorbed solute concentration in equilibrium with an aqueous concentration of unity. Sorbents dominated by "young" SOM are modeled as having sorption sites with identical site energy. In this case, $n = 1$ and the Freundlich isotherm translates to a linear relationship. As sorbent heterogeneity increases, the value of n decreases and the isotherm becomes progressively nonlinear.

(a) monolayer adsorption on mineral surfaces or condensed SOM

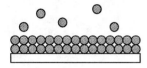

(b) multi-layer adsorption on mineral surfaces or condensed SOM

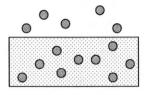

(e) partitioning or absorption in amorphous SOM

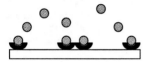

(c) site-specific adsorption on mineral surfaces or condensed SOM

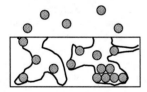

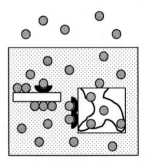

(d) intraparticle or intraorganic pore diffusion and adsorption

(f) solute sorption to geosorbent comprising i) absorption or partitioning, ii) surface adsorption, iii) site-specific adsorption, iv) intraparticle pore diffusion and adsorption

Figure 2.5 A particle-scale representation of sorption mechanisms operational during organic solute-geosorbent interactions. Filled circles represent solute molecules. Reactions shown in figures (a) through (d) occur on mineral surfaces or on "glassy" organic matter domains. Solute-partitioning is prevalent in "young" amorphous organic matter. Figure (f) illustrates the simultaneous occurrence of all sorption mechanisms in geosorbents.

Most geosorbents are inherently heterogeneous and values of n are rarely 1. As illustrated in Figure 2.5(f), natural sorbents can comprise of free mineral surfaces with nonspecific sorption sites, mineral surfaces with specific sites for solute sorption, intraparticle pores, amorphous SOM, condensed SOM, and SOM with microporosity or voids. Ball and Roberts (1991a; 1991b) studied sorption of chlorinated organics in microporous mineral solids from the Borden aquifer and found that sorption was diffusion-controlled and nonlinear (see Case Study 1). Karichkoff (1984) illustrated that a SOM content as small as 0.1% can lead to near complete coverage of the mineral surfaces in soils and sediments. Solid-liquid phase distribution of solute on such sorbents is completely controlled by the amount and type of SOM. In such cases, the dual mode model (Table 2.1) can adequately describe the overall sorption process. This model assumes the SOM to be comprised of two types of "domains" – an *amorphous* domain that consists of a relatively homogeneous, gel-like organic phase, and a *condensed* domain made up of a distribution of heterogeneous voids or microcrystalline sorption-sites. A partitioning-type, absorption behavior is assumed to occur between the organic solute and the amorphous SOM, while Langmuir-type adsorption is believed to control solute phase distribution between the voids or crystalline sites and the aqueous phase (Xing et al., 1996; Xing and Pignatello, 1997). Absorption of the solute into the amorphous SOM regions is rapid, linear and noncompetitive. Adsorption to the microsites is nonlinear and competitive.

Case Study 1. Ball and Roberts (1991a; 1991b) studied the sorption of chlorinated organic chemicals on aquifer material removed from the Borden aquifer in Canada. Tetrachloroethene (PCE) and 1,2,4,5-tetrachlorobenzene (TeCB) were used as model solutes representing chlorinated aliphatic and chlorinated aromatic contaminants, respectively. The aquifer material was pulverized yielding solids with a mean particle size of 0.008 mm. Pulverization resulted in 110 to 400% increase in the specific surface area of aquifer solids. The mean surface area, intraparticle porosity, and organic carbon content of the particles were 1.7 m^2/g, 1.3%, and 0.021% respectively. Contact times of tens to hundreds of days were necessary to achieve sorption equilibrium in unpulverized aquifer material. Pulverization reduced the equilibrium time to 1 day for PCE and 30 to 60 days for TeCB. Sorption of TeCB (K_{OW} = 4.6) to the aquifer solids was approximately four times greater than that of PCE (K_{OW} = 2.6). Although sorption of PCE appeared linear at low aqueous concentrations, significant deviations from linearity occurred when concentrations spanned 4 to 5 orders of magnitude.

Linear, Freundlich and Langmuir coefficients describing PCE and TeCB sorption to the pulverized Borden aquifer material.

	PCE					TeCB				
	K_d	K_F	n	q_o	q_oK_L	K_d	K_F	n	q_o	q_oK_L
Low	0.87	1.0	0.90	189	0.89	122	135	0.80	1900	129
concentrations[a]	(0.024)[b]	(0.013)		(0.018)		(0.094)	(0.051)		(0.057)	
All	0.57	1.3	0.79	3350	0.66	100	135	0.82	15400	102
concentrations	(4.700)	(0.10)		(0.064)		(0.290)	(0.047)		(0.066)	

units for K_d = mL/g, K_F = (μg/kg)/(μg/L)^n, q_o = μg/kg, and q_oK_L = mL/g
[a]< 50 μg/L for PCE and < 12 μg/L for TeCB; [b]mean weighted square error

The amount of solute sorption by the Borden aquifer material exceeded the magnitude predicted from f_{oc} by 10 fold, implying a significant inorganic contribution to sorption. An intraparticle diffusion model was utilized to describe the time-dependent solute uptake. Effective pore diffusion coefficients of PCE and TeCB were estimated to be 2 to 3 orders of magnitude lower than the bulk aqueous diffusivities. The rate of approach to sorption equilibrium was greater for smaller particles and for the less strongly sorbing solute indicating that intragranular diffusion controlled solute mass-transfer. This study demonstrated the importance of slow, particle scale sorption in the context of a natural aquifer material. Understanding such long-term sorption phenomena is important because both contamination and remediation processes in the natural environment can occur over extended periods of times.

The condensed SOM is highly porous at the nanometer scale resulting in a solute phase distribution behavior that is diffusion-controlled and protracted. Lesan and Bhandari (2003) reported that atrazine sorption to surface soils continued to occur throughout the duration of their 84-day study (see Figure 2.3). Xing et al. (1996) noted competition between atrazine and other organic compounds in soils and glassy polymers during their 2-day study. Karapanagioti et al. (2000) related the slow, nonlinear sorption of phenanthrene in an aquifer material to the occurrence of glassy (coal-type) organic matter. Huang and Weber (1998) reported sorption equilibrium times ranging from 90 days to more than one year for phenanthrene sorption on sorbents containing predominantly glassy SOM (shales and kerogen). Johnson et al. (2001) removed the "young" SOM from a peat by subcritical water extraction and observed phenanthrene sorption isotherms become progressively more nonlinear as the residual SOM became more enriched in the condensed SOM.

In a comprehensive study, Xing et al. (1994) reported the sorption behavior of phenol on a variety of geosorbents including mineral surfaces such as goethite, kaolinite and montmorillonite, predominantly amorphous organic matter such as cellulose, heterogeneous organic matter such as chitin, collagen and lignin, and predominantly condensed organic matter such as activated carbon. These authors showed that the biopolymers sorbed 2 to 45 times more phenol than the minerals, with the lignin sorbing 13 to 22 fold more phenol than the cellulose. Nonlinear sorption of sulfonamide antibiotics to soils indicated specific interactions with phenolic, carboxylic and N-heterocyclic regions of the SOM (Thiele-Bruhn et al., 2004).

Palomo and Bhandari (2005; 2006) cautioned against an assumption of true thermodynamic sorption equilibrium based solely on the achievement of constant aqueous- or solid-phase solute concentrations in batch reactors. These researchers indicated that solute-sorbent interactions continue to impact the fate of contaminants long after the apparent sorption equilibrium suggested by measured aqueous or solid-phase concentrations. Huang and Weber (1996) have proposed the three-domain model to explain sorption of hydrophobic organic solutes in geosorbents. According to these researchers, geosorbents consist of three primary sorption domains (i) exposed inorganic mineral; (ii) amorphous SOM; and (iii) condensed SOM. Organic solutes first access the mineral and amorphous SOM when contacted with the sorbent. Sorption to these domains is relatively rapid. The solute may simultaneously or sequentially access the condensed SOM to which it sorbs slowly. Other researchers have proposed a two-compartment model consisting of "rapid" and "slow" sorption domains (Xing et al., 1996; Xing and Pignatello, 1996).

2.3   Desorption

The importance of understanding desorption processes to achieve desired end-points in the remediation of contaminated soils and sediments remains unmatched by the paucity of desorption studies reported in the literature. Desorption, the mass-transfer of a sorbed solute from a surface into the bulk fluid phase, can greatly control the design and economics of clean-up efforts at sites contaminated with organic pollutants. The current state of understanding of desorption processes argues that solute desorption is highly influenced by the mechanisms involved during the sorption process.

2.3.1   Mechanisms and Kinetics

The rate and extent of desorption of organic solutes from natural sorbents is controlled by solute properties, solution chemistry, surface chemistry of the sorbent, and solute-sorbent contact time during sorption. A solute with high aqueous solubility and low K_{OW} resulting from its molecular size, configuration, polarity, or ability to ionize, desorbs more readily than one with a high molecular weight, high lipophilicity and low aqueous solubility. A modification of the solution chemistry by the addition of salts, water miscible organic solvents, or surface-active agents can significantly increase desorption of sorbed organic solutes. Salts can displace ionized organic solutes sorbed on exchange sites, addition of solvents such as methanol or ethanol can increase the solubility of organic solutes in the fluid, and surfactants or detergents can reduce surface tension and enhance desorption by allowing mass-transfer of organics into micelles. It should be noted, however, that altering solution chemistry changes the process from true desorption to "extraction". The remainder of this section will focus on a discussion of true desorption processes.

Sorbent properties can exert a dominating influence on the kinetics and extent of desorption of organic solutes from soils and sediments. Desorption of the sorbed solute from a nonporous and nonreactive mineral surface is often rapid and completely reversible. For example, Thiele-Bruhn and coworkers (2004) observed that sorption of sulfonamide antibiotics to mineral soil colloids was small and the sorbed compounds desorbed readily from clay-size fractions. The presence of microporous regions and reactive sorption sites on mineral surfaces can retard solute desorption and significantly impact sorption reversibility (see Case Study 2). When the relationship between q_e and C_e during desorption is not the same as that during sorption, the desorption process is considered to be hysteretic. This often occurs because the reverse mass-transfer of solute molecules sorbed on intraparticle sites is retarded due to unfavorable pore geometry (constricted or dead-end pores) and the fact that smaller solute concentration gradients operational during desorption exert a significantly diminished driving force for the diffusive flux of the organic contaminant exiting the pores (Ball and Roberts,

1991b). Linearity of sorption-desorption isotherms and the extent of separation between the sorption-desorption isotherms have been used by several researchers as quantitative measures of desorption hysteresis (Table 2.2).

Desorption of organic solutes from geosorbents containing predominantly "young" or amorphous SOM is analogous to solute release from gel-like polymers and is generally rapid, linear and completely reversible (Carroll et al., 1994; LeBoeuf and Weber, 2000b). Palomo and Bhandari (2005) reported little or no hysteresis in the desorption of 2,4-dichlorophenol from surface soils in contact with the solute for 1 day. Minor increases in hysteresis were apparent as the sorption contact period was increased from 1 day to 84 days. Huang and Weber (1997) found little apparent hysteresis for phenanthrene desorption from humic acid and peat, but significantly high hysteresis for desorption of the same solute from shale and kerogen. The fast desorption kinetics and sorption reversibility in amorphous SOM can be attributed to the lack of mass-transfer resistance from the gel-like SOM into the aqueous phase.

Table 2.2. Hysteresis Indices (HIs) - quantitative measures of desorption hysteresis proposed by various researchers.

$HI = \dfrac{q_e^d - q_e^s}{q_e^s}\bigg|_{C_e}$	q_e^d = solid-phase solute concentration after desorption corresponding to an aqueous concentration C_e. q_e^s = solid-phase solute concentration after sorption corresponding to an aqueous concentration C_e.	Huang and Weber (1997)
$HI = \dfrac{n_d}{n_s}$	n_d = Freundlich linearity coefficient obtained from desorption isotherm n_s = Freundlich linearity coefficient obtained from sorption isotherm	Barriuso et al. (1994)
$HI = \left[\dfrac{n_s}{n_d} - 1\right]$	n_d = Freundlich linearity coefficient obtained from desorption isotherm n_s = Freundlich linearity coefficient obtained from sorption isotherm	Ha (2002)

Sorption of organic chemicals on microcrystalline polymers and condensed SOM is analogous to the behavior of these solutes in porous minerals. Solute molecules in contact with condensed SOM domains interact with a distribution of sorption sites with different energies. These microcrystalline sites may occur on the surface of the condensed SOM or in microvoids within the SOM matrix. Desorption from such sites is very slow, nonlinear, and often hysteretic (Carroll et al.; 1994). Bhandari and Lesan (2003) reported that atrazine desorption from two surface soils was more nonlinear than sorption resulting in significant desorption hysteresis (Figure 2.6). These researchers attributed the hysteresis to desorption resistance experienced by solute molecules that sorbed preferentially on high-energy sites at low concentrations. Schlebaum et al. (1999) argued that a lack of high affinity sorption sites on sandy soils resulted in higher pentachlorobenzene desorption rates with increasing initial solute concentrations during sorption. LeBoeuf and Weber (2000b) reported that younger natural sorbents such as cellulose and humic acid exhibited less desorption hysteresis than diagenetically aged sorbents such as kerogen and coal. White and Pignatello (1999) showed that pyrene displaced sorbed phenanthrene during desorption from the condensed SOM domains of two soils.

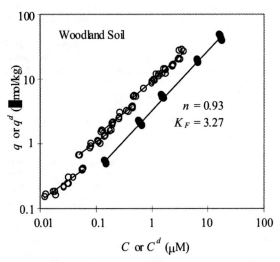

Figure 2.6 Freundlich-type sorption (●) and desorption (○) isotherms for atrazine from a woodland soil reveal the occurrence of significant hysteresis (Bhandari and Lesan, 2003)

Several studies have shown that the rate and extent of desorption of organic solutes decrease with increasing solute-sorbent contact time during sorption. Pavlostathis and coworkers reported that desorption of chlorinated ethenes from contaminated soils followed a biphasic pattern – an initial fast phase of the order of hours followed by a subsequent slower solute desorption rate of the order of days or weeks (Pavlostathis and Jaglal, 1991; Pavlosthathis and Mathavan, 1992). Pignatello et al. (1993) used a two-compartment diffusion sorption model to evaluate the elution of aged and freshly added herbicides from soil. Their results suggested that the slow desorption of 82-92% of the solute resulted from the diffusive efflux of the solute from microparticles or microstructures in the soil. Lesan and Bhandari (2003; 2004) showed that atrazine desorption hysteresis and extractability decreased as sorption contact time increased from 1 hour to 84 days. Biphasic sorption data and time-dependent decreases in desorption rate indicate an initial rapid release of solute from nonporous and nonreactive mineral surfaces and amorphous SOM (Alexander, 1995). The subsequent slow desorption is likely caused by the protracted, diffusion-controlled release of the solute from intraparticle micropores or microvoids associated with condensed SOM (Ball and Roberts, 1991; Alexander, 1995).

Case Study 2. Bhandari and coworkers (Dove et al., 1992; Bhandari et al., 1994; 2001) evaluated the treatability of petroleum hydrocarbon contaminated soils using a bench-scale soil washing system. Three soils (Salem, New River and Eagle Point) were contaminated in the laboratory by exposure to petroleum distillates. The initial total petroleum hydrocarbon (TPH) concentrations in the soils were 7,800, 26,000 and 15,000 mg/kg, respectively. The soil washing system consisted of a high-shear, completely-mixed reactor followed by an upflow column to separate sand-sized particles from the remaining soil, and a sedimentation basin designed to separate silt-sized particles from clays. Approximately 12 L of water was used to wash 0.5 kg of soil for a duration of 4 hours. Residual TPH concentrations after washing were found to be correlated to the particle size. Sands, which comprised the largest weight fraction of the soil, contained the smallest mass fraction of TPH in the washed soil. On the other hand, clays, which comprised the smallest weight fraction of soil, had the largest residual TPH concentrations. Clay particles exhibit large specific surface areas that allow them to retain high concentrations of sorbed contaminants.

Particle size distribution and residual TPH distribution in Salem, New River and Eagle Point soils after soil washing and after biotreatment.

	Weight fraction[a]	Initial TPH (mg/kg)	Fraction of initial TPH in soil fraction after washing[b]	Residual TPH (mg/kg)	Residual TPH after biotreatment[c] (mg/kg)
Salem Soil		7,800		3,500	
Sand	0.61		0.02	300	< 60
Silt	0.20		0.12	4,800	100
Clay	0.15		0.28	15,000	400
New River Soil		26,000		5,000	
Sand	0.31		0.01	900	600
Silt	0.41		0.03	2,000	< 100
Clay	0.26		0.14	15,000	
Eagle Point Soil		15,000		3,000	
Sand	0.60		< 0.01	100	< 50
Silt	0.20		0.04	3,000	100
Clay	0.15		0.14	15,000	900

[a] as determined from particle fractionation during soil washing. 2 to 5% by weight of soil was lost in froth in the high-shear mixer.
[b] the remaining TPH was removed in the washwater, air and froth.
[c] sands were composted for 90 days, silts and clays were treated in completely-mixed biological reactors for 50 and 95 days, respectively.

The washed soil fractions were subjected to biological treatment by the addition of a petroleum degrading bacterial consortium. Biotreatment resulted in significant removal of the residual contamination from all soil fractions except New River sand. Scanning electron microscopy revealed that this sand contained significantly higher concentrations of surface iron oxides than the other sands. Low petroleum removal efficiencies for New River sand were attributed to its high surface iron oxide content and the affinity of these metal oxides to strongly bind natural and petroleum-derived organic carbon.

2.3.2 Contaminant Sequestration and Binding

Desorption and extraction experiments have demonstrated the disappearance of persistent organic compounds applied to soils and sediments during field and laboratory investigations (Clay and Koskinen, 1990; Pignatello and Huang, 1991; Pavlostathis and Mathavan, 1992; Ma et al.; 1993; Roy and Krapac, 1994; Hatzinger and Alexander, 1995; Lesan and Bhandari, 2004). The "aging effect" observed in these studies results in the sequestration or binding of the solute at long solute-sorbent contact times. Formation of "bound" residues in soils has been reported for several organic contaminants and is believed to occur as a result of (i) physical entrapment or sequestration of the chemical in mineral or organic soil matrices (Pignatello et al., 1993; Weber et al., 2001), and (ii) chemical bond formation between the chemical or its principal metabolites and SOM components (Bollag, 1992). Barriuso and Koskinen (1996) evaluated atrazine residue formation in different soil size fractions as a function of time and found that humified SOM fractions associated with soil fines retained the largest quantities of the herbicide. Dankwardt et al. (1996) reported nonextractable residue formation as a result of ligation of atrazine to soil humic and fulvic acids.

Sequestration is defined as the entrapment of solute molecules in nanometer scale pores in the mineral domain or within SOM macromolecules (Ball and Roberts, 1991; Huang and Weber, 1997). Sequestration of organic chemicals may result from the diffusion of these solutes into microvoids in the SOM domain followed by a reconfiguration of the SOM macromolecular structure such that it makes reverse mass-transport more tortuous. A sequestered organic chemical is physically trapped in the mineral or organic domains but not covalently bound to the SOM. Sequestered solutes are believed to be recoverable from geosorbents when subjected to the right combination of solvents and extraction protocol.

Organic solutes such as phenols and anilines can form covalent bonds with SOM resulting in their irreversible binding to geosorbents (Bhandari et al., 1996; 1997). Such processes are analogous to the oxidative polymerization of humus precursors during the synthesis of SOM by a process called humification. The oxidative coupling of phenolic and anilinic contaminants occurs as a free-radical mechanism catalyzed by the action of transition metal oxides or soil enzymes. Oxidative polymerization often leads to the incorporation of natural or synthetic phenols/anilines into SOM (Bollag, 1992). Oxides of iron and manganese can transform phenols and anilines to free radicals or quinones that subsequently undergo oxidative coupling with other free radicals (polymerization) or with reactive moieties in the SOM structure (cross-coupling) (Berry and Boyd, 1985; Stone, 1987; Laha and Luthy, 1990). Polymerized solutes can adhere strongly to geosorbents due to their low aqueous solubility and high lipophilicity. Cross-coupled phenols and anilines are covalently and, therefore, irreversibly bound to the SOM. Kim et al. (2002) reported the formation of bound residues of 8-hydroxybentazon by δ-MnO_2. Selig et al. (2003) observed that δ-MnO_2-catalyzed cross-coupling of phenol was more effective in soils that contained "young" and chemically reactive, amorphous SOM than those that contained diagenetically older and condensed SOM.

Oxidative coupling of phenols and anilines can also be catalyzed by soil enzymes (Bollag, 1992). For example, soil peroxidases or polyphenol oxidases contain active sites capable of oxidation-reduction reactions in the presence of peroxide or molecular oxygen. Phenols and anilines are oxidized to free radicals or quinones, which subsequently polymerize to form heterogeneous oligomers or cross-couple with soil humus (Figure 2.7). Bollag and coworkers reported the cross-coupling of chlorinated phenols and aniline to SOM derived compounds in the presence of a fungal laccase enzyme (Bollag and Liu, 1985; Liu et al., 1987; Tatsumi et al., 1994). Data from laboratory experiments show that the addition of peroxidase enzyme can significantly enhance irreversible binding of phenols and polynuclear aromatic hydrocarbons to soils (Berry and Boyd; 1985; Bhandari and Xu, 2001; Huang et al., 2002; Xu and Bhandari, 2003a; 2003b; Weber and Huang, 2003; Palomo and Bhandari, 2006). Several researchers have demonstrated that enzyme catalyzed polymerization and cross-coupling reactions can result in detoxification of the contaminated media (Bollag et al., 1988; Dec and Bollag, 1990; Wagner and Nicell, 2002; Xu et al., 2004).

2.4   Inorganic Contaminants

The top three inorganic contaminants on the EPA 2003 Hazardous Substances Priority List include heavy metals such as arsenic (As), lead (Pb), and mercury (Hg) (CERCLA, 2003). Other heavy metals such as chromium (Cr), cadmium (Cd), cobalt (Co), nickel (Ni), and zinc (Zn) as well as radioactive contaminants such as uranium (U), radium (Ra), thorium (Th), barium (Ba), strontium (Sr), plutonium (Pu), and cesium (Cs) are also found at many industrial and military hazardous waste sites. The natural processes in soil and sediment affect the chemical forms and the speciation of these contaminants and, therefore, their sorption-desorption behavior with the mineral and organic components of the soil and sediment systems. The major factors that affect the chemical forms and the speciation of these contaminants are pH, redox potential, competing ions, soil minerals, organic matter, and biological activity (Adriano, 2001). The focus of this section will be on how these factors affect the sorption-desorption and, therefore, the fate and transport of representative inorganic contaminants at hazardous waste sites and the impact on contamination treatments.

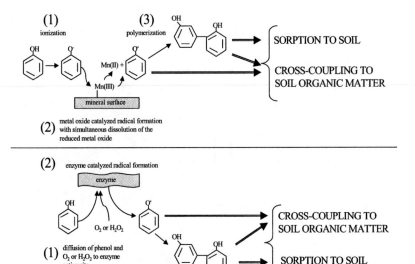

Figure 2.7 Transition metal oxide and enzyme catalyzed polymerization and cross-coupling of phenolic chemicals.

2.4.1 Speciation in the Soil Environment

While some heavy metals such as As, Cr, and Hg have multiple oxidation states and exist in different associated species depending on the redox potential and pH in the soil and sediment systems, others such as Cd and Pb occur in natural and impacted environments mainly in single oxidation states. This section describes the speciation of both categories of metals in contaminated environments.

Figure 2.8 illustrates the speciation of As, Cr and Hg as a function of the redox potential (E_h) and pH conditions in aqueous systems. In aerobic environments (high E_h), arsenic exists in the +5 oxidation state as H_3AsO_4 in acidic conditions (pH < 2), $H_2AsO_4^-$ or $HAsO_4^{2-}$ in the pH range from 2 to 12, and as AsO_4^{3-} in highly alkaline conditions (pH > 12). The acid dissociation constants (pK$_a$s) of H_3AsO_4 are 2.2, 7.0, and 11.5. In more reduced environments, arsenic exists in the +3 oxidation state as H_3AsO_3 at pH < 9, as $H_2AsO_3^-$ or $HAsO_3^{2-}$ within the pH range 9 to 12.7, and as AsO_3^{3-} at extreme alkaline conditions (pH > 12.7). The pK$_a$ values for H_3AsO_3 are 9.2, 12.1, and 13.4, respectively. As illustrated in Figure 2.8a, while $H_2AsO_4^-$ and $HAsO_4^{2-}$ are the two dominant arsenic species in surface soils under normal aerobic conditions, wet soils may also contain H_3AsO_3. In more reduced environments such as water saturated soils, arsenic compounds such as As_2S_3 and AsS_2^- may be formed at pH values > 6 in the presence of sulfur or H_2S. Elemental arsenic (oxidation state of 0) is formed under extremely reduced conditions (E_h < -0.25 V) and pH > 7. Although, at trace levels, As is an essential nutrient for certain animals, at high levels it is toxic in both the +3 and +5 oxidation states (Adriano, 2001).

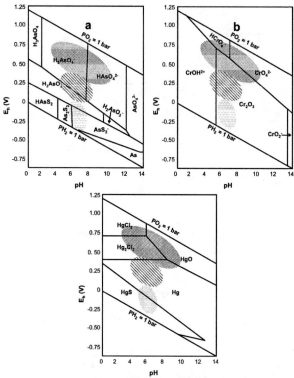

Figure 2.8 E_h-pH diagram for (a) As-O_2-S-H_2O system (a) (modified from Bissen and Frimmel, 2003), (b) for Cr-O-H system (modified from Brookins, 1988), and (c) for Hg-O-S-Cl system (modified from Adriano, 2001). The shaded ovals indicate the range of redox potentials and pH values for normal soils (top oval), wet soils (middle oval), and water logged soils (lower oval), respectively (Bohn et al., 1985).

Chromium exists in the +6 oxidation state as $HCrO_4^-$ in highly aerated normal surface soils at pH < 7, and as CrO_4^{2-} is more dominant at pH > 7 (Figure 2.8b). In reduced environments such as wet and water logged soils, Cr (+6) is reduced to Cr (+3), resulting in Cr_2O_3 as the dominant Cr species. In acidic and reduced environments, Cr_2O_3 dissolves to form $CrOH^{2+}$. While Cr (+3) is an essential element to humans and animals, Cr (+6) is a potent carcinogen (Adriano, 2001).

The dominant mercury species in acidic aerated surface soils are $HgCl_2$ with oxidation state of +2 or Hg_2Cl_2 with oxidation state of +1 (Figure 2.8c). In alkaline aerated surface soils, HgO is the dominant Hg species, which has oxidation state of +2. In wet and water logged soils, elemental Hg is most often the dominant Hg species, however, HgS can be formed in the presence of sulfur or H_2S in acidic and highly reduced environment. Hg^{2+} can be methylated by methane-producing bacteria under aerobic or anaerobic conditions, resulting in monomethyl mercury (CH_3Hg^+) or dimethyl mercury (($CH_3)_2Hg$). HgS can be oxidized to sulfate which can then undergo methylation (Adriano, 2001). Both forms of methyl Hg are highly toxic compared to inorganic Hg and can easily cross the blood-brain barrier and the placenta in humans (Zhang et al., 2003). Methyl mercury is known to biomagnify in the food chain (U.S. EPA, 1996).

Cadmium and lead normally exist in the environment in +2 oxidation states. Their speciation in the solution is affected largely by pH, complexing anions, competing cations, dissolved organic matter, soil minerals, and biological factors. Different soluble species have different sorption and desorption capacity on soil minerals and organic matter and, therefore, different mobility and bioavailability in the environment. Soluble Pb^{2+} and Cd^{2+} are in general considered the most toxic and bioavailable species for organisms (Adriano, 2001).

2.4.2 Transformation of Heavy Metals

Arsenic in soils and sediments can undergo abiotic and/or biotic transformations. Oxides of transition metals such as Mn (+4) and Fe (+3) can catalyze the oxidation of As (+3) to As (+5) through an electron transfer mechanism (Adriano, 2001). On the other hand, As (+5) can co-precipitate with $Fe(OH)_3$ or hydrous Mn oxides to form minerals such as $FeAsO_4 \bullet 2H_2O$ or $Mn_3(AsO_4)_2$. Numerous bacteria, fungi, yeasts and algae (Frankenberger and Arshad, 2002), as well as certain plants (Fitz and Wenzel, 2002), are able to transform As compounds via a variety of processes including oxidation, reduction, methylation, and demethylation. Figure 2.9 summarizes the transformation pathways of arsenic in the environment.

In normal surface soils, reduction of Cr (+6) to Cr (+3) is favored due to the presence of reductants such as Fe (+2), sulfides, and organic matter, while simultaneously Cr (+3) can be oxidized to Cr (+6) by strong oxidants such as Mn oxides (Adriano, 2001). The reduction of Cr (+6) is favored at pH < 6; both reduction and oxidation of Cr may be inhibited under more alkaline conditions (James et al., 1997). The inter-conversion of Cr (+3) and Cr (+6) in the presence of oxidants such as manganese oxides and reductants such as natural organic matter (NOM) is illustrated in Figure 2.10. The direction of this conversion is largely affected by the pH. The abiotic reduction of Cr (+6) by NOM was found to be significant in acidic conditions, but the reduction rates were negligible at neutral or more alkaline conditions. However, in the presence of the bacterial strain *Shewanella putrefaciens* CN32, NOM was able to enhance the reduction of Cr (+6) significantly even at neutral pH (Gu and Chen, 2003). The hydroquinone groups in NOM are believed to participate in the reduction of Cr (+6) (Elovitz and Fish, 1995). Oxidation of Cr (+3) in soils can occur under most field conditions, when high-valency Mn oxides are available as electron acceptors for the reaction (Adriano, 2001).

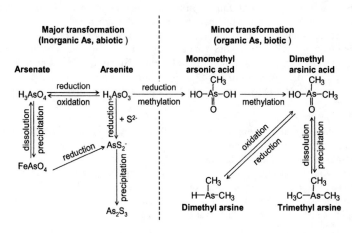

Figure 2.9 Simplified transformation pathways of arsenic in the environment (Adriano, 2001).

In the case of mercury, many of its species can be reduced to elemental Hg, which is highly volatile and can be transported in the atmosphere from one location to another. Mercury ions can be methylated by both aerobic and anaerobic bacteria, resulting in volatile methyl mercury compounds that are more bioavailable and toxic (Figure 2.11). Under anaerobic conditions Hg ions can precipitate to form the insoluble and less bioavailable HgS.

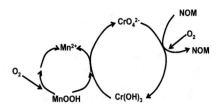

Figure 2.10 Simplified illustration of the interconversion of Cr (+3) and Cr (+6) in typical soils and sediments (modified from James, 1996).

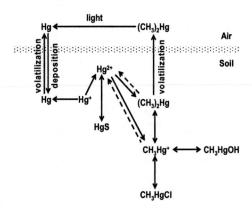

Figure 2.11 Abiotic and biotic transformation of Hg in the soil. Broken arrows denote slow reaction (modified from McBride, 1994).

In a study investigating the transformation of cadmium in calcareous soils, Hirsch and Banin (1990) found that the small percent of the added Cd ($\sim$ 2%) remaining in the soil solution was comprised of Cd^{2+} and $CdHCO_3^+$ (Figure 2.12). Increasing the pH decreased the amount of Cd^{2+} and increased the amount of $CdHCO_3^+$ in the soil solution until the pH reached 8. At pH > 8, $CdCO_3^0$ was the dominant cadmium species. Thus, the high CO_2 and high pH conditions in the rhizosphere of plants grown in calcareous soils can decrease the amount of Cd^{2+} significantly. In soils with high Cl^- concentrations, Cd^{2+} primarily complexes with the Cl^- to form $CdCl_2^0$, $CdCl_3^-$, and $CdCl_4^{2-}$ (Garcia-Miragaya and Page, 1976; Smolders et al., 1996). Cadmium can also form soluble complexes with sulfate anions and dissolved organic matter (Kabata Pendias, and Pendias, 1992).

Like Cd, only a negligible amount of total Pb (< 1%) is associated with the soil solution in Pb contaminated soils (Alloway, 1995; Cances et al., 2003) and the forms of soluble Pb species depend on the soil pH and the presence of complexing anions. Computer-based speciation analysis conducted for aqueous systems indicates that at pH < 7, Pb^{2+} is the dominant soluble Pb species, and its quantity decreases with increasing pH. At pH in the range of 7 to 8, the amount of $PbOH^+$ increases and becomes the dominant species at a pH of 8.5 (Figure 2.13). Complexing anions such as Cl^-, CO_3^{2-}, SO_4^{2-}, and other organic acids can also form soluble species with Pb^{2+} such as $PbCl^+$, $PbCl_2^0$, $PbHCO_3^-$, and $PbCO_3^0$ (Roy et al., 1993; Essington, 2004).

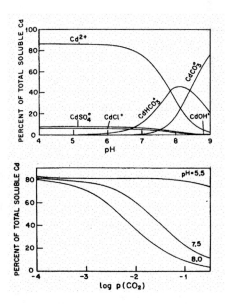

Figure 2.12 Calculated Cd speciation in the pH range of 4 to 9 (left figure) and the effect of CO_2 and pH on Cd^{2+} percentages (right figure) in the solution of a typical calcareous soil (from Hirsch and Banin, 1990).

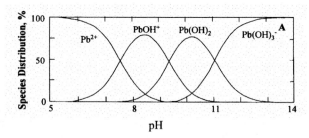

Figure 2.13 Computer prediction of soluble Pb speciation as a function of pH (Ardriano, 2001).

Mineral dissolution is a major factor affecting the amount of trace metals in the solution phase. Mineral dissolution is a function of pH, free cation and anion activities, and redox conditions (Lindsay, 1979). Figure 2.14a indicates that at pH < 7.5, the concentration of Cd^{2+} in the solution phase is controlled by the dissolution of $Cd_3(PO_4)_2$ and decreases with increasing pH. Above pH 7.5, soluble Cd^{2+} continues to decrease with increasing pH and $CdCO_3$ dissolution becomes the controlling factor for the soluble Cd. In the case of Pb, Figure 2.14b suggests that various Pb phosphate minerals control the amount of Pb present in the solution phase and the pH is negatively correlated with amount of Pb in solution.

2.4.3 Sorption-Desorption Mechanisms and Kinetics

The kinetics and mechanisms of sorption-desorption of heavy metals are greatly influenced by the oxidation state of the metal, which in turn is impacted by the pH and the redox potential of the soil system. Due to their pH dependent surface charges, oxide minerals constitute the most important sorption domains for heavy metals. For example, arsenate and arsenite anions sorb or adhere electrostatically to mineral surfaces that are positively charged. In general, arsenite associates less strongly than arsenate to soils and oxide minerals (Smith et al., 1999; Adriano, 2001). Arsenate and phosphate have comparable acid dissociation constants and their salts have similar solubility products resulting in a analogous sorption behaviors on mineral surfaces. Figure 2.15 shows the formation of inner sphere complexes through step-wise ligand exchange reactions during the sorption of arsenate and arsenite on oxide minerals (Sun and Doner, 1996; Fendorf et al., 1997; Grossl et al., 1997). Spectroscopic results from a study by Fendorf et al. (1997) showed that monodentate complexation was favored at low As concentrations, whereas bidentate complexation was favored at higher concentrations.

Sorption of arsenate and arsenite on oxide minerals and soils are relatively fast reactions that reach equilibrium within seconds to hours (Grossl et al., 1997; Smith et al., 1999). These sorption reactions have been observed to follow both Langmuir and Freundlich-type sorption behaviors (Smith et al., 1999; Adriano, 2001). Freundlich K_F values for arsenate sorption on four Australian soils tested by Smith et al. (1999) varied from 3 to 62 L/kg, and were positively correlated with the oxide content of the soils. Sorption of arsenate on the soils tested was higher than that of arsenite. Phosphate and carbonate ions can compete effectively with arsenate for sorption sites on oxide minerals (Manning and Goldberg, 1996; Hiemstra and Riemsdijk, 1999; Gao and Mucci, 2001; Appelo et al., 2002; Dixit and Hering, 2003). Appelo et al. (2002) found that desorption of arsenate from iron oxide surfaces increased about 20 fold when exposed to ~ 80 mg/L carbonate. This phenomenon has been attributed to the high As concentrations reported in the groundwater in Bangladesh (Appelo et al., 2002). Other researchers reported similar reductions in arsenate and arsenite sorption on iron oxides in the presence of phosphate ions (Jain and Loeppert, 2000; Dixit and Hering, 2003). Desorption of arsenate from oxide minerals has been found to occur rapidly, reaching equilibrium within 24 h (O'Reilly et al., 2001).

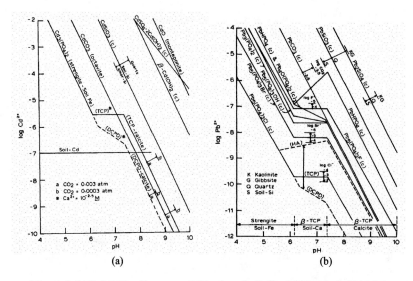

Figure 2.14 Solubility diagrams of Cd minerals (left) and Pb minerals (right) assuming $[CO_2(g)] = 10^{-3.5}$ M, $[Cl^-] = 10^{-3}$ M, and $[SO_4^{2-}] = 10^{-3}$ M (from Lindsay, 1979).

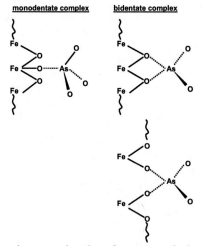

Figure 2.15 Inner sphere complexation of arsenate sorbed on Fe oxide surface (Fendorf et al., 1997)

The chromate anion also associates preferentially with oxide mineral surfaces forming inner-sphere monodentate and/or bidentate complexes (Fendorf et al., 1997). Sorption of chromate on oxide mineral surfaces is rapid, typically reaching equilibrium within a few hours, although higher pHs may reduce sorption rates (Garman et al., 2004). The presence of naturally occurring ligands such as oxalate, silicic acid, and phosphate can compete with chromate for sorption sites on mineral surfaces resulting in enhanced mobility of chromate in natural systems (Tzou et al., 1998; Mesuere and Fish, 1992; Garman et al., 2004). The sorption of chromate on oxide minerals has been observed to follow both Langmuir and Freundlich-type equilibrium behavior (Adriano, 2001).

Trivalent Cr tends to hydrolyze in acidic solutions to species such as $Cr(OH)^{2+}$, $Cr(OH)_4^{2+}$ or $Cr(OH)_{12}^{6+}$. These cations sorb readily through cation exchange reactions to clay minerals that have permanent negative charges. In the case of trivalent Cr, Fendorf and Sparks (1994) found that Cr^{3+} formed a monodentate surface complex on amorphous silica at surface coverages < 20%, while at greater surface coverages discrete chromium hydroxide surface clusters were observed. Cr (+3) also associates with soil organic matter through complexation (Stewart et al., 2003).

Langmuir and Freundlich-type sorption behaviors of Hg (+2) have been observed to occur on clay minerals and mineral oxides (Adriano, 2001). Sorption equilibrium was achieved within hours. Spectroscopic studies by Kim et al. (2004a, 2004b) demonstrated that the formation of monodentate and bidentate inner-sphere complexes was the dominant mode of Hg (+2) sorption on oxide minerals. These studies also suggested that common complexing ligands such as Cl^- and SO_4^{2-} could significantly impact the sorption of the Hg (+2) ions to oxide mineral surfaces. The presence of chloride resulted in reduced sorption of Hg (+2) mainly due to the formation of stable and nonsorbing $HgCl_2$ complexes in solution and, therefore, limiting the amount of free Hg^{2+} available for sorption. Conversely, the presence of sulfate resulted in enhanced Hg (+2) sorption. This was attributed to either direct sorption of sulfate ions on the sorbent or accumulation of SO_4^{2-} at the sorbent-water interface with a consequent reduction of the positive surface charge on the mineral surface that otherwise electrostatically inhibited Hg (+2) sorption.

Among clay minerals, illite has the highest sorption capacity for Hg (+2) and kaolinite the lowest (Adriano, 2001). Research has shown that Hg^{2+} and methyl-Hg have strong affinity for binding sites such as -SH functional groups and weaker affinity for binding sites such as carboxyl and phenol functional groups in natural organic matter (Xia et al., 1999; Drexel et al., 2002; Skylberg et al., 2003). Sorption of methyl-Hg on clay minerals and oxide minerals occurs rapidly and appears to follow the first order kinetics. Surface complexation and surface precipitation were responsible for the methyl-Hg sorption (Desauziers et al., 1997).

Both cadmium and lead associate strongly with clay minerals, mineral oxides, and soil organic matter. Sorption and desorption of Cd and Pb in soils occur rapidly (complete within hours) and follow Langmuir or Freundlich-type equilibrium behaviors (Alloway, 1995; Selim and Iskandar, 1999; Adriano, 2001). While inner-sphere complexes are formed during Cd and Pb sorption at low concentrations (Selim and Iskandar, 1999), surface precipitation is also possible at higher concentrations (Farley et al., 1985). Factors such as pH, ionic strength, complexing ligands, and competing cations affect Cd and Pb sorption in soils. Over a narrow pH range, the sorption of both metals increases rapidly with increasing pH. For example, over a pH range of 5.5 to 6.5 sorption of Cd to a sandy soil was observed to double for every 0.5 unit increase of pH Tran et al (1998). Pb sorption on oxide minerals showed a similar increase over the pH range of 4 to 5 (McKenzie, 1980). Increasing ionic strength decreased Cd sorption, due to competition from calcium ions for in soils (Petruzzelli and Pezzarossa, 2003). Bolan et al. (1999) reported that the presence of phosphate in two soils with variable charge components enhanced Cd sorption by forming metal-ligand complexes. The phosphate-induced Cd sorption was believed to be caused by an increase in the negative charge on the mineral surface due to phosphate sorption. Boekhold et al. (1993) observed that formation of chloro-complexes of Cd significantly reduced Cd sorption in soils.

Desorption of Cd and Pb in soil depends on the properties of the sorbent and the chemistry of the soil solution (Tiller et al., 1984; Backes et al., 1995; Davis and Upadhyaya, 1996; Gray et al., 1998; Ponizovsky and Tsadilas, 2004). Davis and Upadhyaya (1996) observed that increasing pH from 5 to 7 resulted in reduced Cd desorption from an iron oxide surface. The presence of Ca^{2+} increased the desorption of Cd presumably due to competition between Cd^{2+} and Ca^{2+} for the surface sites and the addition of Cl^- suppressed Cd^{2+} desorption probably as a result of re-adsorption of metal-ligand complexes onto the mineral. Similar results were found by Ponizovsky and Tsadilas (2004) for Pb desorption from soils, even though the extent of Pb desorption from soils was much less than Cd desorption.

2.5   Implications for Site Remediation

Sorption and binding affect mass-transfer of contaminants from the aqueous phase to the solid phase during groundwater flow resulting in retardation of the contaminant front. Conversely, the resistance to mass-transfer from the solid phase to the aqueous phase in contaminated aquifers results in the classic "hockey stick" type tailing observed during pump and treat remediation. The natural attenuation phenomenon in contaminated soil and groundwater environments includes mass-transfer processes such as sorption, precipitation and binding of contaminants to soil, sediment or aquifer particles. The selection of monitored natural attenuation (MNA) as a remediation technology for contaminated sites, therefore, necessitates a sound understanding of sorption, desorption and binding phenomena. Sorption and binding processes could contribute significantly to the natural attenuation of contaminants in soils and sediments with high SOM content.

Organic matter is a key component of soils and sediments and plays a critical role in influencing the solid-liquid mass-transfer of organic and inorganic solutes. Both the content and the type of SOM can impact the rates and extents of sorption and desorption. Clay mineral surfaces provide large surfaces areas for contaminant phase-transfer. Mineral oxides can catalyze the oxidation of organic contaminants in soils and aquifer materials. In surface soils, soil enzymes can influence the transformation of organic contaminants through humification-type reactions. Soil-solute contact time can control the rates and extents of sorption and desorption. Organic and inorganic contaminants may be trapped in microporous sites in the mineral or SOM domains and may become irreversibly bound to SOM. Remediation approaches for sorbed organic solutes should focus on increasing the rate of mass-transfer of these contaminants into the aqueous phase followed by biological or chemical oxidation. Alternately, engineered humification processes could be employed for certain types of organic solutes to incorporate these chemicals into SOM.

Unlike organic contaminants, metals are not chemically or biologically transformed into innocuous end products and remediation approaches largely rely on strategies that convert or maintain the metal at an oxidation state and species that is the least mobile, bioavailable, or toxic. Often, this is accomplished by converting the metal into a form that is readily adsorbed on surfaces. Permeable reactive barriers such as those containing zero-valent iron can be used to catalyze the transformation of the metal, provide a surface for adsorption or both.

References

Adamson, A.W. (1990) *Physical Chemistry of Surfaces*, Wiley-Interscience, New York, NY.

Adriano, D. C. (2001) *Trace Elements in Terrestrial Environments: Biogeochemistry, Bioavailability, and Risks of Metals.* 2nd ed. Springer-Verlag, New York.

Alexander, M. (1995) How toxic are toxic chemicals in soil? Environ. Sci. Technol. 29:2713-2717.

Alloway, B. J. (1995). *Heavy Metals in Soils.* 2nd ed. Blackie Academic & Professional.

Appelo, C.A.J., Van Der Weiden, M.J.J., Tournassat, C., Charlet, L. (2002) Surface complexation of ferrous iron and carbonate on ferrihydrite and the mobilization of arsenic. *Environ. Sci. Technol.* 36:3096 – 3103.

Backes, C.A., McLaren, R.G., Rate, A.W., Swift, R.S. (1995) Kinetics of cadmium and cobalt desorption from iron and manganese oxides. Soil Sci. Soc. Am. J. 59:778-785.

Ball, W.P., Roberts, P.V. (1991a) Long-term sorption of halogenated organic chemicals by aquifer material. 1. Equilibrium. *Environ. Sci. Technol.* 25:1221-1237.

Ball, W.P., Roberts, P.V. (1991b) Long-term sorption of halogenated organic chemicals by aquifer material. 2. Intraparticle diffusion. *Environ. Sci. Technol.* 25:1237-1249.

Barriuso, E., Laird, D.A., Koskinen, W.C., Dowdy, R.H. (1994) Atrazine desorption from smectites. *Soil Sci. Soc. Am. J.* 60:150-157.

Barriuso E., Koskinen W.C. (1996) Incorporating nonextractable atrazine residues into soil size fraction as a function of time. *Soil Sci Soc Am J.* 60:150-157.

Berry, D.F., Boyd, S.A (1985) Reaction rates of phenolic humus constituents and anilines during cross-coupling. *Soil. Biol. Biochem.* 17:631-636.

Bhandari, A., Dove, D.C., Novak, J.T. (1994) Soil washing and biotreatment of petroleum contaminated soils. *J. Environ. Engrg.* 120:1151-1169.

Bhandari, A., Lesan, H.M. (2003) Isotherms for atrazine desorption from two surface soils. *Environ. Engrg. Sci.* 20:257-263.

Bhandari, A., Novak, J.T., Berry, D.F. (1996) Binding of 4-monochlorophenol to soil. *Environ. Sci. Technol.* 30:2305-2311.

Bhandari, A., Novak, J.T., Burgos, W.D., Berry, D.F. (1997) Irreversible binding of chlorophenols to soil and its impact on bioavailability. *J Environ Engrg.* 123: 506-513.

Bhandari, A., Novak, J.T., Dove, D.C. (2001) Effect of soil washing on petroleum hydrocarbon distribution on sand surfaces. *J. Haz. Subst. Res.* 2:7-1 to 7-13.

Bhandari, A., Xu, F. (2001) Impact of peroxidase addition on the sorption-desorption behavior of phenolic contaminants in surface soils. *Environ. Sci. Technol.* 35:3163-3168.

Bissen, M., Frimmel, F. H. (2003) Arsenic – a review. Part I. Occurrence, toxicity, speciation, mobility. *Acta Hydrochim. Hydrobiol.* 31:9-18.

Boekhold, A.E., Temminghoff, E.J.M., van der Zee, S.E.A.T.M. (1993) Influence of electrolyte composition and pH on cadmium sorption by an acid sandy soil. J. Soil Sci. 44:85-96.

Bohn, H.L., McNeal, B.L., O'Connor, G.A. (1985) *Soil Chemistry*, 2nd Edition. John Wiley & Sons, NY.

Bolan, N. S., Naidu, R., Khan, M. A. R., Tillman, R. W., Syers, J. K. (1999) The effects of anion sorption on sorption and leaching of cadmium. *Aust. J. Soil Res.* 37:445-460.

Bollag, J. M. (1992) Decontaminating soil with enzymes. *Environ. Sci. Technol.* 26:1876-1881.

Bollag, J.M., Liu, S.Y. (1985) Copolymerization of halogenated phenols and syringic acid. *Pestic. Biochem. Physiol.* 23:261-272.

Bollag, J.M., Shuttleworth, K.L., Anderson, D.H. (1988) Laccase-mediated detoxification of phenolic compounds. *Appl. Environ. Microbiol.* 54:3086-3091.

Brookins, D. G. (1988). *Eh-pH Diagrams for Geochemistry.* Springer-Verlag.

Burgos, W.D., Pisutpaisal, N., Mazzarese, M.C., Chorover, J. (2002) Adsorption of quinoline to kaolinite and montmorillonite. *Environ. Engrg. Sci.* 19:59-68.

Cances, B., Ponthieu, M., Castrec-Rouelle, M., Aubry, E., Benedetti, M. F. (2003) Metal ions speciation in a soil and its solution: experimental data and model results. *Geoderma* 113: 341-355.

Carroll, K.M., Harkness, M.R., Bracco, A.A., Balcarcel, R.R. (1994) Application of a permeant/polymer diffusional model to the desorption of polychlorinated biphenyls from Hudson River sediments. *Environ. Sci. Technol.* 28:253-258.

CERCLA (2003) Priority List of Hazardous Substances. (http://www.atsdr.cdc.gov/clist.html).

Clay S.A., Koskinen W.C. (1990) Characterization of alachlor and atrazine desorption from soils. *Weed Sci.* 38:74-80.

Dankwardt A., Hock B., Simon R., Freitag D., Kettrup A. (1996) Determination of non-extractable triazine residues by enzyme immunoassay: investigation of model compounds and soil fulvic and humic acids. *Environ Sci Technol.* 30:3493-3500.

Davis, A. P., Upadhyaya, M. (1996) Desorption of cadmium from goethite (α-FeOOH). Water Res. 30:1894-1904.

Dec, J., Bollag, J.M. (1990) Detoxification of substituted phenols by oxidoreductive enzymes through polymerization reactions. *Arch. Environ. Contam. Toxicol.* 19:543-550.

Desauziers, V., Castre, N., Le Cloirec, P. (1997) Sorption of methylmercury by clays and mineral oxides. *Environ. Technol.* 18:1009-1018.

Dixit, S., Hering, J. G. (2003) Comparison of arsenic(V) and arsenic(III) sorption onto iron oxide minerals: Implications for arsenic mobility. *Environ. Sci. Technol.* 37:4182 – 4189.

Dove, D.C., Bhandari, A., Novak, J.T. (1992) Soil washing: advantages and pitfalls. *Remediation,* 3(1):55-67.

Drexel, R. T., Haitzer, M., Ryan, J. N., Aiken, G . R., Nagy, K. L. (2002) Mercury(II) sorption to two Florida Everglades peats: Evidence for strong and weak binding and competition by dissolved organic matter released from the peat. *Environ. Sci. Technol.* 36:4058-4064.

Elovitz, M. S., Fish, W. (1995). Redox interactions of Cr(VI) and substituted phenols: Products and mechanism. *Environ. Sci. Technol.* 29:1933-1943.

Essington, M. E. (2004) *Soil and Water Chemistry: An Integrative Approach.* CRC Press.

Farley, K. J., Dzombak, D. A., Morel, F. M. M. (1985) A surface precipitation model for the sorption of cations on metal oxides. *J. Colloid Interface Sci.* 106:226-242.

Fendorf, S. E., Sparks, D. L. (1994) Mechanisms of chromium(III) sorption on silica. 2. Effect of reaction conditions. Environ. Sci. Technol. 28:290-297.

Fendorf, S., Eick, M. J., Grossl, P., Sparks, D. L. (1997). Arsenate and chromate retention mechanisms on goethite. 1. Surface Structure. *Environ. Sci. Technol.* 31:315-320.

Fitz, W. J., Wenzel, W. W. (2002). Arsenic transformations in the soil-rhizosphere-plant system: fundamentals and potential application to phytoremediation. *J. Biotechno.* 99:259-278.

Frankenberger, W. T. Jr., Arshad, M. (2002) Volatilization of arsenic. In *Environmental Chemistry of Arsenic.* Frankenberger, W. T. Jr. (Ed.).Marcel Dekker, New York. pp. 363-380.

Gao, Y., Mucci, A. (2001) Acid base reactions, phosphate and arsenate complexation, and their competitive adsorption at the surface of goethite in 0.7 M NaCl solution. Geochim. Cosmochim. Acta. 65:2361-2378.

Garcia-Miragaya, J., Page, A. L. (1976) Influence of ionic strength and inorganic complex formation on the sorption of trace amounts of cadmium by montmorillonite. Soil Sci. Soc. Am. J. 40:658-663.

Garman, S. M., Luxton, T. P., Eick, M. J. (2004) Kinetics of chromate adsorption on goethite in the presence of sorbed silicic acid. *J. Environ. Qual.* 33:1703-1708.

Gray, C. W., Mclaren, R. G., Roberts, A. H. C., Condron, L. M. (1998) Sorption and desorption of cadmium from some New Zealand soils: effect of pH and contact time. *Aust. J. Soil Res.* 36:199-216.

Grossl, P.R., Eick, M.J., Sparks, D.L., Goldberg, S., Ainsworth, C.C. (1977) Arsenate and chromate retention mechanisms on goethite. 2. Kinetic evaluation using a pressure-jump relaxation technique. *Environ. Sci. Technol.* 31:321-326.

Gu, B., Chen, J. (2003). Enhanced microbial reduction of CR (VI) and U (VI) by different natural organic matter fractions. *Geochimica et Cosmochi. Acta.* 67:3575-3582.

Ha, H.R., Vinitnantharat, S. (2002) Competitive removal of phenol and 2,4-dichlorophenol in biological activated carbon systems. Environ. Technol. 21:398-396.

Hatzinger, P.B., Alexander, M. (1995) Effect of aging of chemicals in soil on their biodegradability and extractability. *Environ. Sci. Technol.* 29:537-545.

Hiemstra, T., Van Riemsdijk, W. H. (1999) Surface structural ion adsorption modeling of competitive binding of oxyanions by metal (hydr)oxides. *J. Colloid Interface Sci.* 210:182-193.

Hirsch, D., Banin, A. (1990) Cadmium speciation in soil solutions. *J. Environ. Qual.* 19:366-372.

Huang, Q., Selig, H., Weber, W.J. (2002) Peroxidase-catalyzed oxidative coupling of phenols in the presence of geosorbents: Rates of nonextractable product formation. *Environ. Sci. Technol.* 36:596-602.

Huang, W., Schlautman, M.A., Weber, W.J. (1996) A distributed reactivity model for sorption by soils and sediments. 5. The influence of near-surface characteristics in mineral domains. *Environ. Sci. Technol.* 30, 2993-3000.

Huang, W., Weber, W.J. (1996) A distributed reactivity model for sorption by soils and sediments. 4. Intraparticle heterogeneity and phase-distribution relationships under nonequilibrium conditions. *Environ. Sci. Technol.* 30, 881-888.

Huang, W., Weber, W.J. (1997) A distributed reactivity model for sorption by soils and sediments. 10. Relationships between desorption, hysteresis, and the chemical characteristics of organic domains. *Environ. Sci. Technol.* 31, 2562-2569.

Huang, W., Weber, W.J. (1998) A distributed reactivity model for sorption by soils and sediments. 11. Slow concentration-dependent sorption rates. *Environ. Sci. Technol.* 32, 3549-3555.

Jain, A., Loeppert, R. H. (2000) Effect of competing anions on the adsorption of arsenate and arsenite by ferrihydrite. *J. Environ. Qual.* 29:1422-1430.

James, B. R. (1996). The challenge of remediating chromium-contaminated soil. *Environ. Sci. Technol.* 30:248A-251A.

James, B. R., Petura, J. C., Vitale, R. J., Mussoline, G. R. (1997) Oxidation-reduction chemistry of chromium: relevance to the regulation and remediation of chromate-contaminated soils. *J. Soil Contam.* 6:569-580.

Johnson, M.D., Huang, W., Weber, W.J. (2001) A distributed reactivity model for sorption by soils and sediments. 13. Simulated diagenesis of natural sediment organic matter and its impact on sorption/desorption equilibria. *Environ. Sci. Technol.* 35, 1680-1687.

Kabata Pendias, A., Pendias, H. (1992). *Trace Elements in Soils and Plants.* 2nd ed. CRC Press, Baton Rouge, FL.

Karapanagioti, H.K., Kleineidam, S., Sabatini, D.A., Grathwohl, P., Ligouis, B. (2000) Impacts of heterogeneous organic matter on phenanthrene sorption: Equilibrium and kinetic studies with aquifer material. *Environ. Sci. Technol.* 34:406-414.

Karickhoff, S.W. (1984) Organic pollutant sorption in aquatic systems. *J Hydraul. Eng.*, 110, 707-734.

Kim, C. S., Rytuba, J. J., Brown, G. E. (2004a) EXAFS study of mercury(II) sorption to Fe- and Al-(hydr)oxides I. Effects of pH. *J. Colloid Interface Sci.* 271:1-15.

Kim, C. S., Rytuba, J. J., Brown, G. E. (2004b) EXAFS study of mercury(II) sorption to Fe- and Al-(hydr)oxides II. Effects of chloride and sulfate. *J. Colloid Interface Sci.* 270:9-20.

Kim, J.S., Park, J.W., Lee, S.E., Kim, J.E. (2002) Formation of bound residues of 8-hydroxybentazon by oxidoreductive catalysts in soil. *J. Agric. Food Chem.* 50:3507-3511.

Laha, S., Luthy, R.G. (1990) Oxidation of aniline and other primary aromatic amines by manganese dioxide. *Environ. Sci. Technol.* 24:363-373.

LeBoeuf, E.J. and Weber, W.J. (1997) A distributed reactivity model for sorption by soils and sediments. 8. Discovery of a humic acid glass transition and an argument for a polymer-based model. *Environ. Sci. Technol.* 31, 1697-1702.

LeBoeuf, E.J. and Weber, W.J. (2000a) Macromolecular characteristics of natural organic matter. 1. Insights from glass transition and enthalpic relaxation behavior. *Environ. Sci. Technol.* 34, 3623-3631.

LeBoeuf, E.J. and Weber, W.J. (2000b) Macromolecular characteristics of natural organic matter. 2. Sorption and desorption behavior. *Environ. Sci. Technol.* 34, 3632-3640.

Lesan, H.M., Bhandari, A. (2004) Contact time dependent atrazine residue formation in surface soils. *Water Res.* 38: 4435-4445.

Lesan. H.M., Bhandari, A. (2003) Atrazine sorption on surface soils: Time-dependent phase distribution and apparent desorption hysteresis. *Water Res.* 37:1644-1654.

Lindsay, W. L. (1979) *Chemical Equilibria in Soils.* John Wiley & Sons, Inc.

Liu, S.Y., Minard, R.D., Bollag, J.M. (1987) Soil-catalyzed complexation of the pollutant 2,6-diethylaniline with syringic acid. *J. Environ. Qual.* 16:48-53.

Ma L., Southwick L.M., Willis G.H., Selim H.M. (1993) Hysteretic characteristics of atrazine adsorption-desorption by a sharkey soil. *Weed Sci.* 41:627-633.

Manning, B.A., Goldberg, S. (1996) Modeling competitive adsorption of arsenate with phosphate and molybdate on oxide minerals. *Soil Sci. Soc. Am. J.* 60:121-131.

McBride, M. B. (1994) *Environmental Chemistry of Soils.* Oxford University Press.

McGinley, P.M., Katz, L.E., Weber, W.J. (1993) A distributed reactivity model for sorption by soils and sediments. 2. Multicomponent systems and competitive effects. *Environ. Sci. Technol.* 27, 1524-1531.

McKenzie, R. M. (1980) The adsorption of lead and other heavy metals on oxides of manganese and iron. *Aust. J. Soil Res.* 18:61-73.

Mesuere, K., Fish, W. (1992) Chromate and oxalate adsorption on goethite. 2. surface complexation modeling of competitive adsorption. *Environ. Sci. Technol.* 26:2365-2370.

O'Reilly, S. E., Strawn, D. G., Sparks, D. L. (2001) Residence time effects on arsenate adsorption/desorption mechanisms on goethite. *Soil Sci. Soc. Am. J.* 65:67-77.

Palomo, M., Bhandari, A. (2005) Time-dependent sorption-desorption behavior of 2,4-dichlorophenol and its polymerization products in surface soils. *Environ. Sci. Technol.* 39:2143-2151.

Palomo, M., Bhandari, A. (2006) Impact of aging on the formation of bound residues after peroxidase-mediated treatment of 2.4-DCP contaminated soils. *Environ. Sci. Technol.* 40:3402-3408.

Pavlostathis, S.G., Jaglal, K. (1991) Desorptive behavior of trichloroethylene in contaminated soil. *Environ. Sci. Technol.* 25:274-279.

Pavlostathis, S.G., Mathavan, G. (1992) Desorption kinetics of selected volatile organic compounds from contaminated soils. *Environ. Sci. Technol.* 26:532-538.

Petruzzelli, G., Pezzarossa, B. (2003) Ionic strength influence on heavy metal sorption processes by soil. *J. de Physique IV.*107:1061-1064.

Pignatello, J.J., Ferrandino, F.J., Huang, L.Q. (1993) Elution of aged and freshly added herbicides form a soil. *Environ. Sci. Technol.* 27:1563-1571.

Pignatello, J.J., Huang, L.Q. (1991) Sorptive reversibility of atrazine and metolachlor residues in field soil samples. *J. Environ. Qual.* 20:222-228.

Pignatello, J.J., Xing, B. (1996) Mechanisms of slow sorption of organic chemicals to natural particles. *Environ. Sci. Technol.* 30: 1-11.

Ponizovsky, A. A., Tsadilas, C. D. (2004) Effect of solution composition and flow rate on kinetics of lead(II) desorption from Alfisol. *Communi. Soil Sci. Plant Anal.* 35:649-664.

Roy W.R., Krapac I.G. (1994) Adsorption and desorption of atrazine and deethylatrazine by low organic carbon geologic materials. *J Environ Qual.* 23:549-556.

Roy, W. R., Krapac, I. G., Steele, J. D. (1993) Sorption of cadmium and lead by clays from municipal incinerator ash-water suspensions. *J. Environ. Qual.* 22:537-543.

Schlebaum, W., Schraa, G., Van Riemsdijk, W.H. (1999) Influence of nonlinear sorption kinetics on the slow-desorbing organic contaminant fraction in soil. *Environ. Sci. Technol.* 33:1413-1417.

Schwarzenbach, R.P., Gschwend, P.M., Imboden, D.M. (1993) *Environmental Organic Chemistry*, Wiley-Interscience, New York, NY.

Selig, H., Keinath, T.M., Weber, W.J. (2003) Sorption and manganese-induced oxidative coupling of hydroxylated aromatic compounds by natural geosorbents. *Environ. Sci. Technol* 37:4122-4127.

Selim, H. M., Iskandar, I. K. (1999). *Fate and Transport of Heavy Metals in the Vandose Zone*. Lewis Pub.

Skylberg, U., Qian, L., Frech, C., Xia, K., Bleam, W. F. (2003) Distribution of mercury, methyl mercury and organic sulphur species in soil, soil solution and stream of a boreal forest catchment. *Biogeochemistry.* 64:53-76.

Smith, E., Naidu, R., Alston, A. M. (1999) Chemistry of arsenic in soils: I. Sorption of arsenate and arsenite by four Australian soils. *J. Environ. Qual.* 28:1719-1726.

Smolders, E., McLaughlin, M. J. (1996) Chloride increases cadmium uptake in Swiss chard in a resin-buffered nutrient solution. *Soil Sci. Soc. Am. J.* 60:1443-1447.

Stewart, M. A., Jardine, P. M., Barnett, M. O., Mehlhorn, T. L., Hyder, L. K., McKay, L. D. (2003) Influence of soil geochemical and physical properties on the sorption and bioaccessibility of chromium(III). *J. of Environ. Qual.* 32:129-137.

Stone, A.T. (1987) Reductive dissolution of manganese (III/IV) oxides by substituted phenols. *Environ. Sci. Technol.* 21:979-988.

Sun, X., Doner, H. E. (1996) An investigation of arsenate and arsenite bonding structures on goethite by FTIR. *Soil Sci.* 161:865-872.

Tatsumi, K., Freyer, A., Minard, R.D., Bollag, J.M. (1995) Enzyme-mediated coupling of 3,4-dichloroaniline and ferulic acid: A model for pollutant binding to humic materials. *Environ. Sci. Technol.* 28:210-215.

Thiele-Bruhn, S., Seibicke, T., Schulten, H.-R., Leinweber, P. (2004) Sorption of sulfonamide pharmaceutical antibiotics on whole soils and particle-size fractions. *J. Environ. Qual.* 33:1331-1342.

Tran, Y. T., Bajracharya, K., Barry, D. A. (1998) Anomalous cadmium adsorption in flow interruption experiments. *Geoderma* 84:169-184.

Tzou, Y. M., Chen, Y. R., Wang, M. K. (1998) Chromate sorption by acidic and alkaline soils. *J. Environ. Sci. Health, Part A: Toxic/Hazardous Substances & Environmental Engineering.* A33:1607-1630.

U.S. EPA. (1996). *Mercury: Study Report to Congress, vol 1: Executive Summary*. U.S. EPA/R-96-001a, NTIS, Springfield, VA.

Wagner, M., Nicell, J.A. (2002) Detoxification of phenolic solutions with horseradish peroxidase and hydrogen peroxide. *Water Res.* 36:4041-4052.

Weber, W.J. (2001) *Environmental Systems and Processes. Principles, Modeling and Design*. Wiley-Interscience, New York, NY.

Weber, W.J., DiGiano, F.A (1996) *Process Dynamics in Environmental Systems*. Wiley-Interscience, New York, NY.

Weber, W.J., Huang, Q. (2003) Inclusion of persistent organic pollutants in humification processes: Direct chemical incorporation of phenanthrene via oxidative coupling. *Environ. Sci. Technol.* 37:4221-4227.

Weber, W.J., LeBoeuf, E.J., Young, T.M., Huang, W. (2001) Contaminant interactions with geosorbent organic matter: Insights drawn from polymer sciences." *Water Res.* 35:853-868.

WEF (2002) *Handbook on Sediment Quality*. Water Environment Federation, Alexandria, VA.

Wershaw, R.L. (1986) A new model for humic materials and their interactions with hydrophobic organic chemicals in soil-water or sediment-water systems. *J. Contamint Hydrol.* 1:29-45.

White, J.J., Pignatello (1999) Influence of bisolute competition on the desorption kinetics of polycyclic aromatic hydrocarbons in soil. *Environ. Sci. Technol.* 33:4292-4298.

Xia, K., Bleam, W. F., Skyllberg, U., Bloom, P. R., Nater, E. A. (1999) XAS study of the binding of mercury (II) to reduced sulfur in soil organic matter. *Environ. Sci. Techol.* 33:257-261.

Xing, B., McGill, W.B., Dudas, M.J., Maham, Y., Hepler, L. (1994) Sorption of phenol by selected biopolymers: Isotherms, energetics, and polarity. *Environ. Sci. Technol.* 28:466-473.

Xing, B., Pignatello, J.J. (1996) Time-dependent isotherm shape of organic compounds in soil organic matter: implications for sorption mechanism. *Environ. Toxicol. Chem.* 15:1282-1288.

Xing, B., Pignatello, J.J. (1997) Dual-mode sorption of low-polarity compounds in glassy poly(vinyl chloride) and soil organic matter. *Environ. Sci. Technol.* 31, 792-799.

Xing, B., Pignatello, J.J., Gigliotti, B. (1996) Competitive sorption between atrazine and other organic compounds in soils and model sorbents. *Environ. Sci. Technol.* 30, 2432-2440.

Xu, F., Bhandari (2003b) Sorption and binding of 1-naphthol and naphthol polymerization products to surface soils. *J. Environ. Engrg.* 129:1041-1050,

Xu, F., Bhandari, A. (2003a) Retention and extractability of phenol, cresol, and dichlorophenol exposed to two surface soils in the presence of horseradish peroxidase enzyme. *J. Agric. Food Chem.* 51:183-188

Xu, F., Koch, D.E., Kong, I.C., Hunter, R.P., Bhandari, A. (2004) Peroxidase mediated oxidative coupling of 1-naphthol: Characterization of polymerization products. (Submitted).

Zhang, W., Aschner, M., Ghersi-Egea, J. F. (2003). Brain barrier systems: a new frontier in metal neurotoxicological research. *Toxic. App. Pharm.* 192:1-11

CHAPTER 3

Couples of Precipitation-Dissolution and Reduction-Oxidation Reactions

IRENE M. C. LO AND KEITH C. K. LAI

3.1   Introduction

Precipitation and dissolution are two main chemical reactions determining the fate of chemicals or contaminants in natural waters and in many water and wastewater treatment processes. Mineral precipitation and dissolution are the prime factors altering the chemical composition of natural water (Stumm and Morgan, 1996). Treatment processes, such as lime-soda softening, iron removal, coagulation with hydrolyzed metal salts and phosphate precipitation, are all based on the precipitation (Snoeyink and Jenkins, 1980). Likewise, the behavior of compounds containing carbon, nitrogen, sulfur, iron and manganese in the treatment processes and natural water is largely governed by redox reactions (Snoeyink and Jenkins, 1980). Hence thorough evaluation of the redox conditions in polluted areas is often a prerequisite in predicting the fate of pollutants and for selecting suitable treatment or mitigation approaches (Christensen et al., 2000). For instance, the redox condition information of a contaminant plume in subsurface allows evaluation of further plume development and potential risks to downgradient groundwater resources, and assessment of the appropriateness of natural attenuation as a remediation option.

In this chapter, some basic theories and definitions about the precipitation and redox processes will be first introduced. Then the mechanisms involving in the removal of dissolved ions such as hexavalent chromium [Cr(VI)], uranium, nitrate (NO_3^-), nitrite (NO_2^-) and perchlorate (ClO_4^-), and organic compounds such as chlorinated aliphatic hydrocarbons (CAHs) and chlorinated organic micropollutants from aqueous environment via the processes of precipitation and/or reduction processes will be discussed. Case studies emphasizing the efficacies and potentials of the chemical precipitation and reduction processes commonly occurring in natural environment for the remediation of hazardous wastes will also be presented. Finally, discussion on the engineering application of these natural processes for hazardous waste remediation will be provided.

3.2   Precipitation and Dissolution

3.2.1   Precipitation

Precipitation only happens in a supersaturated solution. However, it is possible for a slightly supersaturated solution to be stable indefinitely with respect to a solid phase if there is no either continuous increase in the degree of supersaturation or addition of fine particles of substances into the solution. Generally, precipitation involves several steps, which include nucleation, crystal growth, and aging, ripening and agglomeration of solids (Walton, 1967; Nielson, 1964). Nucleation is the initial phase of precipitation, which involves the formation of nuclei from clusters of molecules or ion pairs of the component ions of precipitates, or from fine particles chemically unrelated to the precipitates but with some similarities in crystal lattice structure (Snoeyink and Jerkins, 1980). Spontaneous precipitation of solid phase can only take place on the surface of the nuclei. Precipitation from homogeneous solution (i.e., solution containing no solid phase) requires formation of nuclei from ions in the solution which is called homogenous nucleation. On the other hand, it is defined as heterogeneous nucleation if foreign particles form the nuclei for subsequent precipitation. In natural water, most nucleation is heterogeneous since aqueous solution usually contains fine particles of various types.

Nucleation is an energy-consuming process since an organized structure of the nuclei with defined surfaces is created from a random arrangement of solution constituents. Stable nuclei can only be formed when an activation energy barrier is surmounted. To overcome this energy barrier, supersaturated solution or solution having a concentration greater than that predicted by the equilibrium with precipitate is required. Homogeneous nucleation generally possesses higher activation energy barrier or requires higher degree of supersaturation than heterogeneous nucleation. This is because heterogeneous nucleation involves foreign particles which behave as catalysts in chemical reactions to lower the activated energy barrier. Figure 3.1 illustrates the catalytic influence from foreign particles on the nucleation in which the activation energy barrier of heterogeneous nucleation (ΔG_1) is many times smaller than that of homogeneous nucleation (ΔG_2) (Stumm, 1992). The activation energy for the nucleation under different saturation states is also illustrated (Appelo and Postma, 1993). As evident, the activation energy decreases with increasing the saturation state or degree of supersaturation.

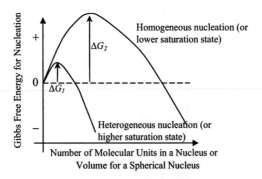

Figure 3.1 Comparison of the activation energy of homogeneous and heterogeneous nucleation at a specific saturation state or comparison of the activation energy of the nucleation under different saturation states

The volume of nuclei showing the activation energy barriers of the nucleation (i.e., ΔG_1 and ΔG_2) is known as critical nucleus. In case the nucleus size passes the size of the critical nucleus, the solids enter the domain of crystal growth in which precipitate constituent ions continuously deposit onto the nucleus (Appelo and Postma, 1993). It is interesting to note that high saturation state generally yields poor crystals. This is because the activation energy barrier is low under this circumstance, thereby allowing many nuclei to form but few grow larger afterwards. Contrarily, at low saturation state, the activation energy barrier is high so that crystal growth dominates over nucleation, which results in large crystals.

Aging of solids is related to the change in solid crystal structure over time. Since the initial solids formed by the precipitation may not thermodynamically be the most stable forms, the crystal structure of the initially formed precipitates may change to more stable phases. Since the stable solids generally have lower solubility than the initially formed solids, this change in the crystal structures may be consequently accompanied by the additional precipitation. Except aging, a phenomenon called ripening may also occur whereby the crystal size of the precipitates increases. Small particles usually have higher solubility than large particles. Therefore, the solution concentration in equilibrium with small particles is higher than that in equilibrium with large particles. In an aqueous system containing both large and small particles, large particles can continue to grow since the solution is still supersaturated with respect to it. However, the continuous decrease in solution concentration owing to the growth of large particles consequently creates an undersaturated condition with respect to small particles, thereby leading to the dissolution of small particles. Accordingly, small particles disappear and the size of large particles increases. In addition to ripening, agglomeration of particles to form larger particles also enhances the conversion of small particles to large particles.

3.2.2 Dissolution

Solubility product (K_{sp}) is the name given to the equilibrium constant, which is related to the dissolution of a precipitate in pure water to its constituent ions. Table 3.1 lists the dissolution reaction of the minerals commonly found in groundwater, and the corresponding K_{sp} and solubility (Seidell, 1958; Freeze and Cherry, 1979; Kiely, 1998). As an example of K_{sp}, the equilibrium constant for the dissolution of calcite ($CaCO_3$), based on the law of mass action, is shown in Eq. (3.1).

$$Equilibrium\ Constant,\ K_{sp} = \{Ca^{2+}\}\{CO_3^{2-}\} \qquad \text{(Eq. 3.1)}$$

Unlike K_{sp} which is the colloquial term for equilibrium constant, solubility refers to the amount of a substance, which can dissolve in solution under certain condition. Temperature is one of the dominant factors affecting the equilibrium position of the dissolution reaction and thereby is capable of influencing the magnitude of K_{sp}. To correct K_{sp} under different temperatures, Van't Hoff equation is applied, which is shown in Eq. (3.2).

$$\ln K_{sp,T_1} - \ln K_{sp,T_2} = \frac{-\Delta H_r^0}{R}\left(\frac{1}{T_1} - \frac{1}{T_2}\right) \qquad \text{(Eq. 3.2)}$$

As noted above, supersaturation is one of the prerequisites for the precipitation. Therefore, understanding the saturation state of solution can indicate the possibility of the occurrence of precipitation and also the quality of crystal growth. To quantify the saturation state, a concept called saturation index (Ω) is applied in which the actual ion activity product (IAP) in water sample is compared to the activity product at equilibrium. For instance, the Ω for the precipitation of calcite can be determined from Eq. (3.3).

$$\Omega = \frac{\{Ca^{2+}\}_{actual}\{CO_3^{2-}\}_{actual}}{\{Ca^{2+}\}_{equilibrium}\{CO_3^{2-}\}_{equilibrium}} = \frac{IAP}{K_{sp}} \qquad \text{(Eq. 3.3)}$$

Solution is said to be in equilibrium if the Ω is one. In case the Ω is larger than one, the solution is supersaturated; whereas it is under subsaturated if the Ω is smaller than one (Appelo and Postma, 1993).

Table 3.1 Mineral dissolution reactions, and the corresponding solubility products (K_{sp}) and/or solubilities at 25 ^{0}C and 1 atm

Mineral	Dissolution reaction	K_{sp}	Solubility at pH 7 (mg/L)
Amorphous silica	$SiO_2(s) + 2H_2O(l) \Longleftrightarrow Si(OH)_4(aq)$	$10^{-2.7}$	120
Calcite	$CaCO_3(s) \Longleftrightarrow Ca^{2+} + CO_3^{2-}$	$10^{-8.4}$	100[a], 500[b]
Copper hydroxide	$Cu(OH)_2(s) \Longleftrightarrow Cu^{2+} + 2OH^-$	$10^{-18.8}$	—
Dolomite	$CaMg(CO_3)_2(s) \Longleftrightarrow Ca^{2+} + Mg^{2+} + 2CO_3^{2-}$	$10^{-17.0}$	90[a], 480[b]
Epsomite	$MgSO_4 \cdot 7H_2O \Longleftrightarrow Mg^{2+} + SO_4^{2-} + 7H_2O(l)$	—	267000
Ferric hydroxide	$Fe(OH)_3(s) \Longleftrightarrow Fe^{3+} + 3OH^-$	$10^{-37.2}$	—
Fluorite	$CaF_2(s) \Longleftrightarrow Ca^{2+} + 2F^-$	$10^{-9.8}$	160
Gibbsite	$Al_2O_3 \cdot 2H_2O(s) + H_2O(l) \Longleftrightarrow 2Al^{3+} + 6OH^-$	10^{-34}	0.001
Gypsum	$CaSO_4 \cdot 2H_2O \Longleftrightarrow Ca^{2+} + SO_4^{2-} + 2H_2O(l)$	$10^{-4.5}$	2100
Halite	$NaCl(s) \Longleftrightarrow Na^+ + Cl^-$	$10^{+1.6}$	360000
Hydroxylapatite	$Ca_5OH(PO_4)_3(s) \Longleftrightarrow 5Ca^{2+} + 3PO_4^{3-} + OH^-$	$10^{-55.6}$	30
Magnesite	$MgCO_3(s) \Longleftrightarrow Mg^{2+} + CO_3^{2-}$	$10^{-4.4}$	—
Mirabillite	$Na_2SO_4(s) \cdot 10H_2O \Longleftrightarrow 2Na^+ + SO_4^{2-} + 10H_2O(l)$	$10^{-1.6}$	280000
Quartz	$SiO_2(s) + 2H_2O(l) \Longleftrightarrow Si(OH)_4(aq)$	$10^{-3.7}$	12
Sylvite	$KCl(s) \Longleftrightarrow K^+ + Cl^-$	$10^{+0.9}$	264000

Source: Seidell (1958); Freeze and Cherry (1979); Kiely (1998)
[a]Partial pressure of $CO_2 = 10^{-3}$ atm.
[b]Partial pressure of $CO_2 = 10^{-1}$ atm.

3.2.3 Factors Affecting Ion Activity

Considering a pure water system only containing a type of dissolved ion such as calcium, all the calcium ions are surrounded by a shield of water molecule dipoles. In this case, all the calcium ions in the system can partake in precipitation and dissolution processes. Hence the activity coefficient of calcium ions (γ_{Ca}^{2+}) is equal to one, and the calcium activity required for the precipitation is exactly equal to its molality as illustrated in Eq. (3.4). However, in the presence of the charged solutes, which are not the constituent ions for calcium precipitation, additional electrostatic shielding of the calcium ions by these charged solutes occurs. Under this circumstance, only certain portion of calcium ions is surrounded by water molecule dipoles for the precipitation and dissolution, thereby resulting in the γ_{Ca}^{2+} of less than unity. As a result, calcium activity is less than its molarity and high molarity of calcium ions is needed to maintain the same calcium activity required for the precipitation. In other words, more calcium ions can dissolve into the solution containing the charged solutes, which results in a higher calcium solubility than the solution without containing the charged solutes. The phenomenon of increasing in the calcium solubility in the presence of the charged solutes or with increasing solution ionic strength is known as ionic strength effect.

$$\{Ca^{2+}\} = \gamma_{Ca^{2+}}(Ca^{2+})$$
(Eq. 3.4)

In contrast to the ionic strength effect, if the electrolyte containing precipitate constituent ions is added into the solution saturated with the constituent ions, the solubility of the ions decreases and precipitation occurs eventually. This is because the activity product [e.g., $\{Ca^{2+}\}\{CO_3^{2-}\}$ for calcite] must adjust to attain a value equal to its K_{sp}. This phenomenon is known as common-ion effect (Freeze and Cherry, 1979).

3.3   Reduction and Oxidation

3.3.1   Basic Concepts

"Redox" is an abbreviation of reduction and oxidation. By definition, oxidation is a reaction in which a substance loses or donates electrons; whereas reduction is a reaction in which a substance gains or accepts electrons. Since redox reactions involve electron transfer from one atom to another, the oxidation state of the reactants and products after the reaction are changed. The oxidation state, sometimes referred to the oxidation number, represents a hypothetical charge that an atom would have if the ion or molecule was to dissociate (Freeze and Cherry, 1979).

As an example of redox reactions, the reaction between dichromate ion ($Cr_2O_7^{2-}$) and ferrous iron (Fe^{2+}) is shown in Eq. (3.5).

$$Cr_2O_7^{2-} + 6Fe^{2+} + 14H^+ \Longleftrightarrow 2Cr^{3+} + 6Fe^{3+} + 7H_2O(l)$$
(Eq. 3.5)

Totally, 6 electrons are transferred from Fe^{2+} to reduce hexavalent chromium [Cr(VI)] in $Cr_2O_7^{2-}$ to trivalent chromium [Cr(III)]. In this redox reaction, Fe^{2+} acts as a reducing agent or reductant since it donates electrons for the reduction of $Cr_2O_7^{2-}$. Contrarily, $Cr_2O_7^{2-}$ is said to be an oxidizing agent or oxidant because it accepts the electrons released from the oxidation of Fe^{2+} (Snoeyink and Jenkins, 1980). To obtain a balanced chemical equation, redox reactions can be viewed as consisting of two half reactions which are used to balance the number of elements involved and electroneutrality. Eqs. (3.6) and (3.7) show the balanced half reactions for the redox reaction between $Cr_2O_7^{2-}$ and Fe^{2+}.

$$6Fe^{2+} \Longleftrightarrow 6Fe^{3+} + 6e^-$$
(Eq. 3.6)
$$Cr_2O_7^{2-} + 14H^+ + 6e^- \Longleftrightarrow 2Cr^{3+} + 7H_2O(l)$$
(Eq. 3.7)

Table 3.2 summarizes the general rules used for balancing redox reactions. Despite the fact that free electrons are used to balance the electroneutrality of half reactions, electrons can only be exchanged and are not found in free states. Based on the conservation of the electrons, a reduction reaction is always coupled with an oxidation reaction (Appelo and Postma, 1993).

Table 3.2 Guides for balancing redox reactions

Step	Description
1.	For each half reaction, write the oxidized and reduced species into the equation and balance the elements at left and right, except hydrogen and oxygen
2.	Balance the number of oxygen atoms by adding H_2O
3.	Balance the number of protons by adding H^+
4.	Balance electroneutrality by adding electrons
5.	Subtract the two half reactions to obtain the complete redox reaction

3.3.2 Redox Equilibria

The redox reaction between $Cr_2O_7^{2-}$ and Fe^{2+} [Eq. (3.5)] can be described in terms of Gibbs free energy, as illustrated in Eq. (3.8). Water does not appear in the equation since by definition, it has a unit activity.

$$\Delta G_r = \Delta G_r^{0} + RT \ln\left(\frac{\{Cr^{3+}\}^2\{Fe^{3+}\}^6}{\{Cr_2O_7^{2-}\}\{Fe^{2+}\}^6\{H^+\}^{14}}\right) \qquad \text{(Eq. 3.8)}$$

This Gibbs free energy of a reaction (ΔG_r) is related to the voltage developed by the redox reaction in an electrochemical cell by Eq. (3.9).

$$\Delta G = -nFE \qquad \text{(Eq. 3.9)}$$

Substitution of Eq. (3.9) into Eq. (3.8) results in the Nernst equation [Eq. (3.10)] in which E^0 is the sum of the standard potentials for the oxidation half reaction (E^0_{oxd}) and reduction half reaction (E^0_{red}) [Eq. (3.11)].

$$E = E^0 - \frac{RT}{nF}\ln\left(\frac{\{Cr^{3+}\}^2\{Fe^{3+}\}^6}{\{Cr_2O_7^{2-}\}\{Fe^{2+}\}^6\{H^+\}^{14}}\right) \qquad \text{(Eq. 3.10)}$$

$$E^0 = E^0_{ox} + E^0_{red} \qquad \text{(Eq. 3.11)}$$

The standard potential for various half reactions, like the Gibbs free energy of formation, cannot be determined absolutely. They are conventionally measured with reference to a particular half reaction which is assigned a standard potential of zero at 25 ^{0}C and 1 atm. This reaction is the reduction of hydrogen ion (H^+) to hydrogen gas (H_2) [Eq. (3.12)].

$$H^+ + e^- \Longleftrightarrow \tfrac{1}{2} H_2(g) \qquad \text{(Eq. 3.12)}$$

The physical setup which defines the standard potential of a half reaction is illustrated in Fig. 3.2 (Snoeyink and Jenkins, 1980). At the left side of the electrochemical cell, a hydrogen gas electrode or standard hydrogen electrode is shown, which consists of a platinum electrode over which hydrogen gas is bubbled in a solution of pH = 0, thereby fulfilling standard state conditions. In the right compartment, an inert platinum electrode is immersed into the solution containing ferrous and ferric irons. The two electrodes are connected to a voltmeter and the electrical circuit is closed by a salt bridge, which helps to maintain the electroneutrality in the compartments. The Nernst equation for the redox reaction shown in Fig. 3.2 is listed as follows.

$$E = E^0 - \frac{RT}{F} \ln\left(\frac{\{Fe^{2+}\}\{H^+\}}{\{Fe^{3+}\}(P_{H_2})^{1/2}} \right) \qquad \text{(Eq. 3.13)}$$

Since both partial pressure of hydrogen gas (P_{H2}) and $\{H^+\}$ are in unity in the standard hydrogen electrode, they are usually omitted from Eq. (3.13) and indicated instead by adding a postscript h to E [Eq. (3.14)].

$$Eh = E^0 - \frac{RT}{F} \ln\left(\frac{\{Fe^{2+}\}}{\{Fe^{3+}\}} \right) \qquad \text{(Eq. 3.14)}$$

Both redox potential (Eh) and E^0 for any half reactions are expressed as the potential with respect to the standard state H^+/H_2 reaction. The value of the standard potential of a half reaction indicates the tendency to release or to accept electrons so that listing of the standard potentials for various half reactions are useful to obtain a first overview

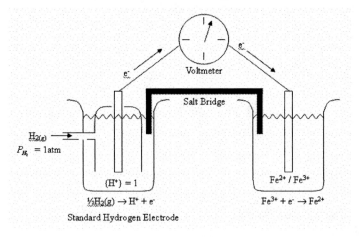

Figure 3.2 Electrochemical cell for the definition of standard potential of a half reaction.

over possible reactions. Table 3.3 lists the standard potentials for some of half reactions (Sillen and Martell, 1964; Stumm and Morgan, 1970; Snoeyink and Jerkins, 1980; Dean, 1985). As seen, the standard potentials for Eq. (3.6) and Eq. (3.7) are -0.77 and +1.33 V, respectively. Therefore, the standard potential for the redox reaction between $Cr_2O_7^{2-}$ and Fe^{2+}, based on Eq. (3.11), is equal to 0.56 V corresponding to ΔG_r^0 of -324 kJ. Positive value of the standard potential or negative value of ΔG_r^0 indicates that Eq. (3.5) proceeds spontaneously to right if all the activities are in unity.

In a similar fashion as pH, which shows the distribution of all acid-base equilibria, Eh determines the distribution of all redox equilibria in aqueous environment. However, in contrast to pH, Eh is difficult to be measured unambiguously in most of natural water. As aforementioned, Eh measurement is made with an inert platinum electrode against the standard hydrogen electrode, which unfortunately is impractical to set up in the field. Therefore, a reference electrode of known potential rather than the standard hydrogen electrode is usually applied instead and the measured potential ($E_{measured}$) is then accordingly corrected with reference to the standard hydrogen electrode using Eq. (3.15). For instance, calomel reference electrode ($KCl_{(sat)}$, $Hg_2Cl_{2(s)}$:$Hg_{(l)}$), which has a potential ($E_{reference}$) of 0.244 V at 25 ^0C, is one of the most common reference electrodes used for Eh measurement.

$$Eh = E_{measured} - E_{reference} \qquad \text{(Eq. 3.15)}$$

3.3.3 Relation of Electron Activity and Redox Potential

Considering the half reaction of the reduction of $Cr_2O_7^{2-}$ to Cr^{3+}, instead of expressing the half reaction in terms of Gibbs free energy, it can also be expressed as Eq. (3.16) based on the law of mass action.

$$K_1 = \frac{\{Cr^{3+}\}^2}{\{Cr_2O_7^{2-}\}\{H^+\}^{14}\{e^-\}^6} \qquad \text{(Eq. 3.16)}$$

Unlike the Nernst equation, the electron activity appears explicitly in the equation. The concept of electron activity is a theoretical treatment of redox reaction, which considerably simplifies the mathematical equations of redox reaction and can manifest the tendency of ions to release or accept electrons. However, it is worth to note that the electron activity should not be interpreted as the concentration of electrons since it can only be exchanged and cannot exist in free states. In analogy to pH, the definition of pε is represented by Eq. (3.17).

$$p\varepsilon = -\log \{e^-\} \qquad \text{(Eq. 3.17)}$$

Rewriting Eq. (3.16) in logarithmic form yields Eq. (3.18).

$$\log K_1 = 2\log \{Cr^{3+}\} - \log \{Cr_2O_7^{2-}\} + 14 \, pH + 6 \, p\varepsilon \qquad \text{(Eq. 3.18)}$$

Table 3.3 Standard potentials for some of half reactions at 25 ^{0}C

Half reaction	E^0 (V)
$H^+ + e^- \Longleftrightarrow \frac{1}{2}H_2(g)$	0
$Na^+ + e^- \Longleftrightarrow Na(s)$	-2.72
$Mg^{2+} + 2e^- \Longleftrightarrow Mg(s)$	-2.37
$Cr_2O_7^{2-} + 14H^+ + 6e^- \Longleftrightarrow 2Cr^{3+} + 7H_2O(l)$	+1.33
$Cr^{3+} + e^- \Longleftrightarrow Cr^{2+}$	-0.41
$MnO_4^- + 2H_2O(l) + 3e^- \Longleftrightarrow MnO_2(s) + 4OH^-$	+0.59
$MnO_4^- + 8H^+ + 5e^- \Longleftrightarrow Mn^{2+} + 4H_2O(l)$	+1.51
$Mn^{4+} + e^- \Longleftrightarrow Mn^{3+}$	+1.65
$MnO_2(s) + 4H^+ + 2e^- \Longleftrightarrow Mn^{2+} + 2H_2O(l)$	+1.23
$Fe^{3+} + e^- \Longleftrightarrow Fe^{2+}$	+0.77
$Fe^{2+} + 2e^- \Longleftrightarrow Fe(s)$	-0.44
$Fe(OH)_3(s) + 3H^+ + e^- \Longleftrightarrow Fe^{2+} + 3H_2O(l)$	+1.06
$3Fe^{3+} + 4H_2O(l) + e^- \Longleftrightarrow Fe_3O_4(s) + 8H^+$	+0.55
$Cu^{2+} + e^- \Longleftrightarrow Cu^+$	+0.16
$Zn^{2+} + 2e^- \Longleftrightarrow Zn(s)$	-0.76
$CO_2(g) + 8H^+ + 8e^- \Longleftrightarrow CH_{4(g)} + 2H_2O(l)$	+0.17
$CO_2(g) + 3H^+ + 4e^- \Longleftrightarrow CH_3COO^-$ (Acetate)	-0.06
$ClO_4^- + 8H^+ + 8e^- \Longleftrightarrow Cl^- + 4H_2O(l)$	+1.39
$NO_3^- + 2H^+ + 2e^- \Longleftrightarrow NO_2^- + H_2O(l)$	+0.84
$NO_3^- + 10H^+ + 8e^- \Longleftrightarrow NH_4^+ + 3H_2O(l)$	+0.88
$2NO_3^- + 12H^+ + 10e^- \Longleftrightarrow N_{2(g)} + 6H_2O(l)$	+1.24
$NO_2^- + 8H^+ + 6e^- \Longleftrightarrow NH_4^+ + 2H_2O(l)$	+0.89
$2NO_2^- + 8H^+ + 6e^- \Longleftrightarrow N_{2(g)} + 4H_2O(l)$	+1.53
$SO_4^{2-} + 3H^+ + 2e^- \Longleftrightarrow HSO_3^- + H_2O(l)$	+0.11
$S(s) + 2e^- \Longleftrightarrow S^{2-}$	-0.44
$SO_4^{2-} + 10H^+ + 8e^- \Longleftrightarrow H_2S(g) + 4H_2O(l)$	+0.34
$SO_4^{2-} + 9H^+ + 8e^- \Longleftrightarrow HS^- + 4H_2O(l)$	+0.24

Similarly, the oxidation half reaction of Fe^{2+} shown in Eq. (3.6) can be written as Eqs. (3.19) and (3.20) using the law of mass action.

$$K_2 = \frac{\{Fe^{3+}\}\{e^-\}}{\{Fe^{2+}\}}$$ (Eq. 3.19)

$$\log K_2 = \log \{Fe^{3+}\} - p\varepsilon - \log \{Fe^{2+}\}$$ (Eq. 3.20)

By applying the pε concept, redox speciation in the redox reaction between $Cr_2O_7^{2-}$ and Fe^{2+} can be simply determined from Eqs. (3.18) and (3.20). This is because the two half reactions in the same solution have equal value of pε if they are in equilibrium.

Both Nernst equation and pε concept are commonly used in the literature for the description of redox equilibria. pε concept is extremely useful in dealing with the redox equilibria in the solution involving both redox and other equilibria such as acid-base and complexation. The reason is that pε concept lets the algebra of redox reaction become similar to other mass action expressions, thereby allowing the same algorithm to be used in the computation procedure. However, the disadvantage of using pε concept is that it is non-measurable quantity. In case the problems such as analytical methods and corrosion involving electrochemical cells are concerned, Nernst equation is more preferable since it can be related to voltage measurements directly. The relationship between Eh and pε is given by Eq. (3.21).

$$Eh = \frac{2.303RT}{F} p\varepsilon \qquad \text{(Eq. 3.21)}$$

3.3.4 Redox in Subsurface

Table 3.4 shows the redox processes commonly occurring in the subsurface environment along with the corresponding reduction and oxidation half reactions, and their standard pε at pH 7 [i.e., $p\varepsilon^0(W)$] (Christensen et al., 2000). Thermodynamically, a reduction half reaction (upper part of Table 3.4) can combine with any oxidation reaction if the $p\varepsilon^0(W)$ of the reduction half reaction is higher than the $p\varepsilon^0(W)$ of the oxidation half reaction (lower part of Table 3.4). In addition, sequences of half reactions can be constructed which range from highly oxidized conditions to highly reduced conditions. Therefore, according to Table 3.4, oxygen gas reduces prior to nitrate followed by the reduction of manganese oxides and then by the reduction of iron oxyhydroxides (Appelo and Postma, 1993).

The full redox reaction created from the combination of the half reactions in accordance with their $p\varepsilon^0(W)$ does not necessarily indicate that the reaction can occur and how the reaction proceeds. It can only show that the full redox reaction is thermodynamically feasible. In fact, most significant redox reactions in aquifers mentioned in Table 3.4 are microbially mediated reactions and involve conversion of organic matter (Christensen et al., 2000). Thus the actual pathway of the full redox reaction may be much more complicated than that proposed by the half reactions. For instance, combination of half reactions (A) and (H) is the typical aerobic respiration reaction, and half reactions (G) and (H) are involved in methane fermentation. Furthermore, combination of (B) and (H) is microbial denitrification; whereas (A) and (M) are microbial nitrification (Stumm and Morgan, 1996).

3.3.5 Eh-pH Diagrams

Eh-pH diagrams are the graphs showing the equilibrium occurrence of ions or minerals as domains relative to *Eh* and pH. The diagrams are also known as Pourbaix diagrams, which were developed by M. J. N. Pourbaix and coworkers in 1950's (Pourbaix et al. 1963). *Eh*-pH diagrams offer a convenient way to describe the major changes expected in speciation in redox-active systems. In such diagrams, areas of predominance in *Eh*-pH or pε-pH coordinate system for various species involving in redox, acid-base, precipitation, and complexation equilibria are established. In this section, construction of the *Eh*-pH diagram for a Fe-H_2O system containing Fe^{2+} and Fe^{3+} solution components, as well as Fe_3O_4 (magnetite), Fe^0 (zero-valent iron) and α-Fe_2O_3 (hematite) solids will be described as an example.

Table 3.4 Redox processes commonly occurring in aquifers, and the corresponding half reactions and standard electron activities [i.e., $p\epsilon^0(W)$] at pH 7

Type of reaction	Half reaction	$p\epsilon^0$ (W)
O_2 reduction (A)	$\frac{1}{4} O_2(g) + H^+ + e^- \Leftrightarrow \frac{1}{2} H_2O(l)$	+13.75
Denitrification (B)	$\frac{1}{5} NO_3^- + \frac{6}{5} H^+ + e^- \Leftrightarrow \frac{1}{10} N_2(g) + \frac{3}{5} H_2O(l)$	+12.62
Mn(IV) reduction (C)	$\frac{1}{2} MnO_2(s) + \frac{1}{2} HCO_3^- (10^{-3}) + \frac{3}{2} H^+ + e^- \Leftrightarrow \frac{1}{2} MnCO_3(s) + H_2O(l)$	+8.9[a]
Fe(III) reduction (D)	$FeOOH(s) + HCO_3^- (10^{-3}) + 2H^+ + e^- \Leftrightarrow FeCO_3(s) + 2H_2O(l)$	-0.8[a]
Organic-C reduction (E)	$\frac{1}{2} CH_2O + H^+ + e^- \Leftrightarrow \frac{1}{2} CH_3OH$	-3.01
SO_4^{2-} reduction (F)	$\frac{1}{8} SO_4^{2-} + \frac{9}{8} H^+ + e^- \Leftrightarrow \frac{1}{8} HS^- + \frac{1}{2} H_2O(l)$	-3.75
CO_2 reduction (G)	$\frac{1}{8} CO_2(g) + H^+ + e^- \Leftrightarrow \frac{1}{8} CH_4(g) + \frac{1}{4} H_2O(l)$	-4.13
Organic-C $\rightarrow$ CO_2 (H)	$\frac{1}{4} CH_2O + \frac{1}{4} H_2O(l) \Leftrightarrow \frac{1}{4} CO_2(g) + H^+ + e^-$	-8.20
Organic-C$\rightarrow$VFA (I)	$\frac{1}{2} CH_2O + \frac{1}{2} H_2O(l) \Leftrightarrow \frac{1}{2} HCOO^- + \frac{3}{2} H^+ + e^-$	-7.68
Sulfide oxidation (J)	$\frac{1}{8} HS^- + \frac{1}{2} H_2O(l) \Leftrightarrow \frac{1}{8} SO_4^{2-} + \frac{9}{8} H^+ + e^-$	-3.75
Fe(II) oxidation (K)	$FeCO_3(s) + 2H_2O(l) \Leftrightarrow FeOOH(s) + HCO_3^- (10^{-3}) + 2H^+ + e^-$	-0.8[a]
CH_4 oxidation (L)	$\frac{1}{2} CH_4(g) + \frac{1}{2} H_2O(l) \Leftrightarrow \frac{1}{2} CH_3OH + H^+ + e^-$	+2.88
Nitrification (M)	$\frac{1}{8} NH_4^+ + \frac{3}{8} H_2O(l) \Leftrightarrow \frac{1}{8} NO_3^- + \frac{5}{4} H^+ + e^-$	+6.16
Mn(II) oxidation (N)	$\frac{1}{2} MnCO_3(s) + H_2O(l) \Leftrightarrow \frac{1}{2} MnO_2(s) + \frac{1}{2} HCO_3^- (10^{-3}) + \frac{3}{2} H^+ + e^-$	+8.9[a]

Note: VFA refers to volatile fatty acid and the source of the data is from Christensen et al. (2000).

[a]These data correspond to bicarbonate ion (HCO_3^-) concentration of 10^{-3} M.

First of all, the stability domain of water in the Fe-H_2O system is considered. For the oxidation of water, the half reaction is

$$O_2(g) + 4H^+ + 4e^- \Longleftrightarrow 2H_2O(l) \qquad (Eq. 3.22)$$

By using the concept of the electron activity, Eq. (3.23) is established for conditions at 25 °C.

$$p\varepsilon = 20.8 + \frac{1}{4} \log P_{O_2} - pH \qquad \text{(Eq. 3.23)}$$

Similarly, Eqs. (3.24) and (3.25) are established for the reduction of water.

$$2H_2O(l) + 2e^- \Longleftrightarrow H_2(g) + 2OH^- \qquad \text{(Eq. 3.24)}$$

$$p\varepsilon = -\frac{1}{2} \log P_{H_2} - pH \qquad \text{(Eq. 3.25)}$$

As illustrated in Figure 3.3, Eqs. (3.23) and (3.25) are plotted as straight lines (1) and (2) respectively. It can be seen that water is converted to oxygen gas if Eh lies above the line (1) and water is reduced to hydrogen gas if Eh is below the line 2. On the other hand, water is stable if Eh lies in the region between the lines (1) and (2). Line (3), which is drawn based on the logarithmic form of the mass-action relation of Eq. (3.26) [i.e., Eq. (3.27)], shows the equilibrium between Fe^{2+} and Fe^{3+} where their concentrations are equal. As seen, Fe^{2+} is the predominant species of dissolved iron at low Eh conditions; whereas Fe^{3+} is the predominant species at high Eh conditions.

$$Fe^{2+} \Longleftrightarrow Fe^{3+} + e^- \qquad \log K_{25} = -13.0 \qquad \text{(Eq. 3.26)}$$

$$p\varepsilon = 13.0 + \log [Fe^{3+}] - \log [Fe^{2+}] \qquad \text{(Eq. 3.27)}$$

Equilibrium between Fe^0 solid and Fe^{2+} [Eq. (3.28)] is represented by line (4) in Figure 3.3 in which it is drawn based on Eq. (3.29) for Fe^{2+} concentration of 10^{-5} M. Below this line, iron only exists as Fe^0 solid in this Fe-H_2O system; while Fe^{2+} is the predominant species above this line.

$$Fe^0 \Longleftrightarrow Fe^{2+} + 2e \qquad \log K_{25} = 13.83 \qquad \text{(Eq. 3.28)}$$

$$p\varepsilon = \frac{1}{2} \log [Fe^{2+}] - 6.915 \qquad \text{(Eq. 3.29)}$$

A boundary showing the equilibrium between Fe^{3+} and α-Fe_2O_3 [Eq. (3.30)] is represented by line (5) drawn from Eq. (3.31) for a Fe^{3+} concentration of 10^{-5} M. The region to the right of this line is the area of predominance of the α-Fe_2O_3 solid. To the left of this line, Fe^{3+} predominates.

$$2Fe^{3+} + 3H_2O(l) \Longleftrightarrow \alpha\text{-}Fe_2O_3(s) + 6H^+ \qquad \log K_{25} = 3.75 \qquad \text{(Eq. 3.30)}$$

$$pH = -0.625 - \frac{1}{3} \log [Fe^{3+}] \qquad \text{(Eq. 3.31)}$$

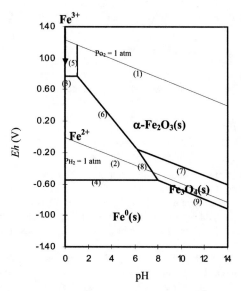

Figure 3.3 Eh-pH diagram of the Fe-H_2O system at 25 ^{0}C containing Fe^{2+} and Fe^{3+} solution components, as well as Fe_3O_4, Fe^0 and α-Fe_2O_3 solids (Assuming total iron concentration = 10^{-5} M)

Line (6) drawn based on Eq. (3.33) indicates the equilibrium between Fe^{2+} and α-Fe_2O_3 [Eq. (3.32)]. As seen, the right side of the line is predominated by α-Fe_2O_3 solid; whereas Fe^{2+} is the predominated species on the left side.

$$2Fe^{2+} + 3H_2O(l) \Longleftrightarrow \alpha\text{-}Fe_2O_3(s) + 6H^+ + 2e^- \quad \log K_{25} = -22.25 \qquad \text{(Eq. 3.32)}$$
$$p\varepsilon = 11.125 - 3pH - \log [Fe^{2+}] \qquad \text{(Eq. 3.33)}$$

Equilibrium between α-Fe_2O_3 and Fe_3O_4 solids [Eq. (3.34)] is represented by line (7) in Figure 3.3 drawn based on Eq. (3.35).

$$2Fe_3O_4(s) + H_2O(l) \Longleftrightarrow 3 \ \alpha\text{-}Fe_2O_3(s) + 2H^+ + 2e^- \quad \log K_{25} = -7.25 \qquad \text{(Eq. 3.34)}$$
$$p\varepsilon = 3.62 - pH \qquad \text{(Eq. 3.35)}$$

Line (8) drawn from Eq. (3.37) indicates the equilibrium between Fe^{2+} and Fe_3O_4 [Eq. (3.36)]. In addition, a boundary showing the equilibrium between Fe^0 and Fe_3O_4 [Eq. (3.38)] is represented by line (9) drawn from Eq. (3.39)

$$3Fe^{2+} + 4H_2O(l) \Longleftrightarrow Fe_3O_4(s) + 8H^+ + 2e^- \quad \log K_{25} = -29.76 \qquad \text{(Eq. 3.36)}$$
$$p\varepsilon = 14.88 - 4pH - \frac{3}{2} \log [Fe^{2+}] \qquad \text{(Eq. 3.37)}$$

$$3Fe^0 + 4H_2O(l) \iff Fe_3O_4(s) + 8H^+ + 8e^- \qquad \log K_{25} = 11.72 \qquad \text{(Eq. 3.38)}$$
$$p\epsilon = -1.47 - pH \qquad \text{(Eq. 3.39)}$$

From the constructed Eh-pH diagram for the Fe-H$_2$O system, it is readily to observe that Fe^0 only exists at very low Eh conditions. Moreover, at low pH conditions, iron species preferably exist as Fe^{2+} and Fe^{3+} rather than Fe$_3$O$_4$ and α-Fe$_2$O$_3$. For constructing the Eh-pH diagram for unknown or more complex systems, identification of the solids and solution components in the systems are important. This can be achieved by searching the literature for relevant reactions and equilibrium data.

3.4 Removal of Contaminants by Reduction and Precipitation Processes

3.4.1 Chromium

In the United States, chromium is one of the inorganic pollutants frequently detected in groundwater at hazardous sites (National Research Council, 1994). Its ubiquity in contaminated sites is due to its wide application in industries, such as steel production, leather tanning, electroplating, pigment and chemical manufacturing as well as in corrosion prevention (Buerge and Hug, 1997; Sedlak and Chan, 1997). Generally, chromium exists in trivalent [Cr(III)] and hexavalent [Cr(VI)] forms in natural environment in which the former chromium species dominate under anoxic or suboxic conditions; while the latter dominate under oxic condition. Cr(III) is an essential human nutrient for glucidic metabolism. It has relatively low mobility in aqueous environment since its oxides and hydroxides possess very low K_{sp} (Loyaux-Lawniczak et al., 2000). As shown in Fig. 3.4, at pH less than 3.6, chromium ion (Cr^{3+}) is the predominant Cr(III) species. At pH above 3.6, chromium hydroxyl species, such as Cr(OH)$^{2+}$, Cr(OH)$_2^+$, Cr(OH)$_3$ and Cr(OH)$_4^-$, are the predominant Cr(III) species (Richard and Bourg, 1991; Palmer and Wittbrodt, 1991; USEPA, 2000). Moreover, it can be readily observed that the area of predominance of Cr(III) species mainly exists in low Eh conditions in comparison with that of Cr(VI) species.

In marked contrast to Cr(III), Cr(VI) is highly toxic and a potential carcinogen (Fendorf and Li, 1996). Cr(VI) oxyanions including chromate (HCrO$_4^-$), bichromate ions (CrO$_4^{2-}$) and Cr$_2$O$_7^{2-}$ are more soluble than Cr(III) hydroxyl species and thereby spread easily in aquifers (Nriagu and Nieboer, 1988). Because of the marked contrast of the toxicity and mobility between Cr(VI) and Cr(III), reduction of Cr(VI) to Cr(III) followed by Cr(III) precipitation is the major principles applied for the remediation of Cr(VI)-contaminated groundwater (Powell et al., 1995; Blowes et al., 1997; Lo et al., 2003a). This remedial principle is also applied for remediating various inorganic contaminants, such as uranium, selenium and technetium (Powell et al., 1998).

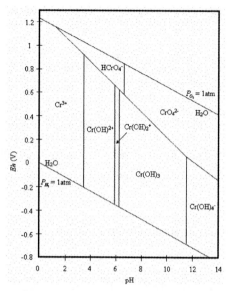

Figure 3.4 *Eh*-pH diagram for chromium (Adapted from US EPA, 2000)

Thermodynamically, reduction of Cr(VI) to Cr(III) requires the presence of electron donors or reductants. In natural aquifers, Fe^{2+} in minerals and solution, reduced sulfur compounds and soil organic carbon are reductants commonly found. Ferrous iron-bearing minerals include carbonate- $[Fe_4^{2+}Fe_2^{3+}(OH)_{12}]^{2+}[CO_3 \cdot 2H_2O]^{2-}$, sulfate- $[Fe_4^{2+}Fe_2^{3+}(OH)_{12}]^{2+}[SO_4 \cdot nH_2O]^{2-}$ and chloride-containing green rusts $[Fe_3^{2+}Fe^{3+}(OH)_8]^+[Cl]^-$, and magnetite (Fe_3O_4) (Loyaux-Lawniczak et al., 2000; Peterson et al., 1997; Williams and Scherer, 2001). As an example of the Cr(VI) reduction by the green rust precipitates, Eq. (3.40) shows the possible chemical equation involving in the redox reaction between CrO_4^{2-} and sulfate-containing green rust (Loyaux-Lawniczak et al., 2000). As seen, CrO_4^{2-} is reduced to Cr(III) substituted ferrihydrite $(Fe_{45/11}Cr_{10/11}HO_8)$ with concomitant production of hydroxide ion (OH^-), which subsequently provides thermodynamically favorable condition for further precipitation of Cr(III) (Palmer and Wittbrodt, 1991).

$$15Fe_4^{2+}Fe_2^{3+}(OH)_{12}SO_4(s)+20CrO_4^{2-} \Longleftrightarrow 22Fe_{45/11}Cr_{10/11}HO_8(s)+15SO_4^{2-}+10OH^-+74H_2O(l)$$
$$\text{(Eq. 3.40)}$$

Eq. (3.41) describes the redox reaction between CrO_4^{2-} and Fe_3O_4 in which Fe_3O_4 is oxidized to Fe_2O_3 while CrO_4^{2-} is reduced to Cr^{3+}. After the reduction, the Cr(III) subsequently precipitates as Cr(OH)$_3$ via Eq. (3.42).

$$6Fe_3O_4(s) + 2CrO_4^{2-} + 5H^+ \Longleftrightarrow 9Fe_2O_3(s) + 2Cr^{3+} + 5OH^- \qquad \text{(Eq. 3.41)}$$

$$2Cr^{3+} + 5OH^- <==> Cr(OH)_3(s) + Cr(OH)_2^+ \qquad \text{(Eq. 3.42)}$$

In addition to ferrous iron-bearing minerals, the efficacy of aqueous Fe^{2+} in reducing Cr(VI) has been well recognized (Fendorf and Li, 1996). In treating industrially generated chromium wastes, aqueous Fe^{2+} is one of the main reductants currently being used (Eary and Rai, 1988). The general chemical equation for the redox reaction between Cr(VI) and aqueous Fe^{2+} is described in Eq. (3.43)

$$H_xCrO_4^{x-2} + 3Fe^{2+} + (y+3z)H_2O(l) <==> Cr(OH)_y^{3-y} + 3Fe(OH)_z^{3-z} + (x+y+3z)H^+$$
$$\text{(Eq. 3.43)}$$

Reduced sulfur compounds including sulfide (S^{-2}) and sulfite (SO_3^{2-}) are also effective reductants with respect to Cr(VI) (Palmer and Wittbrodt, 1991). In case SO_3^{2-} is in excess, the reduction of Cr(VI) follows Eq. (3.44); whereas the redox reaction follows Eq. (3.45) in the presence of excess Cr(VI). Although S^{2-} can thermodynamically reduce Cr(VI) (see Table 3.3), studies indicate that Fe^{2+} must be present in order to obtain both Cr(VI) reduction and precipitation [Eq. (3.46)] (Palmer and Wittbrodt, 1991; Simon et al., 2002).

$$2HCrO_4^- + 6H^+ + 4HSO_3^- <==> 2Cr^{3+} + 2SO_4^{2-} + S_2O_6^{2-} + 6H_2O(l) \qquad \text{(Eq. 3.44)}$$

$$2HCrO_4^- + 5H^+ + 3HSO_3^- <==> 2Cr^{3+} + 3SO_4^{2-} + 5H_2O(l) \qquad \text{(Eq. 3.45)}$$

$$Cr_2O_7^{2-} + 2FeS(s) + 7H_2O(l) <==> 2Cr(OH)_3(s) + 2Fe(OH)_3(s) + 2S(s) + 2OH^- \qquad \text{(Eq. 3.46)}$$

The other important soil constituents, which can contribute to Cr(VI) reduction, is soil organic carbon. In fact, the amount of $Cr_2O_7^{2-}$ being reduced by soil is the recognized method widely used for measuring the content of soil organic carbon (Nelson and Sommers, 1982). Eq. (3.47) depicts the idealized chemical equation involved.

$$2Cr_2O_7^{2-} + 3C + 16H^+ <==> 4Cr^{3+} + 3CO_2(g) + 8H_2O(l) \qquad \text{(Eq. 3.47)}$$

In general, soil organic carbon mainly exists in the form of humic substances, including humic acid, fulvic acid and humin. It is well known that the functional groups, such as carboxylic acids, alcohols and phenols, in these humic substances act as the main reductants for the Cr(VI) reduction (Wiberg, 1965; Hayes, 1985). Bloomfield and Pruden (1980) found that water soluble soil organic matter is effective in reducing Cr(VI) at pH less than 4 but not effective at pH greater than 5. Furthermore, Stollenwerk and Grove (1985) reported a significant reduction in Cr(VI) concentration in groundwater samples by the spiked fulvic acid particularly in acidified samples.

In addition to the natural reductants, zero-valent metals such as zero-valent iron (Fe^0) are another thermodynamically favorable reductants with respect to Cr(VI).

It is effective under both anoxic and oxic conditions (Cantrell et al., 1995), and usually more reactive than most natural reductants in aquifers (Rai and Zachara, 1988; Henderson, 1994; Cantrell et al., 1995; Fendorf and Li, 1996; Sedlak and Chan, 1997). Theoretically, Fe^0 donates electrons to reduce Cr(VI) to Cr(III) and simultaneously oxidizes to ferric iron (Fe^{3+}) as shown in Eq. (3.48). The reduced Cr(III) is then removed from aqueous solution by precipitation of chromium hydroxide [Eq. (3.42)], or co-precipitation of mixed chromium-iron hydroxide solids as shown in Eq. (3.49) (Powell et al., 1995; Blowes et al., 1997) or mixed chromium-iron oxyhydroxide solids as depicted in Eq. (3.50) (Eary and Rai, 1988; Schwertmann et al., 1989). The reduced Cr(III) usually forms precipitates on the surface of Fe^0 or reductant solids rather than in bulk solution since heterogeneous nucleation usually has lower activation energy barrier than homogeneous nucleation (Appelo and Postma, 1993).

$$CrO_4^{2-} + Fe^0(s) + 4H_2O(l) \Longleftrightarrow Cr^{3+} + Fe^{3+} + 8OH^- \qquad \text{(Eq. 3.48)}$$

$$(1-x)Fe^{3+} + xCr^{3+} + 3H_2O(l) \Longleftrightarrow Cr_xFe_{1-x}(OH)_3(s) + 3H^+ \qquad \text{(Eq. 3.49)}$$

$$(1-x)Fe^{3+} + xCr^{3+} + 2H_2O(l) \Longleftrightarrow Cr_xFe_{1-x}OOH(s) + 3H^+ \qquad \text{(Eq. 3.50)}$$

The decrease in the Cr(VI) concentration by the reduction and precipitation processes generally can be described by a pseudo first-order kinetic model as shown in Eq. (3.51) (Cantrell et al., 1995; Alowitz and Scherer, 2002). Since the rate constant of the Cr(VI) reduction was found to be directly proportional to the ratio of the surface area of reductant to the solution volume or the surface area concentration (λ_s), Eq. (3.52) is also used for describing the rate of the Cr(VI) reduction (Williams and Scherer, 2001).

$$\frac{d[C]}{dt} = -k_{obs}[C] \qquad \text{(Eq. 3.51)}$$

$$\frac{d[C]}{dt} = -k_{sa}\lambda_s[C] \qquad \text{(Eq. 3.52)}$$

Table 3.5 summarizes the observed pseudo first-order rate constants (k_{obs}) or half-lives ($t_{1/2}$) of the Cr(VI) reductions by various types of the reductants. As seen, Fe^0 and carbonate-containing green rust have the highest reactivity in reducing Cr(VI) with the $t_{1/2}$ ranging from 2.4 to 8.7 minutes. Lo et al. (2006) also found that each gram of Fe^0 could remove approximately 2.3 to 4.2 mg of Cr(VI) from groundwater at 23 °C. Fe^{2+} in aqueous solution and soil has comparatively low reactivity. The aquifer materials in Odessa, Texas containing Fe^{2+} and soil organic carbon possess the lowest reactivity in which the corresponding $t_{1/2}$ is about 2.5 yrs (Henderson, 1994). In comparison to SO_3^{2-} ($E^0 = -0.04$ V), hydrogen sulfite ($E^0 = +0.11$ V), fulvic acid ($E^0 = +0.5$ V), Fe_3O_4 ($E^0 = +0.55$ V), humic acid ($E^0 = +0.7$ V) and Fe^{2+} ($E^0 = +0.77$ V), Fe^0 has the lowest standard potential for the reduction half reaction ($E^0 = -0.44$ V). From a thermodynamical point of view, Fe^0 has the highest reducing power for reducing Cr(VI) (Palmer and Wittbrodt, 1991; Snoeyink and Jenkins, 1980). Therefore, in the latter part of this chapter, the mechanisms of the precipitation and/or redox reactions among contaminants and Fe^0 will primarily be discussed.

Table 3.5 Observed pseudo first-order rate constants (k_{obs}) and half-lives ($t_{1/2}$) for the Cr(VI) reduction

Reactive material	Pseudo first-order rate constant (k_{obs})	Half-life ($t_{1/2}$)	Experiment	Literature cited	Remarks
Trinity Sandy Aquifer in Odessa, TX	3.2×10^{-5} 1/hr	2.5 yrs	Field experiment	Henderson (1994)	Nearly neutral in pH in the groundwater
Ferrous iron in soil	7×10^{-4} 1/hr 1.1×10^{-3} 1/hr	990 hrs 630 hrs	Batch experiment	Rai and Zachara (1988)	Laboratory experiments
Aqueous ferrous iron	2.0×10^{-6} 1/s [a]	96.3 hrs	Batch experiment	Sedlak and Chan (1997)	Initial Fe^{2+} conc of 0.9 μM (50.2 μg/L) and at pH 7.2
	2.1×10^{-4} 1/min [b] 2.4×10^{-4} 1/min [b]	55.2 hrs 48.4 hrs	Initial rate and stopped-flow methods	Fendorf and Li (1996)	Initial Fe^{2+} conc of 100 μM (5580 μg/L) at pH 6.67
Carbonate containing green rust	2.6×10^{-3} 1/s 4.0×10^{-3} 1/s	4.4 min 2.9 min	Batch experiment	Williams and Scherer (2001)	In the presence of 0.5 g/L of carbonate containing green rust
Zero-valent iron (Fe^0)	6.5 1/hr [c] (0.13 1/hr)*	6.4 min (5.2 hrs)*	Batch experiment	Cantrell et al. (1995)	Initial Cr(VI) conc of 500 μg/L and at pH 8.4
	7.7 1/hr [c] (0.16 1/hr)*	5.4 min (4.4 hrs)*			Initial Cr(VI) conc of 10,000 μg/L and at pH 8.4
	4.8 to 17.2 1/hr [d] (0.25 to 0.9 1/hr)*	2.4 to 8.7 min (0.77 to 2.7 hrs)	Batch experiment	Alowitz and Scherer (2002)	Initial Cr(VI) conc of 10000μg/L and at pH 7.0
	2.3 to 4.2 mg Cr(VI)/g Fe^0 [e]		Column experiment	Lo et al. (2006)	Initial Cr(VI) conc of 25,000 μg/L and at pH 7.1

*The figures listed in parentheses are the k_{obs} and $t_{1/2}$ at 1 m^2/L of the surface area concentration (λ_s).

[a] $\frac{d[Cr(VI)]}{dt} = -k[Fe^{2+}][Cr(VI)] = -k_{obs}[Cr(VI)]$; [b] $\frac{d[Cr(VI)]}{dt} = -k'[Fe^{2+}]^{0.6}[Cr(VI)] = -k_{obs}[Cr(VI)]$

[c] 2.43 m^2/g of Fe^0 specific surface area and 48.6 m^2/L of the surface area concentration (λ_s)

[d] 2.30 m^2/g of Fe^0 specific surface area and 19 m^2/L of the surface area concentration (λ_s)

[e] Cr(VI) removal capacity (mg of Cr(VI) removed per gram of Fe^0). 1.8 m^2/g of Fe^0 specific surface area and 7716 m^2/L of the surface area concentration (λ_s)

3.4.2 Uranium

Uranium is customarily detected in the groundwater within mine waste piles and leachate derived from mine wastes (Dubrovsky et al., 1984; Olsen et al., 1986; Morin et al., 1988). It is the heaviest naturally occurring element and all uranium isotopes (i.e., 238uranium and 235uranium) are radioactive in nature with a decay t1/2 between 108 and 109 yrs (Simon et al., 2002). Uranium is dangerous to human not just because of its radioactivity but also due to its toxicity as a heavy metal. The maximum contaminant level (MCL) of uranium in water is 300 μg/L based on the radiation limit, but down to 20 μg/L with regard to its toxicity (Simon et al., 2002). Naturally, uranium exists in the oxidation states +4 [U(IV)] and +6 [U(VI)]. U(VI), such as uranyl ion ($UO_2{2+}$), is more mobile than U(IV) and the solubility of U(IV) oxides, such as uraninite (UO_2), is of the order of 10-3 mg/L in a pH between 4 and 14. The substantial drop in the uranium solubility after reducing from U(VI) to U(IV) indicates that reduction and precipitation is one of the possible approaches for the remediation of U(VI)-contaminated water and groundwater. Actually, nearly 100% removal of U(VI) at an initial concentration up to 18000 mg/L through precipitation and reduction by Fe^0 has been reported.

As seen from Table 3.6, Gu et al. (1998) reported that half of 1000 mg/L of U(VI) concentration can be removed in a glass vial containing Fe^0 and U(VI) spiked solution in 5.7 min or less. Despite the fact that adsorption of UO_2^{2+} onto the corrosion products of Fe^0 such as iron oxides and oxyhydroxides is known to take place particularly at a high pH (Langmuir, 1978; Hsi and Langmuir, 1985; Ho and Miller, 1986), it was found that less than 4% of UO_2^{2+} was adsorbed by the Fe^0 corrosion products, and the reduction and precipitation with Fe^0 is the major pathway for the removal of UO_2^{2+} (Gu et al. 1998). As illustrated in Eq. (3.53), by using the electrons released from Fe^0 oxidation, UO_2^{2+} is first reduced to UO_2, which subsequently precipitates on the Fe^0 surface (Powell et al., 1998; Ott, 2000). Since the standard potential of U(VI)/U(IV) is always higher than that of Fe^{2+}/Fe^0, the reduction of U(VI) by Fe^0 occurs spontaneously (Simon et al., 2002). However, it is important to note that the reduced U(IV) species on the Fe^0 surface may be reoxidized and remobilized when the reduced system becomes more oxidized.

$$Fe^0(s) + UO_2^{2+} \Longleftrightarrow Fe^{2+} + UO_2(s) \qquad \text{(Eq. 3.53)}$$

3.4.3 Nitrate and Nitrite

In the United States, 18% of private wells were found containing NO_3^- level above the drinking water standard of 10 mg/L nitrate-nitrogen (NO_3^--N). Another 37% of the wells have the levels greater than 3 mg/L NO_3^--N (Till et al., 1998). Although excess amount of NO_3^- can cause eutrophication in water bodies, NO_3^- itself is relatively non-toxic to human, but nitrite (NO_2^-) derived by microbially mediated reduction of NO_3^- can cause human health problems such as methemoglobinemia, liver damage and even cancers (Cabel et al., 1982; Choe et al., 2004). Anthropogenic sources such as nitrogen fertilizers, nitrogen pesticides, animal wastes and septic systems account for most NO_3^- contamination of groundwater (Choe et al., 2000). Storm and irrigation runoffs typically from farmlands always bring high concentration of NO_3^- to aquifers and cause pollution (Huang et al., 1998). Currently, ion exchange, reverse osmosis, biological denitrification and chemical reduction are the technologies commonly applied for the removal of NO_3^- in aqueous environments. However, ion exchange and reverse osmosis require frequent regeneration of the media and generate secondary brine wastes. Biological denitrification also requires intensive maintenance and a constant supply of organic substrate. In comparison to chemical reduction approaches, microbial processes are relatively slow and incomplete (Choe et al., 2000).

Table 3.6 Pseudo first-order rate constants (k_{obs}) and the half-lives ($t_{1/2}$) of U(VI) reduction using Fe^0 as a reductant

Fe^0 type	Fe^0 specific surface area (m^2/g)	Observed pseudo first-order rate constant, k_{obs} (1/min)	Half-life, $t_{1/2}$, (min)
Master-builder $Fe^{0 a}$	0.98	0.39 (0.002)	1.79 (348)
Peerless Fe^0, medium [a]	0.10	0.29 (0.015)	2.41 (47.8)
Peerless Fe^0, coarse [a]	0.08	0.12 (0.008)	5.68 (92.4)
Cercona cast $Fe^{0 a}$	0.02	0.19 (0.048)	3.61 (14.6)
Cercona Fe^0-palladized [a]	na	0.31	2.22
Fe^0 in natural GW [b]	na	(0.023)	(30.1)
Fe^0 in deionized water [b]	na	(0.035)	(19.8)
Fe^0 in 0.4 M NaCl soln. [b]	na	(0.079)	(8.77)
Fe^0 in 0.4 M $NaNO_3$ soln[b].	na	(0.0072)	(96.3)

Note: na refers to not available. The figures shown in parentheses are the k_{obs} and $t_{1/2}$ at 1 m^2/L of λ_s.
[a]The initial uranyl ion (UO_2^{2+}) concentration is 1000 mg/L. All the results are obtained from batch experiments containing 2 g of Fe^0 and 10 mL of solution. The surface area concentration (λ_s) for Master-builder Fe^0, medium size Peerless Fe^0, coarse size Peerless Fe^0 and Cercona cast Fe^0 are 196, 20, 16 and 4 m^2/L, respectively. The source of the data is from Gu et al. (1998).
[b]Iron coupons punched from mild steel plate measuring 1.43 cm in diameter and 0.15 cm in thickness were used as Fe^0. The pH of the natural groundwater is 6.84, and its calcium, magnesium and carbonate concentrations are 52, 5.9 and 96 mg/L, respectively. The source of the data is from Farrell et al. (1999).

Chemical reduction processes using Fe^0 as a reductant can reduce NO_3^- to NO_2^- [Eq. (3.54)], ammonium ion [Eq. (3.55)] or nitrogen gas Eq. [(3.56)], and simultaneously NO_2^- can also be reduced to ammonium ion Eq. [(3.57)] or nitrogen gas Eq. [(3.58)] depending on the reaction conditions. Similar to chromium and uranium, the NO_3^- and NO_2^- reduction by Fe^0 can also be described by a pseudo first-order reaction with respect to NO_3^- or NO_2^- concentration (Choe et al., 2000; Cheng et al., 1997; Alowitz and Scherer, 2002). As shown in Table 3.3, the standard potentials for the NO_3^- and NO_2^- reduction half reactions (i.e., between +0.84 and +1.53 V) are much higher than that of Fe^{2+}/Fe^0 (i.e., -0.44 V) (Dean, 1985). Therefore, Eqs. (3.54) to (3.58) are thermodynamically favorable and the reduction of NO_3^- and NO_2^- by Fe^0 can occur spontaneously. Finally, ammonium ion, which is one of the end products produced from the redox reactions [Eqs. (3.55) and (3.57)], can be removed from aqueous environments by either raising the solution pH to release ammonia gas (NH_3) from the remediated solution (Cheng et al., 1997) or using absorbents such as zeolite (Lee et al. 2007).

$$Fe^0(s) + NO_3^- + 2H^+ \Longleftrightarrow Fe^{2+} + NO_2^- + H_2O(l) \qquad \text{(Eq. 3.54)}$$

$$4Fe^0(s) + NO_3^- + 10H^+ \Longleftrightarrow 4Fe^{2+} + NH_4^+ + 3H_2O(l) \qquad \text{(Eq. 3.55)}$$

$$5Fe^0(s) + 2NO_3^- + 12H^+ \Longleftrightarrow 5Fe^{2+} + N_2(g) + 6H_2O(l) \qquad \text{(Eq. 3.56)}$$

$$3Fe^0(s) + NO_2^- + 8H^+ \Longleftrightarrow 3Fe^{2+} + NH_4^+ + 2H_2O(l) \qquad \text{(Eq. 3.57)}$$

$$3Fe^0(s) + 2NO_2^- + 8H^+ \Longleftrightarrow 3Fe^{2+} + N_2(g) + 4H_2O(l) \qquad \text{(Eq. 3.58)}$$

Previous studies of the temporal change of the concentration of NO_3^-, NH_3 and Fe^{2+} in a batch containing 0.3 g of Fe^0 and 15 mL of solution spiked with 500 mg/L of NO_3^- shows simultaneous decrease in NO_3^- concentration, and increase in the concentration of NH_3 and Fe^{2+} (Huang et al. 1998). This observation conclusively indicates the reduction of NO_3^- to NH_3 and oxidation of Fe^0 to Fe^{2+}. The similarity of the maximum molarity between NO_3^- and NH_3 shows that NH_3 is the main end product of the redox reaction under those experimental conditions. Since reduction of 1 mole of NO_3^- to NH_3 stochiometrically requires 4 mole of Fe^0 [see Eq. (3.55)], the molarity of Fe^{2+} at the end of the experiment was about 4 times higher than the molarity of NH_3.

As evident, all the redox reactions between Fe^0, and NO_3^- and NO_2^- involve the consumption of H^+. Therefore, pH is believed to be one of the most important factors affecting the rate of the NO_3^- and NO_2^- reductions (Choe et al., 2004). Table 3.7 summarizes some of the k_{obs} of the NO_3^- and NO_2^- reductions by Fe^0 under various solution pH (Alowitz and Scherer, 2002; Cheng et al., 1997). It is readily observed that the rate of the NO_3^- and NO_2^- reductions decreases with increasing solution pH. Cheng et al. (1997) also reported 73% decrease in the k_{obs} of the NO_3^- reduction with increasing the solution pH from 5.0 to 7.0. Therefore, it is expected that acidic environment or addition of H^+ is the prerequisite for the effective reduction of NO_3^- and NO_2^- by Fe^0. Choe et al. (2004) found that in an unbuffered solution with initial pH setting with acids, the rise of pH caused by the NO_3^- reduction by Fe^0 and Fe^0 oxidation could be stabilized. This is because the pH was buffered at pH 6.5 by the consumption of OH^- through formation of green rusts. Therefore, the NO_3^- reduction by Fe^0 could be carried out continuously and completely in an unbuffered solution.

Microbially mediated reduction of NO_3^- or biological denitrification is another approach for removing NO_3^- from water and groundwater. Chemoautotrophs using the

H_2 gas produced from the anaerobic oxidation of Fe^0 with water [Eq. (3.59)] can denitrify NO_3^- to nitrogen gas [Eq. (3.60)] (Till et al., 1998). Microbially catalyzed NO_3^--dependent Fe^{2+} oxidation [Eq. (3.61)] involves oxidation of solid phase Fe^{2+} compounds such as α-FeOOH and $FeCO_3$ into $Fe(OH)_3$ with concomitant denitrification of NO_3^- (Weber et al., 2001). Recently, public concerns focus on the enhancement from anaerobic ammonium-oxidizing bacteria (AMMONOX) to the abiotic NO3- reduction by Fe0. As aforesaid, NH_3 is a primary product of the abiotic reductions of NO3- and NO2- by Fe0 [Eqs. (3.55) and (3.57)], which has an adverse

Table 3.7 Pseudo first-order rate constants (k_{obs}) and the corresponding half-lives ($t_{1/2}$) of nitrate (NO_3^-) and nitrite (NO_2^-) reductions using Fe^0 as a reductant

Fe^0 type	pH	Observed pseudo first-order Rate constant, k_{obs} (1/hr)	Half-life, $t_{1/2}$, (hr)
Nitrate			
	5.5	0.95±0.027 (0.0099±0.0003)	0.73±0.02 (70.1±2.0)
	6.0	0.58±0.010 (0.0060±0.0001)	1.20±0.02 (114.7±2.0)
	6.5	0.20±0.086 (0.0021±0.0009)	4.25±1.83 (408.1±175.5)
	7.0	0.13±0.024 (0.0014±0.0003)	5.52±1.02 (529.8±97.8)
Fisher $Fe^{0\,a}$	7.5	0.15±0.057 (0.0016±0.0006)	5.40±2.05 (518.4±197.0)
	8.0	0.13±0.042 (0.0014±0.0004)	5.95±1.92 (571.4±184.6)
	8.5	0.025±0.0008 (0.0003±0.00001)	27.75±0.89 (2663.8±85.2)
	9.0	0.008±0.0001 (0.0001±0.000001)	90.03±1.52 (8642.5±145.9)
	2.75	0.0571±0.0027 (0.0026±0.0001)	12.2±0.6 (267.6±12.7)
	2.89	0.0913±0.0028 (0.0042±0.0001)	7.6±0.23 (167.1±5.1)
Aldrich $Fe^{0\,b}$	3.19	0.0256±0.0033 (0.0012±0.0002)	27.3±3.5 (605.6±78.1)
	3.36	0.0592±0.0117 (0.0027±0.0005)	12.0±2.4 (268.0±53.0)
	6.16	0.00904±0.00148 (0.0004±0.0001)	77.7±6.3 (1733.0±283.7)
	8.00	0.00886±0.00082 (0.0004±0.00004)	78.5±7.2 (1735.6±160.6)
Nitrite			
	5.5	2.6±0.22 (0.137±0.012)	0.27±0.02 (5.1±0.4)
	6.0	2.0±0.37 (0.105±0.019)	0.36±0.07 (6.8±1.3)
	6.5	1.0±0.07 (0.053±0.004)	0.70±0.05 (13.2±0.9)
Fisher $Fe^{0\,c}$	7.0	0.61±0.12 (0.032±0.006)	1.2±0.23 (22.5±4.4)
	7.5	0.23±0.023 (0.012±0.001)	3.0±0.30 (57.8±5.8)
	8.0	0.10±0.020 (0.005±0.001)	7.2±1.44 (137.2±27.4)
	8.5	0.035±0.003 (0.002±0.0002)	20.0±1.77 (379.2±33.6)

Note: All the studies are conducted in batch experiments. The figures shown in parentheses are the k_{obs} and $t_{1/2}$ at 1 m^2/L of the surface area concentration (λ_s).
[a]Initial nitrate concentration 50 mg/L, 2.30±0.18 m^2/g of the Fe^0 specific surface area and 96 m^2/L of the λ_s (Alowitz and Scherer, 2002).
[b]Initial nitrate concentration 20 mg/L NO_3^--N, 0.55±0.05 m^2/g of the Fe^0 specific surface area and 22 m^2/L of the λ_s (Su and Puls, 2004).
[c]Initial nitrite concentration 40 mg/L, 2.30±0.18 m^2/g of the Fe^0 specific surface area and 19 m^2/L of the λ_s (Alowitz and Scherer, 2002).

aesthetic impact on drinking water (Jafvert and Valentine, 1992). According to Eq. (3.62), AMMONOX can oxidize ammonium ion to nitrogen gas by utilizing NO3- as an electron acceptor. Therefore, the growth of AMMONOX on the Fe0 surface does not only help reduce the NO3- concentration; it can also help remove ammonium, thereby allowing elimination of the subsequent process for the treatment of NH3 in remediated solution (Westerhoff and James, 2003).

$$Fe0(s) + 2H_2O(l) \Longleftrightarrow H_2(g) + Fe^{2+} + 2OH^- \qquad \text{(Eq. 3.59)}$$

$$2NO_3^- + 5H_2(g) \Longleftrightarrow N_2(g) + 4H_2O(l) + 2OH^- \qquad \text{(Eq. 3.60)}$$

$$10Fe^{2+} + 2NO_3^- + 24H_2O(l) \Longleftrightarrow 10Fe(OH)_3(s) + N_2(g) + 18H^+ \qquad \text{(Eq. 3.61)}$$

$$5NH_4^+ + 3NO_3^- \Longleftrightarrow 4N_2(g) + 9H_2O(l) + 2H^+ \qquad \text{(Eq. 3.62)}$$

3.4.4 Perchlorate

Perchlorate salt (ClO_4^-) is an unusually stable ion in aqueous solution. It is mainly used as oxidizers in solid rocket fuels, in the manufacture of automotive airbag inflators, as mordants for fabrics and dyes, as additives in lubricant oils, in tanning and finishing leather, and in chemical fertilizers, as well as in the production of paints and enamels (Moore et al., 2003). Upon the recent development of a sensitive ion chromatography procedure lowering the ClO_4^- minimum detection limit from 100 ppb to 4 ppb, ClO_4^- contamination was widely detected in groundwater which may threaten the drinking water supplies of at least 1.2 million people in the United States (Miller and Logan, 2000). Atmospheric sources of ClO_4^- have also been identified leading to its deposition in areas far from anthropogenic sources. Further concern has been developed from the detection of ClO_4^- in milk and other dairy products (Kirk et al., 2005). Since ClO_4^- can affect the thyroid hormone production of human and possibly cause mental retardation in fetuses and infants, it has been added to the United States Environmental Protection Agency's drinking water contaminant candidate list and an action level of 4 ppb was set by California Department of Health Services in 2002 (Zhang et al., 2002). On account of the ubiquity of ClO_4^- in groundwater and its health impact on human, a cost-effective approach for remediation of groundwater containing only the part per billion levels of ClO_4^- is urgently required.

Remediation of ClO_4^- in groundwater is a difficult undertaking because it is highly soluble, mobile and recalcitrant in natural environments. Although ClO_4^- is thermodynamically susceptible to be reduced ($E^0 = +1.39V$), its large activation energy barrier makes the reduction not kinetically labile. Currently, anion exchange, membrane filtration, electrochemical reduction, chemical reduction and microbially mediated reduction are the methods available to treat ClO_4^- in the ppb range in contaminated groundwater. However, anion exchange and membrane filtration require costly resin regeneration and membrane replacement, respectively, and electrochemical reduction is only applicable to laboratory-scale issues (Urbansky and Schock 1999).

Nanoscale Fe^0 was found to be able to chemically reduce ClO_4^- to chlorate (ClO_3^-), chlorite (ClO_2^-), hypochlorite (ClO^-) ions and finally to chloride ions as illustrated in Eqs. (3.63) to (3.66). Cao et al. (2005) reported that in a 150mL batch reactor containing 100mL of ClO_4^--contaminated water, each gram of nanoscale Fe^0 could hourly reduce 0.013, 0.10, 0.64 and 1.52 mg of ClO_4^- at 25, 40, 60 and 75 °C, respectively. However, in comparison to chromium and nitrate, the ClO_4^- reduction rate by nanoscale Fe^0 is still low. Furthermore, initially formed nanoscale Fe^0 tends to agglomerate to form large flocs easily, which leads to markedly decrease in the specific surface areas of the nanoparticles and thereby may result in a rapid loss of its efficiency on the ClO_4^- reduction (He and Zhao, 2005).

$$ClO_4^- + Fe^0(s) + 2H^+ <==> ClO_3^- + Fe^{2+} + H_2O(l) \qquad \text{(Eq. 3.63)}$$

$$ClO_3^- + Fe^0(s) + 2H^+ <==> ClO_2^- + Fe^{2+} + H_2O(l) \qquad \text{(Eq. 3.64)}$$

$$ClO_2^- + Fe^0(s) + 2H^+ <==> ClO^- + Fe^{2+} + H_2O(l) \qquad \text{(Eq. 3.65)}$$

$$ClO^- + Fe^0(s) + 2H^+ <==> Cl^- + Fe^{2+} + H_2O(l) \qquad \text{(Eq. 3.66)}$$

In analogy to nitrate, autotrophic bacteria can sequentially reduce ClO_4^- to chloride ion using H_2 released from the anaerobic Fe^0 oxidation (Sanchez 3003; Yu et al. 2006). As shown in Eq. (3.67), autotrophic bacteria use the released H_2 as an energy source, bicarbonate ion as a carbon source and ammonium ion as a nitrogen source for the reduction of ClO_4^-. Yu et al (2006) reported that in the presence of Fe^0, each gram of autotrophic bacteria (dry weight) could utilize 9.2 mg of ClO_4^- per hour. However, favorable pH conditions between 7 and 8 were found being required for the autotrophic bacteria to establish ClO_4^- reduction pathway. Moreover, in the presence of nitrate, the microbially mediated reduction rate of ClO_4^- was significantly inhibited.

$$0.5H_2(g) + 0.075ClO_4^- + 0.1HCO_3^- + 0.08H^+ + 0.02NH_4^+$$
$$==> 0.02C_5H_7O_2N + 0.56H_2O(l) + 0.075Cl^- \qquad \text{(Eq. 3.67)}$$

3.4.5 Chlorinated Aliphatic Hydrocarbons

CAH-contaminated groundwater has drawn a great concern from publics or environmentalists for the necessity of remediation because of the persistence of CAHs in aquifers and also their toxicity to human body (Montgomery, 2000). Rügge et al. (1999) reported that the $t_{1/2}$ of the natural attenuation of CAHs may range from 128 to 2310 days. Morevoer, epoxidation of CAHs by liver oxidation enzyme was proven producing carcinogenic compounds, thereby resulting in drinking water limits in the order of a few micrograms per litre (Gotpagar et al., 1997; Montgomery, 2000). In fact, of the 25 most frequently detected groundwater contaminants at hazardous waste sites in the United States, 10 are CAHs, such as 1,1,1-trichloroethane (1,1,1-TCA), tetrachloroethylene (PCE), trichloroethylene (TCE) and dichloroethylene (DCE) isomers (National Research Council, 1994). The ubiquity of CAHs in groundwater is due to their extensive applications as dry cleaning and metal degreasing agents, refrigerants, and propellants over the past 50 years (Orth, 1992; Fetter, 1999; Montgomery, 2000; Lai and Lo, 2002).

Abiotic reduction or reductive dechlorination of CAHs by reductants such as Fe^0 is one of the effective approaches for removing CAHs from groundwater (Boronina et al., 1995; Warren et al., 1995; Arnold and Roberts, 1998; Boronina et al., 1998; Fennelly and Roberts, 1998; Lo and Lai, 2002; Lo et al., 2003b; Andrea et al., 2005). Under both aerobic [Eqs. (3.68) and (3.69)] and anaerobic conditions [Eq. (3.59)], Fe^0 oxidation provides electron sources and thermodynamically favorable conditions for the reduction of CAHs through reductive-β-elimination [Eq. (3.70)] and hydrogenolysis [Eq. (3.71)] pathways (Matheson and Tratnyek, 1994; Roberts et al., 1996; Farrell et al., 2000; Liang et al., 2000; Lai and Lo, 2003; Lai et al., 2004).

$$2Fe^0(s) + O_2(g) + 2H_2O(l) \Longleftrightarrow 2Fe^{2+} + 4OH^- \qquad \text{(Eq. 3.68)}$$

$$4Fe^{2+} + 4H^+ + O_2(g) \Longleftrightarrow 4Fe^{3+} + 2H_2O(l) \qquad \text{(Eq. 3.69)}$$

$$RHCl{=}RHCl + 2e^- \Longleftrightarrow RH{\equiv}RH + 2Cl^- \qquad \text{(Eq. 3.70)}$$

$$RHCl{=}RHCl + H^+ + 2e^- \Longleftrightarrow RH_2{=}RHCl + Cl^- \qquad \text{(Eq. 3.71)}$$

In addition, direct hydrogenolysis of CAHs by the H_2 [Eq. (3.72)] released from the anaerobic Fe^0 oxidation can also occur in the presence of catalysts such as Fe_3O_4 (Matheson and Tratnyek, 1994; Ritter et al., 2002).

$$RHCl{=}RHCl + H_2(g) \xrightarrow{\ Catalyst\ } RH_2{=}RHCl + H^+ + Cl^- \qquad \text{(Eq. 3.72)}$$

Unlike chromium and uranium reduction by Fe^0, completely reductive dechlorination of CAHs by Fe^0 is not a one-step process. By following the pseudo first-order kinetics [Eqs. (3.51) or (3.52)], CAHs are sequentially reduced to less chlorinated forms and finally transformed to non-chlorinated end products. As reported by Arnold and Roberts (1998), chlorinated ethenes such as PCE and TCE are sequentially dechlorinated into DCE isomers, vinyl chloride (VC), dichloroacetylene, chloroacetylene, acetylene and finally to ethene and ethane through the pathways of hydrogenolysis, reductive-β-elimination and hydrogenation (Fig. 3.5). Since the standard potentials for various reductive dechlorination reactions (see Table 3.9) are higher than that of Fe^{2+}/Fe^0 (-0.44 V), reductive dechlorination of CAHs by Fe^0 can occur spontaneously.

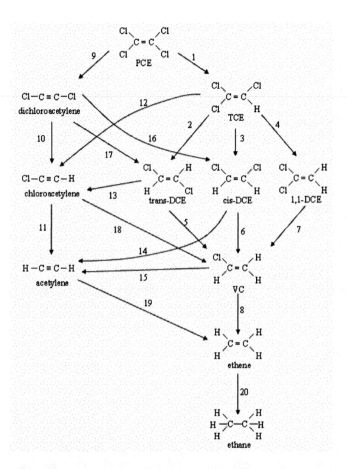

Figure 3.5 Hypothesized reductive dechlorination of chlorinated ethenes and other intermediates by Fe^0. Reactions 1 to 8, 10 to 11 correspond to hydrogenolysis pathway; reactions 9, 12 to 15 are pathways of reductive-β-elimination. Reactions 16 to 20 are hydrogenation reactions (Adapted from Blowes et al., 1999a).

In applying Fe^0 for the remediation of CAH-contaminated groundwater, the reaction time should be carefully determined based on the k_{obs} of different CAHs to ensure that Fe^0 can completely dechlorinate CAHs. This is because the chlorinated intermediates generated from the pathways of hydrogenolysis, reductive-β-elimination and hydrogenation such as cis-DCE and VC are also toxic to human. Therefore, incomplete reduction of CAHs by Fe^0 may potentially generate more toxic contaminants in groundwater. Table 3.9 summarizes the representative kinetic data for the reductive dechlorination by Fe^0.

Table 3.8 Standard potentials for various reductive dechlorination reactions of chlorinated aliphatic hydrocarbons (CAHs) by Fe^0

Contaminants	Pathway	Reaction	E^0 (W) (V)
PCE	Hydrogenolysis [a]	PCE $\rightarrow$ TCE	0.592
TCE	Hydrogenolysis [a]	TCE $\rightarrow$ cis-DCE	0.530
TCE	Hydrogenolysis [a]	TCE $\rightarrow$ 1,1-DCE	0.513
TCE	Hydrogenolysis [a]	TCE $\rightarrow$ trans-DCE	0.509
1,1-DCE	Hydrogenolysis [a]	1,1-DCE $\rightarrow$ VC	0.423
cis-DCE	Hydrogenolysis [a]	cis-DCE $\rightarrow$ VC	0.407
trans-DCE	Hydrogenolysis [a]	trans-DCE $\rightarrow$ VC	0.428
VC	Hydrogenolysis [a]	VC $\rightarrow$ ethene	0.481
Dichloroacetylene	Hydrogenolysis [a]	Dichloroacetylene $\rightarrow$ Chloroacetylene	0.560
Chloroacetylene	Hydrogenolysis [a]	Chloroacetylene $\rightarrow$ acetylene	0.500
PCE	Reductive-β-elimination[b]	PCE $\rightarrow$ dichloroacetylene	0.631
TCE	Reductive-β-elimination[b]	TCE $\rightarrow$ chloroacetylene	0.599
cis-DCE	Reductive-β-elimination[b]	cis-DCE $\rightarrow$ acetylene	0.568
trans-DCE	Reductive-β-elimination[b]	trans-DCE $\rightarrow$ acetylene	0.589
Dichloroacetylene	Hydrogenation [c]	Dichloroacetylene $\rightarrow$ cis-DCE	0.490
Dichloroacetylene	Hydrogenation [c]	Dichloroacetylene $\rightarrow$ trans-DCE	0.470
Chloroacetylene	Hydrogenation [c]	Chloroacetylene $\rightarrow$ VC	0.370

Source: Roberts et al. (1996)

[a]Standard potential corresponding to hydrogenolysis [Eq. (3.71)]: RHCl=RHCl+H$^+$+2e$^-$ <==> RH$_2$=RHCl+Cl$^-$.

[b]Standard potential corresponding to reductive-β-elimination [Eq. (3.70)]: RHCl=RHCl+2e$^-$ <==> RH≡RH+2Cl$^-$

[c]Standard potential corresponding to reduction of triple bond to double bond (hydrogenation).

In addition to Fe^0, reductive dechlorination of chlorinated ethenes by ferrous iron- or sulfur-containing minerals has also been reported (Butler and Hayes, 1999; Lee and Batchelor, 2002). Lee and Batchelor (2002) reported that dichloroacetylene, chloroacetylene, acetylene and ethene are the main dechlorination products detected in the dechlorination process of PCE and TCE by green rust, thereby proposing the domination of reductive-β-elimination over hydrogenolysis as the major dechlorination pathway. However, the dechlorination products of both hydrogenolysis and reductive-β-elimination were detected in the dechlorination process of PCE and TCE by FeS (Butler and Hayes, 1999). Table 3.10 illustrates the typical k_{obs} of the reductive dechlorination of PCE, TCE, cis-DCE and VC by green rust and mackinawite (FeS).

Table 3.9 representative kinetic data for the reductive dechlorination of chlorinated aliphatic hydrocarbons (CAHs) by Fe^0.

CAH	Observed pseudo first-order rate constant, k_{obs} (1/hr)	Half-life, $t_{1/2}$, (hr)
Tetrachloroethylene	2.1^a, $0.104 - 0.779^b$	0.33^a, $0.89 - 6.64^b$
Trichloroethylene	0.39^a, $0.104 - 0.786^b$	1.78^a, $0.882 - 6.69^b$
cis-Dichloroethylene	0.041^a, $0.011 - 0.022^b$	16.9^a, $31.8 - 63.6^b$
trans-Dichloroethylene	0.12^a, $0.374 - 0.419^b$	5.78^a, $1.65 - 1.85^b$
1,1-Dichloroethane	$0.015 - 0.262^b$	$2.65 - 46.8^b$
1,1-Dichloroethylene	0.064^a, $0.033 - 0.466^b$	10.8^a, $1.49 - 20.7^b$
Vinyl Chloride	0.05^a, $0.026 - 0.074^b$	13.9^a, $9.37 - 26.2^b$
Hexachloroethane	31^a	0.022^a
1,1,2,2-Tetrachloroethane	13^a	0.053^a
1,1,1,2-Tetrachloroethane	14^a	0.050^a
1,1,1-Trichloroethane	11^a, $0.087 - 0.579^b$	0.063^a, $1.20 - 7.95^b$
Chloroform	0.92^a, $0.071 - 0.364^b$	0.753^a, $1.91 - 9.78^b$

[a]All k_{obs} and $t_{1/2}$ are determined based on 1000 m^2/L of the surface area concentration (λ_s). The source of the data is from Johnson et al. (1996).
[b]All k_{obs} and $t_{1/2}$ are determined based on 7790 m^2/L of λ_s. The source of the data is from Lai et al. (2006).

Table 3.10 kinetic data for the reductive dechlorination of chlorinated aliphatic hydrocarbons (CAHs) by green rust and iron sulfide

CAH	Reduced by green rust[a]		Reduced by Iron Sulfide[b]	
	Observed pseudo first-order rate constant, k_{obs} (1/hr)	Half-life, $t_{1/2}$, (hr)	Observed pseudo first-order rate constant, k_{obs} (1/hr)	Half-life, $t_{1/2}$, (hr)
PCE	0.0978 (0.000162)	7.09 (4278)	0.00057 (0.00114)	1216 (607.9)
TCE	0.0516 (0.0000854)	13.4 (8115)	0.00149 (0.00298)	465 (232.6)
cis-DCE	0.0313 (0.0000518)	22.1 (13378)		
VC	0.0469 (0.0000776)	14.8 (8925)		

Note: The figures shown in parentheses are the k_{obs} and $t_{1/2}$ at 1 m^2/L of the surface area concentration (λ_s).
[a]The experimental results are obtained from batch kinetic experiments conducted at 23^0C and pH 7 with 604 m^2/L of the λ_s (Lee and Batchelor, 2002).
[b]The experimental results are obtained from batch kinetic experiments conducted at 25^0C and pH 8.3 with 0.5 m^2/L of the λ_s (Butler and Hayes, 1999).

3.4.6 Polychlorinated Biphenyls

Polychlorinated biphenyls (PCBs) having 1 to 10 chlorine atoms bounding to a biphenyl molecule are a kind of chlorinated organic micropollutants (Figs. 3.6a and

3.6b). Of the 209 possible congeners, only about 130 of the structures are energetically favorable in the manufacturing processes. Like many other aromatic hydrocarbons, PCBs are highly lipophilic and chemically stable. Because of their environmental persistence and bioconcentration in the food chain, PCBs have drawn much public concern recently. It was reported that accumulation of PCBs in the human body through food chain may cause neuron-developmental and neurobehavioral deficits in children and cancer (Hester and Harrison, 1996). Despite the fact that production of PCBs has been ceased since 1970s, heavy uses of PCBs as organic diluents, plasticizers, pesticide extenders, adhesives, dust-reducing agents, cutting oils, flame retardants, heat transfer

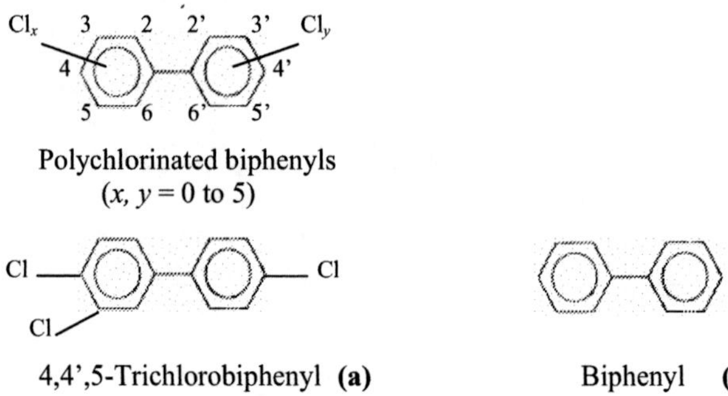

Polychlorinated biphenyls
$(x, y = 0 \text{ to } 5)$

4,4',5-Trichlorobiphenyl **(a)**   Biphenyl   **(b)**

Figure 3.6 Basic structure and nomenclature of (a) polychlorinated biphenyls (PCBs) and (b) biphenyl.

fluids and dielectric fluids for transformers and capacitors since 1930s have resulted in a direct or an indirect release of PCBs into water and soil environments (Kim et al., 2004). It is estimated that 31% of the world PCB produced has been released into the environment and another 65% are currently in use or has been disposed in landfills and hazardous waste dumps, thereby having a possibility of release to the environment (Hester and Harrison, 1996).

Current technologies commonly used to destruct PCBs in water and soil include biodegradation, incineration and reductively chemical dechlorination (Grittini et al., 1995). Aerobic and anaerobic biodegradation of PCBs may take several months for complete decomposition. Incineration of PCB-contaminated soil is effective but it may produce undesirable products such as dioxin (Kim et al., 2004). Reductive dechlorination of PCBs using Fe^0 as a reductant was found to occur only at 400 ^{0}C (Chuang et al., 1995); while there was no dechlorination at room temperature. However, rapidly reductive dechlorination of PCBs was observed using bimetallic materials such as palladized Fe^0 (Pd/Fe^0) as a reductant (Grittini et al., 1995; Korte et al., 2002; Doyle et al., 1998). Similar to the CAH dechlorination by Fe^0, reductive dechlorination of PCBs by Pd/Fe^0 is also a stepwise process in which PCBs are sequentially dechlorinated to less chlorinated biphenyls and finally reduced to biphenyl using the electrons released from the Fe^0 oxidation, as depicted in Eq. (3.73).

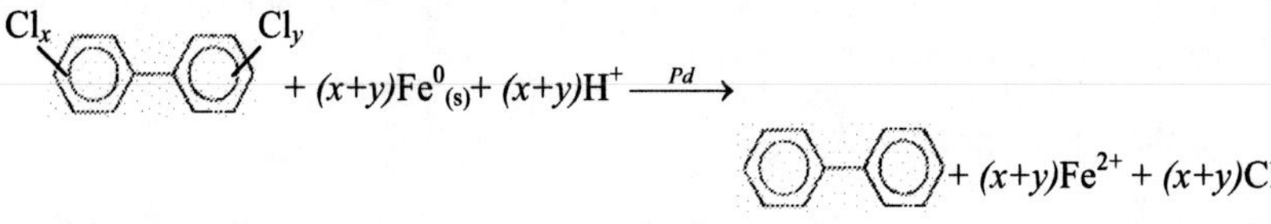

$$Cl_x \text{—⬡—⬡—} Cl_y + (x+y)Fe^0_{(s)} + (x+y)H^+ \xrightarrow{Pd}$$

$$\text{⬡—⬡} + (x+y)Fe^{2+} + (x+y)Cl^-$$

(Eq. 3.73)

Palladium in this redox reaction serves as a catalyst rather than participating into the reaction. As an example of the reductive dechlorination of PCBs by Pd/Fe^0, 4,4',5,5'-tetrachlorobiphenyl (4,4',5,5'-TeCB) on the Pd/Fe^0 surface is first reduced to trichlorobiphenyls (TCBs) and subsequently to dichlorobiphenyls (DCBs). Afterwards, DCBs are reductively dechlorinated to monochlorobiphenyls (MCBs) and finally to biphenyl, as described in Fig. 3.7. Each dechlorination step involves direct transfer of two electrons and the release of a chlorine atom (Yak et al., 2000).

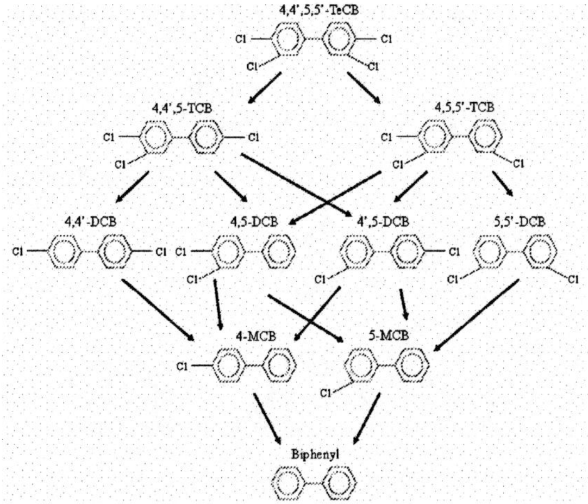

Figure 3.7 Possible dechlorination pathways of 4,4',5,5'-tetrachlorobiphenyl (4,4',5,5'-TeCB) on the Pd/Fe surface.

The reductive dechlorination of PCBs by Pd/Fe^0 is a pseudo first-order reaction (Kim et al. (2004). It is interesting to note that the rate of dechlorination was found in an order of 4-MCB > 3-MCB > 2-MCB. As seen in Table 3.11, the $t_{1/2}$ of the dechlorination of 4-MCB by Pd/Fe^0 is about 41.8 hrs; whereas it is 70.7 and 87.7 hrs for 3-MCB and 2-MCB, respectively. Similar trend of variation in the dechlorination rate constant of MCBs can also be observed for palladized zinc (Pd/Zn). Generally, PCBs with chlorine atom at carbon 4 are preferentially dechlorinated and PCBs with chlorine atom at positions 3 or 5 are followed. Finally, PCBs with chlorine atom at positions 2 or 6 have the least priority to be dechlorinated by bimetallic materials (Kim et al., 2004). In addition, based on the k_{obs} normalized at 1 m^2/L of the λ_s shown in Table 3.11, it is clear to see that MCBs are more susceptible to be dechlorinated by Pd/Fe^0 than DCBs including 3,4-DCB, 2,4-DCB and 2,3-DCB.

Thermodynamically, Zn has a more negative standard potential (E^0 = - 0.76 V) than that of Fe^0 (E^0 = -0.44 V) so that Zn theoretically can provide a higher reducing force for the reductive dechlorination of MCBs. This may be the reason why Pd/Zn shows higher reactivity than Pd/Fe^0 in reductively dechlorinating MCBs (Table 3.11). In a similar manner, palladized sodium (Pd/Na) and magnesium (Pd/Mg) may also possess higher reactivity than Pd/Fe^0 on PCB dechlorination because both sodium and magnesium have more negative standard potentials, which are -2.72 and -2.37 V,

Table 3.11 Kinetic data for the reductive dechlorination of polychlorinated bphenyls (PCBs) by Pd/Fe^0 and Pd/Zn

	Reduced by Pd/Fe$^{0\,a}$			Reduced by Pd/Zn ^{b}	
PCB	**Observed pseudo first-order rate constant, k_{obs} (1/hr)**	**Half-life, $t_{1/2}$, (hr)**	**PCB**	**Observed pseudo first-order rate constant, k_{obs} (1/hr)**	**Half-life, $t_{1/2}$, (hr)**
4-MCB	0.0166 (0.0060)	41.8 (115.5)	4-MCB	0.144 (0.048)	4.8 (14.6)
3-MCB	0.0098 (0.0045)	70.7 (154.0)	3-MCB	0.080 (0.028)	8.7 (25.2)
2-MCB	0.0079 (0.0040)	87.7 (173.3)	2-MCB	0.069 (0.024)	10.0 (28.9)
3,4-DCB	0.0118 (0.0020)	58.7 (355.4)			
2,4-DCB	0.0094 (0.0016)	73.7 (447.1)			
2,3-DCB	0.0086 (0.0015)	80.6 (477.9)			

Note: The experimental results were obtained from batch reactors containing 20 mL of solution and 1 to 2 g of bimetallic materials. The figures shown in parentheses are the k_{obs} and $t_{1/2}$ at 1 m^2/L of the surface area concentration (λ_s).
[a]Pd/Fe specific surface area 0.062±0.002 m^2/g.
[b]Pd/Zn specific surface area 0.059±0.002 m^2/g.

respectively (Doyle et al. 1998). During the reductive dechlorination of PCBs by Pd/Fe^0, water is the dominant hydrogen donor for this redox reaction [Eq. (3.73)]. Unfortunately, extraction of waste or soil for PCBs usually does not employ water. Korte et al. (2002) mentioned that solvent with lower dielectric constant (i.e., less polar) exhibited greater inhibition to the palladium-catalyzed reductive dechlorination reactions. Hence reductive dechlorination of PCBs by Pd/Fe^0 in ethanol was observed to be better than in isopropanol since the former possesses a higher dielectric constant.

3.5   Natural Occurring of Reduction and Precipitation Processes

3.5.1 Hexavalent Chromium Reduction in Trinity Sand Aquifer, Odessa, Texas

Chromium I Superfund site located in Odessa, Texas consists of an unconfined aquifer ranging in the thickness from 55 to 70 ft extended to a depth of about 140 ft below the ground surface (Fig. 3.8). The aquifer is a Cretaceous Trinity Formation (Fm) in which the sand is moderately well-sorted, ferruginous to calcareous and considered to be hydraulically continuous. Between 1969 and 1978, several chrome plating factories were

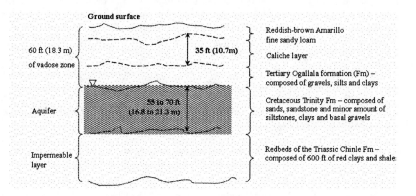

Figure 3.8 Hydrogeologic conditions of Trinity Sand Aquifer, Odessa, Texas.

located at Odessa. Probably because of the direct discharge of wastewater and rinse water containing primarily dissolved Cr(VI) into the aquifer, a maximum Cr(VI) concentration of 72 mg/L was detected in groundwater.

Site assessment of the Cr(VI) contamination in the Trinity Sand aquifer from 1986 to 1991 indicated a noticeable decrease in the Cr(VI) concentration in the groundwater. The maximum Cr(VI) concentration was detected decreasing from 72 mg/L in 1986 to 5.3 mg/L in 1991. The contribution of groundwater advection and dispersion to the drop of the Cr(VI) concentration was believed to be insignificant since the overall areal extent of the Cr(VI) plume remained relatively constant between 1986 and 1991. By taking both dissolved and adsorbed masses of Cr(VI) into account, Henderson (1994) found that only 30% of the total Cr(VI) present in the aquifer in 1986 remained in 1991. Fitting pseudo first-order kinetic model to the mass data of total Cr(VI) resulted in $t_{1/2}$ of approximately 2.5 years, with a corresponding k_{obs} of 3.2×10^{-5} 1/hr. Further analyses of the sediment samples collected from the Trinity Sand aquifer showed that each kilogram of the soils contained approximately

67 mg of total chromium [i.e., Cr(III) and Cr(VI)] in which about 60 to 90% of the total chromium was inferred existing as Cr(III) solids. Hence it is believed that natural reduction of Cr(VI) to Cr(III) followed by heterogeneous precipitation onto the soils should be the main reason leading to the 70% loss of Cr(VI) in the aquifer between 1986 and 1991.

Based on the Eh-pH diagram shown in Fig. 3.4, Cr(III) mainly existed as Cr(OH)$_3$ solids is the thermodynamically stable valence state of chromium in the Trinity Sand aquifer with the pH of 6.7 to 7.4 and the Eh of 0.26 to 0.53 V. It is believed that the Fe^{2+} in the ferruginous soils, and dissolved organic carbon with a concentration as high as 8 mg/L in both groundwater and the soils of the Trinity Sand aquifer acted as major reductants for the Cr(VI) reduction as described in Eqs. (3.43) and (3.47). The abundance of iron oxyhydroxides in the Trinity Sand sediments also provided favorable conditions for the co-precipitation of mixed chromium-iron hydroxide [Eq. (3.49)] or oxyhydroxide solids [Eq. (3.50)]. It is expected that even there is no engineering interventions or remediation, the Cr(VI) concentration in the Trinity Sand aquifer can drop below the MCL of 0.1 mg/L within 10 years by the natural reduction and precipitation processes.

3.5.2 Dechlorination of Chlorinated Solvents at a Superfund Site in St. Joseph, Michigan

The geologic formation of the Superfund site in St. Joseph, Michigan is an unconfined aquifer consisting of a layer of unconsolidated fine sand with some silt and overlaying a lacustrine clay unit. The water table at the Superfund site generally varies from 23 ft (7 m) to 52 ft (16 m) below ground surface. The thickness of the whole sand layer (i.e., unsaturated and saturated layers) ranges from 39 ft (12 m) to 105 ft (32 m) (Semprini et al., 1995). Most likely owing to the disposal of wastewater from an automotive brake manufacturer into an unlined lagoon close to the site from 1950s to 1970s, TCE concentration as high as 100 mg/L was detected in the groundwater.

In 1991, high concentration of TCE, DCE isomers, VC and ethene was detected within 66 ft (20 m) of the center of the plume. The detection of high concentration of the hydrogenolysis products of TCE indicated the occurrence of the naturally reductive dechlorination of TCE in the aquifer (National Research Council, 2000). At a depth of about 65 ft (19.8 m), TCE concentration was detected to be the highest, but methane concentration was nearly the least. At a depth where there were intermediate increase in methane concentration and decrease in sulfate concentration, reductive dechlorination of TCE to cis-DCE was observed, thereby indicating fast transformation of TCE to cis-DCE in a transition zone between the sulfate-reducing zone and methanogenesis zone. At a depth of 75 ft (23 m) at which maximum concentration of VC, ethene and methane, and minimum concentration of sulfate were detected, there were complete disappearance of TCE and cis-DCE. This observation showed that more reducing conditions (i.e., methanogenic conditions) rather than sulfate-reducing conditions are required for further dechlorination of TCE and cis-DCE to VC and ethene.

Two-dimensional chemical concentration contour over the aquifer further verifies the naturally reductive dechlorination of TCE to DCE isomers in less reducing conditions of sulfate reduction, and naturally reductive dechlorination of DCE isomers to VC and ethene under the redox conditions of methanogenesis. High TCE concentration zone was associated an area lacking methane or highly active methanogenic condition. Raised cis-DCE level occurred in the transition zone from low to high methane concentration, thereby showing the dechlorination of TCE to cis-DCE under less reducing conditions. Decrease in cis-DCE concentration was associated with low sulfate values. In the region with elevated methane concentration, elevated VC and ethene concentrations were observed. Hence the methanogenic or more reducing conditions was believed being required for the further dechlorination of TCE and cis-DCE to VC and ethene.

The naturally reductive dechlorination reactions occurring in St. Joseph Michigan certainly had to be coupled with oxidation reactions for the anaerobic transformation of TCE. At the Superfund site, the organic matter leaching from the disposal lagoon containing chemical oxygen demand (COD) as high as 400 mg/L was the main electron donor for the naturally reductive dechlorination of TCE, sulfate reduction and methane production (National Research Council, 2000). It was reported that up to 15% of the drop in the COD in the aquifer was associated with the reductive dechlorination of TCE and 8 to 25% of the TCE in the groundwater was finally converted to ethene. As described in Fig. 3.9, microorganisms in the aquifer first converted complex organic materials or COD to sugars, amino acids and organic acids, which subsequently were fermented to alcohols and fatty acids for energy. Then alcohols and fatty acids were oxidized by another type of microorganisms to produce acetate and H_2. Finally, other sets of microorganisms used TCE, sulfate and carbon dioxide as electron acceptors, and utilized acetate and H_2 as electron donors, which resulted in the naturally reductive dechlorination of TCE in the aquifer (McCarty, 1997; McCarty, 1999).

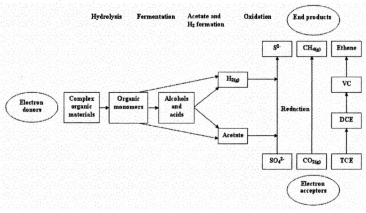

Figure 3.9 Possible steps involved in the naturally reductive dechlorination of TCE in the aquifer at St. Joseph, Michigan.

3.6   Engineering Application

The noticeable decrease in the Cr(VI) concentration in the Trinity Sand aquifer at Odessa, Texas, and the dechlorination of TCE to less chlorinated and non-chlorinated forms in St. Joseph, Michigan show the potential of the redox and precipitation processes for hazardous waste remediation. However, in natural systems, their efficiencies are extremely low. For instance, natural reduction and precipitation processes took nearly 2.5 years to decrease half of the Cr(VI) mass in the Trinity Sand aquifer. Engineering application of the redox and precipitation processes in the technology called permeable reactive barriers (PRBs) can significantly shorten the $t_{1/2}$ of reduction of hazardous wastes in few minutes or few days.

3.6.1   Permeable Reactive Barriers

PRBs are a promising technology for *in situ* remediation of hazardous wastes dissolved in groundwater (Gillham and O'Hannesin, 1992). As shown in Fig. 3.10, PRBs involve an emplacement of reactive materials in the subsurface designed to intercept a contaminant plume, provide a flow path through the reactive media, and transform the contaminants into environmentally acceptable forms via various chemical processes. Thus remediation concentration goals can be attained in the downgradient of the PRBs (Powell et al., 1998; Lai et al. 2006a). The processes that can reduce the contaminant concentration in groundwater include oxidation, reduction, precipitation, chemical transformation and a combination of these processes (Blowes et al., 1999a). The reactive materials packed inside PRBs are generally selected based on the type of hazardous wastes to be treated. Since PRBs require no pumping and extraction systems, no installation of invasive surface structures and equipment, and no further treatment and disposal of waste materials, there is hardly any annual operation and maintenance costs other than site monitoring. Moreover, the land can be put to productive uses while it is being remediated.

Currently, zero-valent metals, such as Fe^0, are commonly used in PRB technology for groundwater remediation because of their relatively high reactivity (USEPA, 2002; Blowes et al., 1999a). The reduction reactions of dissolved hazardous wastes in groundwater by Fe^0 is a heterogeneous reaction which first involves diffusion of dissolved contaminants through the stagnant water layer to the Fe^0 surface and formation of a precursor complex between Fe^0/iron oxides and the surface contaminants (contaminants$_{(surface)}$). During the reduction, the electrons released from the oxidation of Fe^0 or the oxidation of surface-bound ferrous ion ($Fe^{2+}_{surface}$) are transferred to the contaminants$_{(surface)}$. Ideally, in the absence of an oxide layer on Fe^0 surface, the reduction of groundwater contaminants involves primarily electron transfer from Fe^0. Under environmental conditions, however, Fe^0 surface is usually covered by a layer of iron oxides containing $Fe^{2+}_{(surface)}$. In case the iron oxide layer is a non-conductive physical barrier, electron transfer from Fe^0 to the contaminants$_{(surface)}$may only occur through the defects such as pits or grain boundaries where there is rupture of the iron oxide layer.

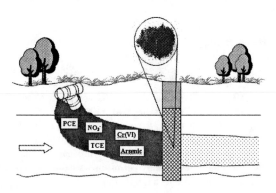

Figure 3.10 Schematic diagram of a PRB in a subsurface environment (from Lo et al., 2007, with permission).

If the iron oxide layer behaves as a semiconductor, the reduction of contaminants occurs indirectly by transport of electron from Fe^0 to the contaminants$_{(surface)}$ through the oxide layer via the oxide conduction band, impurity bands or localized states. In case the iron oxide layer acts as a coordinating surface, the reduction of contaminants such as CAHs may occur by metal-to-ligand electron transfer. The electrons may be donated from surface-bound Fe(II) sites created by the adsorption of aqueous Fe^{2+} [Eq. (3.73)] or originated from the underlying Fe^0 metal creating Fe(II) sites within the oxide lattice [Eq. (3.74)] (Scherer et al., 1998).

$$Fe^0 \quad Oxide$$

$$O\!\!>\!\!Fe^{III}\text{-}OH\text{-}Fe^{II}\cdots\cdots RCl_{(surface)} \rightarrow O\!\!>\!\!Fe^{II}\text{-}OH\text{-}Fe^{III}\cdots\cdots RH_{(surface)} \qquad \text{(Eq. 3.73)}$$

$$Fe^0 \quad O\!\!>\!\!Fe^{II}\text{-}OH\cdots\cdots RCl_{(surface)} \rightarrow O\!\!>\!\!Fe^{III}\text{-}OH\cdots\cdots RH_{(surface)}$$
$$Fe^{2+} \qquad\qquad\qquad\qquad\qquad\qquad\qquad\qquad\qquad\qquad\qquad\qquad \text{(Eq. 3.74)}$$

To date, funnel-and-gate PRB and continuous PRB are two basic configurations commonly applied in contaminated sites for groundwater remediation (Barker, 1998; O'Hannesin, 1998). Funnel-and-gate PRB mainly consists of a low permeability funnel and permeable reactive gate (see Figure 3.11a). Generally, the funnel is composed of low permeability materials, such as sheet pilings, slurry wall and grout curtains (Bedient et al., 1999), and preferably keyed into an impermeable layer, such as clay or bedrock, so as to prevent the underflow of contaminated groundwater. By emplacing the funnel, contaminated groundwater is directed towards the reactive gate containing reactive materials for the reduction or reduction and precipitation. Because of directing large cross-sectional area of water to a much small cross-sectional area of the gate, the groundwater velocity within the reactive gate of a funnel-and-gate PRB should be higher than the natural groundwater flow (Gavaskar et al., 1997; Powell et al. 1998; Gavaskar et al., 2000). Continuous PRB is the most prevalent configuration being used at the moment, as it exerts little disturbance to groundwater flow. As shown in Figure 3.11b, within continuous PRB, reactive materials (e.g., Fe^0) are distributed across the entire path of dissolved contaminant plume. It incorporates no funnel, thereby permitting contaminated groundwater to pass through the reactive barrier under its natural gradient.

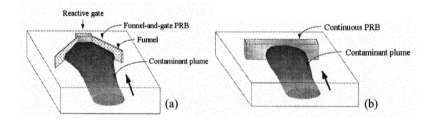

Figure 3.11 Schematic diagrams of (a) funnel-and-gate PRB and (b) continuous PRB (adapted from Powell et al., 1998).

3.6.2 In Situ Treatment of Cr(VI) and TCE Plumes by PRB Installed at Elizabeth City, North Carolina

Extent of Contamination
    In this section, the design, installation, operation and performance of a full-scale PRB installed at the United States Coast Guard (USCG) Support Center near Elizabeth City, North Carolina will be discussed to allow readers to get more familiar with the engineering application of the redox and precipitation processes through the PRB technology for groundwater remediation (Blowes et al., 1999a; Blowes et al., 1999b; Puls et al., 1999a; Puls et al., 1999b; Wilkin et al., 2001). As is evident in Figures 3.12a and 3.12b, Cr(VI) and TCE plumes extending approximately 213 ft (65 m) from an electroplating shop with a width between 115 ft (35 m) and 197 ft (60 m) were detected in USCG Support Center. The maximum observed Cr(VI) and TCE concentrations exceeded 10 and 19 mg/L, respectively (Blowes et al., 1999b). In 1996, a continuous PRB packed with Fe^0 was installed to treat the overlapping Cr(VI) and TCE plumes (Puls et al., 1999a).

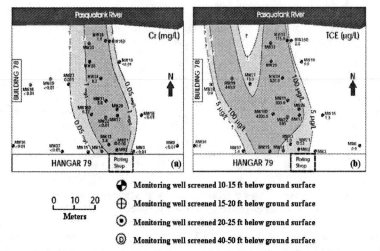

0   10   20
Meters

⬥ Monitoring well screened 10-15 ft below ground surface

⊕ Monitoring well screened 15-20 ft below ground surface

◉ Monitoring well screened 20-25 ft below ground surface

◎ Monitoring well screened 40-50 ft below ground surface

Figure 3.12 Plan view of (a) Cr(VI) plume and (b) TCE plume in the aquifer at USCG Support Center measured in 1994 (adapted from Blowes et al., 1999b).

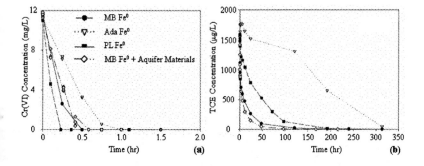

Figure 3.13 Temporal variation of (a) Cr(VI) and (b) TCE concentrations in each of the four batch tests (adapted from Blowes et al., 1999b).

Treatability Studies

Prior to the PRB installation, several laboratory treatability studies were conducted so as to determine the treatability of Cr(VI) and TCE by various reactive materials and also the required thickness of the PRB. Treatability studies involved four batch tests conducted using three different sources of commercially available Fe^0; Master-Builder (MB), Ada, Peerless (PL), and a combination of MB Fe^0 and the aquifer materials collected from the Support Center. According to Figs. 3.13a and 3.13b, rapid reduction of Cr(VI) and TCE concentrations by Fe0 were observed, thereby verifying the effectiveness of Fe^0 on treating the Cr(VI) and TCE overlapping plume. The rate of Cr(VI) reduction was found to be many times higher than the TCE. Complete removal of Cr(VI) from aqueous solution was achieved between 0.25 and 1.0 hrs; whereas complete TCE reduction required 100 to 300 hrs.

For the sake of determining the required thickness of the full-scale PRB in USCG Support Center before the installation, two sets of laboratory column experiments using various types of the commercially available Fe^0 as reactive materials were implemented. The first set of the column experiment was operated at a flow velocity of about 2 ft/d (61 cm/d), which corresponded to the velocity expected in a funnel-and-gate configuration. Another set of the column experiment was operated at a lower flow velocity of approximately 1 ft/d (30.5 cm/d) corresponding to the natural groundwater velocity at the site or the expected velocity within a continuous configuration of PRB. Figures 3.14a to 3.14d illustrate the concentration profiles of TCE, chloroform (TCM), cis-DCE and VC along different laboratory Fe^0 packed columns at 1 ft/d (30.5 cm/d) of seepage velocity. Since the reduction rate of Cr(VI) by Fe^0 was many times higher than those of CAHs (see Figures 3.13a and 3.13b), the required thickness of the full-scale PRB, therefore, was mainly dependent on the reduction rate of CAHs. As evident, complete reduction of TCE, TCM, cis-DCE and VC could be attained at a longitudinal distance of about 20 in (50 cm) from the influent even though there were intermediate production of cis-DCE and VC within

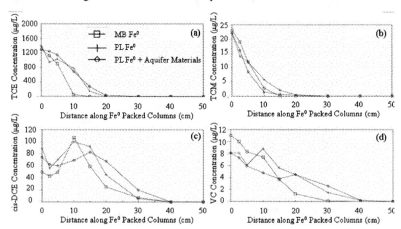

Figure 3.14 Concentration profiles of (a) TCE, (b) TCM, (c) cis-DCE and (d) VC along the laboratory Fe^0 packed columns at a seepage velocity of 1 ft/d (30.5 cm/d) (adapted from Blowes et al., 1999b).

the Fe^0 packed columns. Complete reduction of CAHs at a seepage velocity corresponding to that within funnel-and-gate PRB (i.e., 2 ft/d or 61 cm/d), however, could not be achieved at a longitudinal distance less than 20 in (50 cm). After considering the k_{obs} of the contaminants, factor of safety and the differences of the performance between the continuous and funnel-and-gate configurations, a continuous PRB with a 151 ft (46 m) long, 24 ft (7.3 m) deep and 2 ft (0.6 m) thick composing of 100% Peerless Fe^0 was finally installed at USCG Support Center (Figures 3.15a and 3.15b).

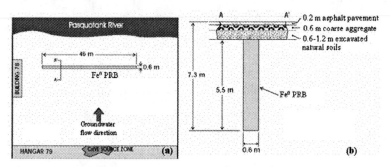

Figure 3.15 (a) Plan view and (b) cross-sectional view of the continuous Fe^0 PRB installed at USCG Support Center near Elizabeth City, North Carolina in 1996 (adapted from Blowes et al., 1999b).

Field Performance of Fe^0 PRB

PRBs are a passive treatment system in which natural flow of dissolved contaminants through the PRBs under the natural hydraulic gradient is required (Lai et al. 2006b and 2006c). Measurement of the hydraulic conductivity of the continuous Fe^0 PRB installed at USCG Support Center and its surrounding aquifer showed that the continuous Fe^0 PRB possessed a relatively higher hydraulic conductivity than the aquifer (Fig. 3.16). Thus dissolved contaminants conclusively passed through the Fe^0 PRB for the reduction or reduction and precipitation. Immediately after the continuous Fe^0 PRB installation (in 1996), rapid decrease in the Cr(VI) concentration across the Fe^0 PRB was detected, as shown in Fig. 3.17a (Blowes et al., 1999a). Upgradient Cr(VI) concentration as high as 5.1 mg/L could be reduced to less than the MCL of 0.1 mg/L within the Fe^0 PRB. All chromium detected in the groundwater was in hexavalent form and no Cr(III) was measured in the collected groundwater samples. The disappearance of both Cr(VI) and Cr(III) in the downgradient aquifer showed the occurrence of reduction of Cr(VI) to Cr(III) [Eq. (3.48)] and subsequent precipitation of Cr(III) within the Fe^0 PRB [Eqs. (3.49) and (3.50)].

In a similar manner, significant decrease in the TCE, cis-DCE and VC concentrations could also be observed across the continuous Fe^0 PRB (see Figs. 3.17b to 3.17d) (Blowes et al., 1999a). Upgradient TCE concentration of up to 5600 µg/L dropped to close to or less than MCL of 5 µg/L within the Fe^0 PRB. cis-DCE and VC concentrations of less than MCLs could also be maintained in most of the downgradient monitoring wells. Because of the reduction of TCE, cis-DCE and VC by Fe^0 to ethene

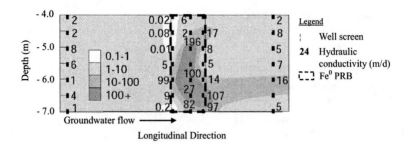

Figure 3.16 Distribution of the hydraulic conductivity over the continuous Fe⁰ PRB installed at USCG Support Center and the surrounding aquifer in 1996 (Adapted from Blowes et al., 1999a).

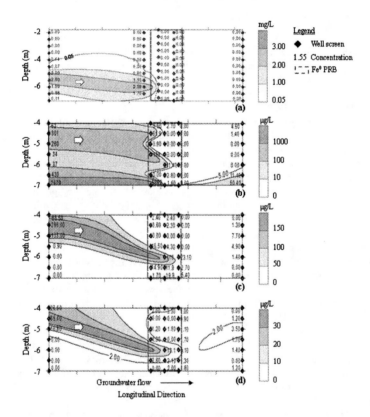

Figure 3.17 Cross-sectional profiles showing (a) total chromium or Cr(VI), (b) TCE, (c) cis-DCE and (d) VC concentrations across the continuous Fe⁰ PRB in 1996 (adapted from Blowes et al., 1999a).

and ethane via hydrogenolysis [Eqs. (3.71) and (3.72)] and reductive-β-elimination [Eq. (3.70)] pathways, simultaneous increases in the ethene and ethane concentrations were observed across the continuous Fe^0 PRB (Figs. 3.18a and 3.18b). In addition, groundwater pH was detected rising from the background values between 6 and 8 to approximately 11.7 within the Fe^0 PRB. *Eh* of groundwater declined from the background values between +0.1 and +0.5 V to the values as low as –0.58 V across the Fe^0 PRB. Low *Eh* environment within the continuous Fe^0 PRB also lowered the background concentrations of some redox sensitive species, such as dissolved oxygen, sulfate and nitrate significantly. Although passivation from mineral precipitates, such as calcite, and the growth of biofilm including sulfate- and metal-reducing bacteria, to the Fe^0 reactivity existed within the Fe^0 PRB (Wilkin et al., 2001; Wilkin and Puls, 2003), there was no pronounced deterioration in the remediation effectiveness 4 and 5 years after the installation. In fact, groundwater samplings along the Fe^0 PRB conducted in 2000 and 2001 still showed significantly reductive dechlorination of CAHs, and effective reduction and precipitation of Cr(VI) within the Fe^0 PRB.

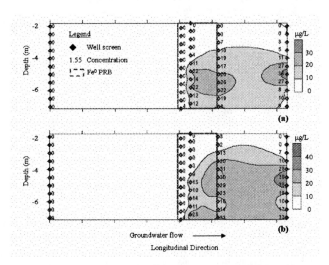

Figure 3.18 Cross-sectional profiles showing (a) ethene and (b) ethane concentrations across the continuous Fe^0 PRB in 1996 (adapted from Blowes et al., 1999a).

3.7 Summary

Chemical precipitation and reduction are effective processes for removing hazardous wastes in groundwater. The occurrence of precipitation of hazardous wastes requires supersaturated conditions in aqueous environments; while the happening of reduction of hazardous wastes requires the presence of reductants possessing lower standard potentials than the hazardous wastes. In the Trinity Sand aquifer at Odessa,

Texas, natural reduction and precipitation processes were able to decrease half of the Cr(VI) mass in the aquifer in 2.5 years. At the Superfund site in St. Joseph, Michigan, naturally reductive dechlorination of TCE occurred, which could completely transform TCE to ethene. Through the precipitation and redox processes, hazardous wastes including Cr(VI) and U(VI) can be first reduced to immobile Cr(III) and U(IV), respectively. They were then removed from aqueous environments by precipitating as oxide, hydroxide and/or oxyhydroxide solids on the reductant surface. Remediation of NO_3^-- or NO_2^--contaminated groundwater can be achieved through the reduction of NO_3^- or NO_2^- to ammonium and nitrogen gas. ClO_4^- in groundwater can be chemically reduced to chlorate (ClO_3^-), chlorite (ClO_2^-), hypochlorite (ClO^-) ions and finally to chloride ions. Stepwise reductive dechlorination of CAHs and PCBs into less chlorinated or non-chlorinated forms is the main approach used to remediate chlorinated aliphatic and aromatic organics from groundwater. With the engineering application of the precipitation and redox processes in the PRB technology, the remediation efficiency of these processes can be significantly enhanced in which the $t_{1/2}$ of the reduction of hazardous waste concentration decreases from few years in purely natural systems to only few days in PRB systems.

Notation

$[C]$	=	concentration of contaminant or hazardous waste (mg/L or M)
E	=	potential developing the electromotive force (emf) of the electrochemical cell (V)
E^0	=	standard potential or the potential with all species present at unit activity at 25 ^{0}C and 1 atm (V)
E^0 (W)	=	standard potential in neutral water (V)
Eh	=	redox potential (V)
$E_{measured}$	=	potential measured by reference electrode (V)
E^0_{ox}	=	standard potential for the oxidation half reaction (V)
E^0_{re}	=	standard potential for the reduction half reaction (V)
$E_{reference}$	=	potential of reference electrode with respect to standard hydrogen electrode (V)
F	=	Faraday's constant (96.42 kJ/V mol or 23.06 kcal/V mol)
ΔG	=	change in Gibbs free energy (kJ/mol)
ΔG_r	=	change in Gibbs free energy of the reaction (kJ/mol)
ΔG_r^0	=	change in standard Gibbs free energy of the reaction (kJ/mol), which is equal to ΔG_r when products or reactants are present at unit activity at a specified standard state (i.e., 25 ^{0}C and 1 atm)
ΔH_r^0	=	enthalpy change of chemical reaction (kJ/mol)
IAP	=	ion activity product
(i)	=	molality of species i (mol/kg)
$\{i\}$	=	activity of species i
$\{i\}_{actual}$	=	activity of species i in actual water sample
$\{i\}_{equilibrium}$	=	activity of species i at equilibrium
K_1, K_2	=	equilibrium constants

K_{25}	=	equilibrium constants at 25°C
K_{sp}	=	solubility product
K_{sp,T_1}	=	solubility product at temperature T_1
K_{sp,T_2}	=	solubility product at temperature T_2
K_W	=	equilibrium constant for water dissociation
k	=	rate constant (1/M s)
k'	=	rate constant ($M^{-0.6}$ 1/min)
k_{obs}	=	observed pseudo first-order rate constant (1/s, 1/min, 1/hr or 1/yr)
k_{sa}	=	k_{obs} normalized to the surface area concentration (L/m hr)
n	=	number of electron transferred in the reaction
n_H	=	number of moles of protons exchanged per mole of electrons
P_{H_2}	=	partial pressure of hydrogen (atm)
P_{O_2}	=	partial pressure of oxygen (atm)
$p\varepsilon^0$	=	standard pε at 25 °C and 1 atm
$p\varepsilon^0$ (W)	=	$p\varepsilon^0$ in neutral water
	=	$p\varepsilon^0 + \dfrac{n_H}{2}\log K_W$
P_{H_2}	=	partial pressure of hydrogen gas (atm)
R	=	ideal gas constant (8.314×10^{-3} kJ/mol K)
T, T_1, T_2	=	absolute temperature (K)
$t_{1/2}$	=	half-life (sec, min, hr or yr)
γ_i	=	activity coefficient of species i (kg/mol)
λ_s	=	surface area concentration (m^2/L)
Ω	=	saturation index

References

Alowitz, M.J., and Scherer, M.M. (2002) Kinetics of nitrate, nitrite, and Cr(VI) reduction by iron metal. *Environ. Sci. Technol.*, 36(3), 299-306.

Andrea, P.D., Lai, K.C.K., Kjeldsen, P., and Lo, I.M.C. (2005) Effect of groundwater inorganics on the reductive dechlorination of TCE by zero-valent iron. *Water, Air, and Soil Pollut.*, 162, 401-420.

Appelo, C.A.J., and Postma, D. (1993) Geochemistry, groundwater and pollution, Rotterdam, Brookfield, VT..

Arnold, W.A., and Roberts, A.L. (1998) Pathways of chlorinated ethylene and chlorinated acetylene reaction with Zn(0). *Environ. Sci. Technol.*, 32(19), 3017-3025.

Arnold, W.A., and Roberts, A.L. (2000) Pathways and kinetics of chlorinated ethylene and chlorinated acetylene reaction with Fe(0) particles. *Environ. Sci. Technol.*, 34(9), 1794-1805.

Barker, J. (1998) Permeable reactive walls – An example for treating a mixed plume. Advances in Innovative Ground-Water Remediation Technologies, Dec 15, Atlanta, GA.

Bedient, P.B., Rifai, H.S., and Newell, C.J. (1999) Ground water contamination transport and remediation, 2nd Ed., Prentice Hall PTR, Upper Saddle River, NJ.

Bloomfield, C., and Pruden, G. (1980) The behavior of Cr(VI) in soil under aerobic and anaerobic conditions. *Environ. Pollut.*, 23, 103-114.

Blowes, D.W., Ptacek, C.J., and Jambor, J.L. (1997) In-situ remediation of Cr(VI)-contaminated groundwater using permeable reactive walls: Laboratory studies. *Environ. Sci. Technol.*, 31(12), 3348-3357.

Blowes, D.W., Puls, R.W., Gillham, R.W., Ptacek, C.J., Bennett, T.A., Bain, J.G., Hanton-Fong, C.J., and Paul, C.J. (1999a) An in situ permeable reactive barrier for the treatment of hexavalent chromium and trichloroethylene in ground water: Volume 2 performance monitoring, EPA/600/R-99/095b, Office of Research and Development, United States Environmental Protection Agency, Washington, D.C.

Blowes, D.W., Gillham, R.W., Ptacek, C.J., Puls, R.W., Bennett, T.A., O'Hannesin, S.F., Hanton-Fong, C.J., and Bain, J.G. (1999b) An in situ permeable reactive barrier for the treatment of hexavalent chromium and trichloroethylene in ground water Volume 1 design and installation, EPA/600/R-99/095a, Office of Research and Development, United States Environmental Protection Agency, Washington, D.C.

Boronina, T., Klabunde, K.J., and Sergeev, G. (1995) Destruction of organohalides in water using metal particles: Carbon tetrachloride/water reactions with magnesium, tin and zinc. *Environ. Sci. Technol.*, 29(6), 1511-1517.

Boronina, T., Lagadic, I., and Klabunde, K.J. (1998) Activated and nonactivated forms of zinc posers: Reactivity toward chlorocarbons in water and AFM studies of surface morphologies. *Environ. Sci. Technol.*, 32(17), 2614-2622.

Buerge, I.J., and Hug, S.J. (1997) Kinetics and pH dependence of chromium(VI) reduction by iron(II). *Environ. Sci. Technol.*, 31(5), 1426-1432.

Butler, E.C., and Hayes, K.F. (1999) Kinetics of the transformation of trichloroethylene and tetrachloroethylene by iron sulfide. *Environ. Sci. Technol.*, 33(12), 2021-2027.

Cabel, B., Kozicki, R., Lahl, U., Podbielski, A., Stachel, B., and Struss, S. (1982) Pollution of drinking water with nitrate. *Chemosphere*, 11(11), 1147-1154.

Cantrell, K.J., Kaplan, D.I., and Wietsma, T.W. (1995) Zero-valent iron for the in situ remediation of selected metals in groundwater. *J. Hazard. Mater.*, 42(2), 201-212.

Cheng, I.F., Muftikian, R., Fernando, Q., and Korte, N. (1997) Reduction of nitrate to ammonia by zero-valent iron. *Chemosphere*, 35(11), 2689-2695.

Choe, S., Chang, Y.Y., Hwang, K.Y., and Khim, J. (2000) Kinetics of reductive denitrification by nanoscale zero-valent iron. *Chemsophere*, 41(8), 1307-1311.

Choe, S., Liljestrand, H.M. and Khim, J. (2004) Nitrate reduction by zero-valent iron under different pH regimes. *Applied Geochemistry*, 19(3), 335-342.

Christensen, T.H., Bjerg, P.L., Banwart, S.A., Jakobsen, R., Heron, G., and Albrechtsen, H.J. (2000) Characterization of redox conditions in groundwater contaminant plumes. J. Contam. Hydrol., 45(3-4), 165-241.

Chuang, F.W., Larson, R.A., and Wessman, M.S. (1995) Zero-valent iron-promoted dechlorination of polychlorinated biphenyls. Environ. Sci. Technol., 29(9), 2460-2463.

Dean, J.A. (1985) Lange's handbook of chemistry, 13th Ed., McGraw-Hill, New York.

Doyle, J.G., Miles, T.A., Parker, E., and Cheng, I.F. (1998) Quantification of total polychlorinated biphenyl by dechlorination to biphenyl by Pd/Fe and Pd/Mg bimetallic particles. Microchem. J., 60(3), 290-295.

Dubrovsky, N.M., Morin, K.A., Cherry, J.A., and Smyth, D.J.A. (1984) Uranium tailings acidification and subsurface contaminant migration in a sand aquifer. Water Poll. Res. J. Canada, 19, 55-89.

Eary, L.E., and Rai, D. (1988) Chromate removal from aqueous wastes by reduction with ferrous ion. Environ. Sci. Technol., 22(8), 972-977.

Farrell, J., Kason, M., Melitas, N., and Li, T. (2000) Investigation of the long-term performance of zero-valent iron for reductive dechlorination of trichloroethylene. Environ. Sci. Technol., 34(3), 514-521.

Fendorf, S.E., and Li, G.C. (1996) Kinetics of chromate reduction by ferrous iron. Environ. Sci. Technol., 30(5), 1614-1617.

Fennelly, J.P., and Roberts, A.L. (1998) Reaction of 1,1,1-trichloroethane with zero-valent metals and bimetallic reductants. Environ. Sci. Technol., 32(13), 1980-1988.

Fetter, C.W. (1999). Contaminant hydrogeology, 2nd Ed., Prentice-Hall, Inc., Simon & Schuster/A Viacom Company, Upper Saddle River, NJ.

Freeze, R.A., and Cherry, J.A. (1979) Groundwater, Prentice-Hall, Inc., Englewood Cliffs, NJ.

Gavaskar, A., Gupta, N., Sass, B., Fox, T., Janosy, R., Cantrell, K., and Olfenbuttel, R. (1997) Design guidance for application of permeable reactive barriers to remediate dissolved chlorinated solvents, Battelle, Columbus, OH.

Gavaskar, A., Gupta, N., Sass, B., Janosy, R., and Hicks, J. (2000) Design guidance for application of permeable reactive barriers for groundwater remediation, Strategic Environmental Research and Development Program (SERDP), Battelle, Columbus, OH.

Gillham, R.W., and O'Hannesin, S.F. (1992) Metal-catalysed abiotic degradation of halogenated organic compounds, IAH Conference-Modern trends in hydrogeology, Hamilton, Ontario, 94-103.

Grittini, C., Malcomson, M., Fernando, Q., and Korte, N. (1995) Rapid dechlorination of polychlorinated biphenyls on the surface of a Pd/Fe bimetallic system. Environ. Sci. Technol., 29(11), 2898-2156.

Gotpagar, J., Grulke, E., Tsang, T., and Bhattacharyya, D. (1997) Reductive dehalogenation of trichloroethylene using zero-valent iron. Environ. Prog., 16(2), 137-143.

Gu, B., Liang, L., Dickey, M.J., Yin, X., and Dai, S. (1998) Reductive precipitation of uranium(VI) by zero-valent iron. Environ. Sci. Technol., 32(21), 3366-3373.

Hayes, M.H.B. (1985) Humic substances in soil, sediment, and water, John Wiley and Sons, New York, NY.

He, F., and Zhao, D. (2005) Preparation and characterization of a new class of starch-stabilized bimetallic nanoparticles for degradation of chlorinated hydrocarbons in water. *Environ. Sci. & Technol.*, 39(9), 3314-3320.

Henderson, T. (1994) Geochemical reduction of hexavalent chromium in the Trinity Sand Aquifer. *Ground Water*, 32(3), 477-486.

Hester, R.E., and Harrison, R.M. (1996) Chlorinated organic micropollutants, Royal Society of Chemistry, Cambridge, UK.

Hsi, C.K.D., and Langmuir, D. (1985) Adsorption of uranyl onto ferric oxyhydroxides: Application of the surface complexation site-binding model. *Geochim. Cosmochim. Acta.*, 49(9), 1931-1941.

Ho, C H., and Miller, N.H. (1986) Adsorption of uranyl species from bicarbonate solution onto hematite particles. *J. Colloid Interface Sci.*, 110(1), 165-171.

Huang, C.P., Wang, H.W., and Chiu, P.C. (1998) Nitrate reduction by metallic iron. *Wat. Res.*, 32(8), 2257-2264.

Jafvert, C.T., and Valentine, R.L. (1992) Reaction scheme for the chlorination of ammoniacal water. *Environ. Sci. Technol.*, 26(3), 577-586.

Johnson, T.L., Scherer, M.M., and Tratnyek, P.G. (1996). Kinetics of halogenated organic compound degradation by iron metal. *Environ. Sci. Technol.*, 30(8), 2634-2640.

Kim, Y.H., Shin, W.S., and Ko, S.O. (2004) Reductive dechlorination of chlorinated biphenyls by palladized zero-valent metals. *Journal of Environmental Science & Health Part A-Toxic/Hazardous Substances & Environmental Engineering*, 39(5), 1177-1188.

Kirk, A.B., Martinelango, P.K., Tian, K., Dutta, A., Smith, E.E., and Dasgupta, P.K. (2005) Perchlorate and iodide in dairy and breast milk. *Environ. Sci. & Technol.*, 39(7), 2011-2017.

Korte, N.E., West, O.R., Liang, L., Gu, B., Zutman, J.L., and Fernando, Q. (2002) The effect of solvent concentration on the use of palladized-iron for the step-wise dechlorination of polychlorinated biphenyls in soil extracts. *Waste Management*, 22(3), 343-349.

Lai, K.C.K., and Lo, I.M.C. (2002) Bench-scale study of the effects of seepage velocity on the dechlorination of TCE and PCE by zero-valent iron. The 6th International Symposium on Environmental Geotechnology, July 2-5, Seoul, Korea.

Lai, K.C.K., and Lo, I.M.C. (2003) Field monitoring of the performance of a PRB at the Vapokon site, Denmark. Permeable Reactive Barriers (PRBs) Action Team Meeting, RTDF, October 15-16, Niagara Falls, NY.

Lai, K.C.K., Lo, I.M.C., and Kjeldsen, P. (2004) Remediation of chlorinated aliphatic hydrocarbons in groundwater using Fe0 PRB. 1st International Symposium on Permeable Reactive Barriers, PRB-Net, March 14-16, Belfast, Northern Ireland.

Lai, K.C.K., Lo, I.M.C., Birkelund, V., and Kjeldsen, P. (2006a) Field monitoring of a permeable reactive barrier for removal of chlorinated organics. *J. of Environmental Engrg., ASCE*, 132(2), 199-210.

Lai, K.C.K., Lo, I.M.C., and Kjeldsen, P. (2006b) Natural gradient tracer test for the permeable reactive barrier in Denmark 1. Field study of tracer movement. *Hazardous, Toxic and Radioactive Waste Management Practice Periodical, ASCE*, 10(4), 231-244.

Lai, K.C.K., Lo, I.M.C., and Kjeldsen, P. (2006c) Natural gradient tracer test for the permeable reactive barrier in Denmark 2. Spatial moments analysis and dispersion of conservative tracer. *Hazardous, Toxic and Radioactive Waste Management Practice Periodical, ASCE*, 10(4), 245-255.

Langmuir, D. (1978) Uranium solution-mineral equilibria at low temperatures with applications to sedimentary ore deposits. *Geochim. Cosmochim. Acta.*, 42(6), 547-569.

Lee, W.J., and Batchelor, B. (2002) Abiotic reductive dechlorination of chlorinated ethylenes by iron-bearing soil minerals. 2. Green rust. *Environ. Sci. Technol.*, 36(24), 5348-5354.

Lee, S., Lee, K., Rhee, S., and Park, J. (2007) Development of a new zero-valent iron zeolite material to reduce nitrate without ammonium release. *J. of Environmental Engrg., ASCE,* 133(1), 6-12.

Liang, L.Y., Korte, N., Gu, B., Puls, R., and Reeter, C. (2000) Geochemical and microbial reactions affecting the long-term performance of in situ "iron barriers. *Adv. Environ. Res.*, 4(4), 273-286.

Lo, I.M.C., and Lai, K.C.K. (2002) Preliminary design of permeable reactive wall based on column studies. The 6th International Congress on Environmental Geotechnics, ISEG, August 11-15, Rio de Janeiro, Brazil.

Lo, I.M.C., Lam, S.C., and Chan, K.K. (2003a) Effects of groundwater characteristics on chromate removal by permeable reactive barriers. International Conference on Contaminated Land, 12-16 May, Gent, Belgium.

Lo, I.M.C., Lai, K.C.K., and Kjeldsen, P. (2003b) Performance of permeable reactive barrier on remedying aliphatic chlorinated organic contaminated groundwater. International Conference on Soil and Groundwater Contamination & Cleanup in Arid Countries, ICSGCC, January 20-23, Muscat, Sultanate of Oman.

Lo, I.M.C., Lam, C.S.C., and Lai, K.C.K. (2006) Hardness and carbonate effects on the reactivity of zero-valent iron for Cr(VI) removal. *Wat. Res.*, 40(3), 595-605.

Lo, I.M.C., Lai, K.C.K., and Surampalli, R. (2007) Zero-Valent iron reactive materials for hazardous waste and inorganics removal, *American Society of Civil Engineers.*

Loyaux-Lawniczak, S., Refait, P., Ehrhardt, J.J., Lecomte, P., and Génin, J.M.R. (2000) Trapping of Cr by formation of ferrihydrite during the reduction of chromate ions by Fe(II)-Fe(III) hydroxysalt green rusts. *Environ. Sci. Technol.,* 34(3), 438-443.

Matheson, L.J., and Tratnyek, P.G. (1994) Reductive dehalogenation of chlorinated methanes by iron metal. *Environ. Sci. Technol.*, 28(12), 2045-2053.

McCarty, P.L. (1997) Breathing with chlorinated solvents. *Science*, 276(5318), 1521-1522.

McCarty, P.L. (1999). Chlorinated organics, American Academy of Environmental Engineers, Annapolis, MD.

Miller, J.P., and Logan, B.E. (2000) Sustained perchlorate degradation in an autotropic, gas-phase, packed-bed bioreactor. *Environ. Sci. & Technol.*, 34(14), 3018-3022.

Montgomery, J.H. (2000) Groundwater chemicals desk reference, 3rd Ed., Lewis Publishers, CRC Press, Boca Raton, FL.

Moore, A.M., DeLeon, C.H., and Young, T.M. (2003) Rate and extent of aqueous perchlorate removal by iron surfaces. *Environ. Sci. & Technol.*, 37(14), 3189-3198.

Morin, K.A., Cherry, J.A., Dave, N.K., Lim, T.P., and Vivyurka, A.J. (1988) Migration of acidic groundwater seepage from uranium-tailings impoundments, 1. Field study and conceptual hydrogeochemical model. *J. Contam. Hydrol.*, 2(4), 271-303.

National Research Council (1994) Alternatives for ground water cleanup, National Academy Press, Washington, D.C.

National Research Council (2000) Natural attenuation for groundwater remediation, Committee on Intrinsic Remediation, Water Science and Technology Board, Board on Radioactive Waste Management, as well as Commission on Geosciences, Environment, and Resources, National Academy Press, Washington, D.C.

Nelson, D.W., and Sommers, L.E. (1982) Methods of soil analysis, part 2, American Society of Agronomy, Inc. and Soil Science Society of America, Inc., Madison WI.

Nielson, A.E. (1964) Kinetics of precipitation, Pergamon Press, New York, NY.

Nriagu, J.O., and Nieboer, E. (1988) Chromium in the natural and human environments, A Wiley-Interscience Publication, John Wiley and Sons, New York, NY.

Olsen, C.R., Lowry, P.D., Lee, S.Y., Larsen, I.L., and Cutshall, N.H. (1986) Geochemical and environmental processes affecting radionuclide migration from a formerly used seepage trench. *Geochim. Cosmochim. Acta*, 50(4), 593-607.

O'Hannesin, S.F. (1998) Groundwater remediation using in-situ treatment wall. Advances in Innovative Ground-Water Remediation Technologies, San Francisco, CA.

Orth, W.S. (1992) Mass balance of the degradation of trichloroethylene in the presence of iron filings, M. Sc. Thesis, Department of Earth Sciences, University of Waterloo, Ontario, Canada.

Ott, N. (2000) Permeable reactive barriers for inorganics, National Network of Environmental Management Studies Fellow, Office of Solid Waste and Emergency Response, United States Environmental Protection Agency, Washington, D.C.

Palmer, C.D., and Wittbrodt, P.R. (1991) Processes affecting the remediation of chromium-contaminated sites. *Environmental Health Perspectives*, 92, 25-40.

Peterson, M.L., White, A.F., Brown, G.E., and Parks, G.A. (1997) Surface passivation of magnetite by reaction with aqueous Cr(VI): XAFS and TEM results. *Environ. Sci. Technol.*, 31(5), 1573-1576.

Pourbaix, M. (1974) Atlas of electrochemical equilibria in aqueous solutions, 2nd English Ed., National Association of Corrosion Engineers, Houston, TX.

Powell, R.M., Puls, R.W., Hightower, S.K., and Sabatini, D.A. (1995) Coupled iron corrosion and chromate reduction: Mechanisms for subsurface remediation. *Environ. Sci. Technol.*, 29(8), 1913-1922.

Powell, R.M., Blowes, D.W., Gillham, R.W., Schultz, D., Sivavec, T., Puls, R.W., Vogan, J.L., Powell, P.D., and Landis, R. (1998) Permeable reactive barrier technologies for contaminant remediation, EPA/600/R-98/125, Office of Research and Development, United States Environmental Protection Agency, Washington, D.C.

Puls, R.W., Blowes, D.W., and Gillham, R.W. (1999a) Long-term performance monitoring for a permeable reactive barrier at the U. S. Coast Guard Support Center, Elizabeth City North Carolina. J. Hazard. Mater., 68(1-2), 109-124.

Puls, R.W., Paul, C.J., and Powell, R.M. (1999b) The application of in situ permeable reactive (zero-valent iron) barrier technology for the remediation of chromate-contaminated groundwater: A field test. *Applied Geochemistry*, 14(8), 989-1000.

Rai, D., and Zachara, J.M. (1988) Chromium reactions in geologic materials. EA-5741, Electric Power Research Institute, Palo Alto, CA.

Richard, F.C., and Bourg, A.C. M. (1991) Aqueous geochemistry of chromium: A review. *Wat. Res.*, 25(7), 807-816.

Ritter, K., Odziemkowski, M.S., and Gillham, R.W. (2002) An in situ study of the role of surface films on granular iron in the permeable iron wall technology. *J. Contam. Hydrol.*, 55(1-2), 87-111.

Roberts, A.L., Totten, L.A., Arnold, W.A., Burris, D.R., and Campbell, T.J. (1996) Reductive elimination of chlorinated ethylenes by zero-valent metals. *Environ. Sci. Technol.*, 30(8), 2654-2659.

Rügge, K., Bjerg, P.L., Pedersen, J.K., Mosbæk, H., and Christensen, T.H. (1999) An anaerobic field injection experiment in a landfill leachate plume, Grindsted, Denmark 1. Experimental setup, tracer movement, and fate of aromatic and chlorinated compounds. *Water Resour. Res.*, 35(4), 1231-1246.

Sanchez, A.F.J. (2003) Fe0-enhanced bioremediation for the treatment of perchlorate in groundwater, Ph.D Thesis, Civil, Architectural and Environmental Engineering, The University of Texas at Austin, Texas.

Scherer, M.M., Balko, B.A., and Tratnyek, P.G. (1998) Chapter 15 The role of oxides in reduction reactions at the metal-water interface. Kinetics and Mechanisms of Reactions at the Mineral-Water Interface., American Chemical Society, Washington, D.C.

Schwertmann, U., Gasser, U., and Sticher, H. (1989) Chromium-for-iron substitution in synthetic goethites, *Geochim. Cosmochim. Acta*, 53(6), 1293-1297.

Sedlak, D.L., and Chan, P.G. (1997) Reduction of hexavalent chromium by ferrous iron. *Geochim. Cosmochim. Acta*, 61(11), 2185-2192.

Semprini, L., Kitanidis, P.K., Kampbell, D.H., and Wilson, J.T. (1995) Anaerobic transformation of chlorinated aliphatic hydrocarbons in a sand aquifer based on spatial chemical distribution. *Water Resour. Res.*, 31(4), 1051-1062.

Shriver, D., and Atkins, P. (1999) Inorganic chemistry, 3rd Ed., W. H. Freeman and Company, New York, NY.

Sillen, L.G., and Martell, A.E. (1964) Stability constants of metal ion complexes, Special Publication No. 16, The Chemical Society, London.

Simon, F.G., Meggyes, T., and McDonald, C. (2002) Advanced groundwater remediation: Active and passive technologies, Thomas Telford, London, UK.

Snoeyink, V.L., and Jenkins, D. (1980) Water chemistry, John Wiley & Sons, New York, NY.

Stollenwerk, K.G., and Grove, D.B. (1985) Reduction of hexavalent chromium in water samples acidified for preservation. *J. Environ. Qual.*, 14(3), 395-399.

Stumm, W., and Morgan, J.J. (1970) Aquatic chemistry, Wiley-Interscience, New York. NY.

Stumm, W. (1992) Chemistry of the solid-water interface: Processes at the mineral-water and particle-water interface in natural systems, Wiley-Interscience, New York.

Stumm, W., and Morgan, J.J. (1996) Aquatic chemistry: Chemical equilibria and rates in natural waters, 3rd Ed., John Wiley & Sons, New York, NY.

Su, C., and Puls, R.W. (2004) Nitrate reduction by zerovalent iron: Effects of formate, oxalalte, citrate, chloride, sulfate, borate, and phosphate. *Environ. Sci. Technol.*, 38(9), 2715-2720.

Till, B.A., Weathers, L.J., and Alvarez, P.J.J. (1998) Fe(0)-supported autotrophic denitrification. *Environ. Sci. Technol.*, 32(5), 634-639.

United States Environmental Protection Agency (2000) In site treatment of soil and groundwater contaminated with chromium, technical resource guide, EPA-625-R-00-005, Office of Research and Development, Washington, D.C.

United States Environmental Protection Agency (2002) Field applications of in situ remediation technologies: Permeable reactive barriers, EPA-68-W-00-084, Technology Innovation Office, Office of Solid Waste and Emergency Response, Washington, D.C.

Urbansky, E.T., and Schock, M.R. (1999) Issues in managing the risks associated with perchlorate in drinking water. *J. Environ. Management.*, 56, 79-95.

Vidumsky, J.E. (2000) Permeable reactive barriers update. Advances in Innovative Ground-water Remediation Technologies Conference, June, Boston, MA.

Walton, A.G. (1967) The formation and properties of precipitates, Wiley-Interscience, New York, NY.

Warren, K.D., Arnolod, R.G., Bishop, T.L., Lindholm, L.C., and Betterton, E.A. (1995) Kinetics and mechanism of reductive dehalogenation of carbon tetrachloride using zero-valence metals. *J. Hazard. Mater.*, 41(2-3), 217-227.

Weber, K.A., Picardal, F.W., and Roden, E.E. (2001) Microbially catalyzed nitrate-dependent oxidation of biogenic solid-phase Fe(II) compounds. *Environ. Sci. Technol.*, 35(8), 1644-1650.

Westerhoff, P., and James, J. (2003) Nitrate removal in zero-valent iron packed columns. *Wat. Res.*, 37(8), 1818-1830.

Wiberg, K.B. (1965) Oxidation in organic chemistry, Academic Press, New York, NY.

Wilkin, R.T., Puls, R.W., and Sewell, G.W. (2001) Long-term performance of permeable reactive barriers using zero-valent iron: An evaluation at two sites, Office of Research and Development, Subsurface Protection and Remediation Division, United States Environmental Protection Agency, Ada, OK.

Wilkin, R.T., and Puls, R.W. (2003) Capstone report on the application, monitoring, and performance of permeable reactive barriers for ground-water remediation: Volume 1 – performance evaluations at two sites, EPA/600/R-03/045a, Office of Research and Development, United States Environmental Protection Agency, Cincinnati, OH.

Williams, A.G.B., and Scherer, M.M. (2001) Kinetics of Cr(VI) reduction by carbonate green rust. *Environ. Sci. Technol.*, 35(17), 3488-3494.

Yak, H.K., Lang, Q., and Wai, C.M. (2000) Relative resistance of positional isomers of polychlorinated biphenyls toward reductive dechlorination by zerovalent iron in subcritical water. *Environ. Sci. Technol.*, 34(13), 2792 – 2798.

Yu X.Y., Amrhein, C., Deshusses, M.A., and Matsumoto, M.R. (2006) Perchlorate reducing by autotrophic bacteria in the presence of zero-valent iron. *Environ. Sci. & Technol.*, 40(4), 1328-1334.

Zhang, H., Bruns, M.A., and Logan B.E. (2002) Perchlorate reduction by a novel chemolithoautotropic hydrogen-oxidizing bacterium. *Environ. Microbiol.*, 4(10), 570-576.

CHAPTER 4

Biological Assimilation and Degradation

S. BALA SUBRAMANIAN, SONG YAN, R. D. TYAGI AND R. Y. SURAMPALLI

4.1   Introduction

Many xenobiotic compounds are highly carcinogenic and mutagenic, causing major health disorders such as acute, chronic, and systemic allergic reactions. They, in turn, affect the nervous, respiratory and circulatory systems, liver, organ, kidney and reproductive system of human beings and impact the surrounding ecosystem as well (Steven et al., 2000). Xenobiotic pollutants may be broadly categorized into organic, inorganic and biological classes. Organic compounds mainly include aliphatic and aromatic hydrocarbons, chlorinated biphenyls (PCBs), and polycyclic aromatic hydrocarbons (PAHs). Inorganic pollutants include heavy metal ions and their complexes (Lajoie et al., 1993). Most of these xenobiotic pollutants have been found to be persistent and recalcitrant. Though physical and chemical treatments are the conventional methods for pollutants removal, the former only renders a temporary solution and the latter may further contaminate the environment by adding chemical reagents or producing new pollutants. Therefore, there is a need to develop innovative biological and natural processes to eliminate these hazardous persistent organic and inorganic compounds from the environment.

Microorganisms play a major role in the breakdown and mineralization of these pollutants. Many researchers have reported that microorganisms in the environment can efficiently adapt to use xenobiotic chemicals as novel growth and energy substrates. Their specialized enzyme metabolic pathways for the degradation of xenobiotic pollutants have been found in microbial strains that metabolize them completely and at considerable rates. These type of microbial strains have been isolated from natural systems, such as soil and wastewater. The degradative enzymes include intracellular and extracellular, which are responsible for bioaccumulation, biotransformation, and co-metabolism, and can degrade low molecular weight compounds and a variety of macromolecules (Ju, 1997; Davis, 2002; Evans, 2003). Among these microorganisms, *Pseudomonas* species are the most important for biodegradation of toxic pollutants. Many aromatic compounds such as benzoate, p-hydroxybenzoate, maldelate, tryptophan, phthalate, and salicylate have been transformed into a common intermediate beta-ketoadipate by 1,2-dioxygenase produced by *Pseudomonas* species (Cafaro et al., 2004). Furthermore, some recalcitrant compounds, such as PAHs, can also be utilized by microbes and synthesize useful byproducts such as bio-surfactants (Cooper et al., 1984). Aliphatic hydrocarbons are assimilated by a wide variety of microorganisms, but not all microbial species are capable of utilizing them as a growth substrate. Many details of microbial hydrocarbon metabolism have been elucidated (Ratledge, 1984) including enzymology (Providenti, 1993), regulation and genetics (Witholt et al., 1990). The biodegradation process may be affected by physicochemical conditions such as temperature, pH, redox potential, salinity, and oxygen concentration or by the availability of the substrates such as solubility, and dissociation from adsorbed materials. Little attention has been paid to the primary interaction of microorganisms and the pollutants involved.

Various microorganisms such as bacteria, cyanobacteria, fungi and algae are able to remove efficiently heavy metals from the environment via bioaccumulation and biosorption processes. Microorganisms can concentrate metals to levels that are largely higher than those existed in the environment. Biosorption is an important approach for the bioremediation of metal-contaminated environments by interactions between metals and live or dead microbial biomass. Microbes adsorb the toxic metals on their cell wall and recover them from polluted sites (Nandakumar et al., 1995). *Chlamydomonas reinhardtii* grown in laboratory media were found to fractionate radioisotope selenium compound (Thomas, 2004). Microorganisms also are found to produce metalothiones, a metal accumulating protein secreted by bacterial species to adsorb metal ions from the surrounding environment.

Similarly, various plant species are also known to remove toxic metals from the soil and water through biosorption and bioassimilation (Hong et al., 2000; Zumriye, 2005). Researchers have found that plants also have the capability to degrade the toxic pollutants through enzymatic degradation. The use of certain plants to clean-up contaminated sites with a variety of pollutants such as heavy metals, radionuclides, chlorinated solvents and pesticides, petroleum hydrocarbons, polychlorinated biphenyls, polynuclear aromatic hydrocarbons, and explosives, is known as phytoremediation (Lasat, 2002; Singh et al., 2003). Elegant transgenic approaches have been designed for the development of mercury or arsenic phytoremediation technologies (Krämer, 2005).

This chapter discusses types of pollutants, their degradative pathways mechanisms along with microorganisms responsible for assimilation and degradation factors affecting their rate of biodegradation and bioavailability of microorganisms. Case studies showing the natural degradation of xenobiotic pollutants and biodegradation of petroleum hydrocarbons in the marine ecosystem are presented.

4.2   Definitions

4.2.1 Biodegradation

Biodegradation is the process by which organic substances are broken down into non-toxic or less harmful by-products by other living organisms and enzymes. Reactions can take place *in vitro* or *in vivo*. Generally, biodegradation process is affected by several important factors, such as temperature, pH, nutrients and oxygen availability. The process may also be evaluated for purposes of hazard assessment such as primary alteration of the chemical structure of the compound resulting in loss of a specific property of that compound and removal of the toxic properties of the compound. This often corresponds to primary biodegradation but it depends on the circumstances under which the products are discharged into the environment and ultimate breakdown of the compound to either fully oxidized or reduced simple molecules such as carbon dioxide/methane, nitrate/ammonium, and water (Martin et al., 2003).

4.2.2 Mineralization

Mineralization is a process where a substance is converted or oxidized from an organic substance to an inorganic substance and derivatives, such as nitrate, carbon dioxide, thereby becoming mineralized. This is also called biomineralization (Schwartz et al., 1997), which takes place during the life of an organism such as the formation of bone tissue or egg shells, largely with calcium. Alternatively, it may be a process which begins after death and burial within sediments by the total replacement of the organic material with various minerals known as fossilization. Frequently this involves either calcite or quartz, but many other minerals such as pyrite may be involved.

4.2.3 Biotransformation

Biotransformation is the process whereby a substance is changed from one chemical form to another (transformed) by a chemical reaction such as oxidation, reduction or hydrolysis of the compounds, conjugation with products of intermediately metabolism such as acetic acid, cysteine, glucuronic acid, glutamate, glycine and ornithine, alkylation or dealkylation, deamination and decarboxylation within the body or any chemical conversion of substances that is mediated by living organisms or enzyme preparations derived therefrom (Ornston et al., 1990). Biotransformation is vital to survival in that it transforms absorbed nutrients into substances required for normal body functions. Biotransformation could serve as an important defense mechanism in which toxic xenobiotics and body wastes are converted into less harmful substances and substances that can be excreted from the body.

4.2.4 Bioaccumulation/Bioassimilation

Bioaccumulation or bioassimilation is a process by which a contaminant is taken up by organisms directly through the physical exposure pathway or through consumption of food containing the contaminated substance. It incorporates the concepts of bioconcentration and biomagnification, Depending on the composition of the food, concentration factors can vary from one environment to another for specific organisms corresponding to a given chemical (Hong et al., 2000). Bioaccumulation is a normal process that can result in injury to an organism only when the equilibrium between exposure and bioaccumulation is overwhelmed, relative to the harmfulness of the chemical. The extent of bioaccumulation depends on concentration of a chemical in the environment, amount of chemical coming into an organism or the capacity of the organism to transfer the chemical from outside to inside the cell, and duration of exposure, and the time necessary for the organism to acquire the chemical and then degrade it.

4.2.5 Bioavailability

Bioavailability of pollutants is an important parameter to determine the quantity of chemical compounds absorbed or degraded into systemic circulation by microbial or ecological receptor that is exposed to soil at a particular site (Davis et al., 2002).

4.3   Common Degradation Pathways and Mechanisms

Bacterial metabolic pathways are a series of chemical reactions occurring within a cell. These reactions are catalyzed by enzymes and controlled by degradative genes. Apart from the usefulness of microbes in the degradation of toxic pollutants, some bacterial species can act as a pathogen in the environment. Such pathogens have to be treated with appropriate methods and via proper waste disposal. Microbes play an important role in the biodegradation and bioremediation of toxic pollutant using their effective metabolic pathways and degradative enzymes, which is eco-friendly and cost effective physical and chemical process. Recent advances in bio-degradation of selected pollutants, mechanisms of degradative pathways and microbes responsible for degradation are presented in Table 1. Mechanism and pathways of biodegradation of organic compounds such as aromatic hydrocarbon and PAHs in petroleum, coal tar and shale oil wastes and polychlorinated biphenyls (PCBs) are discussed. Commonly degradative pathways and catalytic mechanisms of bacteria are given below.

4.3.1 Aromatic hydrocarbons

Degradation of aromatic hydrocarbon, especially PAHs, takes place by "ortho"- and "meta-" cleavage of the dioxygenate ring structure (Dagely, 1986). It is an aerobic degradation process which is catalyzed by inherent microbial populations and could aid in the complete removal of aromatic hydrocarbons in the environment. The aromatics degradation is affected by hydrocarbon concentration, bacterial categories and metabolites, and other chemical and physical parameters. Generally, the ring cleavage pathways consist of thousands of chemical reactions catalyzed by intracellular or extracellular enzymes produced by the microbial populations. These two alternative approaches of ring cleavage pathways clearly indicate comprehensive understanding of catalytic mechanisms preferentially used by different organisms (Fig. 4.1 and 4.2) (Dagely, 1986; Ornston, 1982).

4.3.1.1 Central Metabolic Pathways

Catechol and protocatechuate are the principal catabolites of aromatic metabolism as shown in Fig. 4.1 and 4.2. Central metabolic pathways mainly contain lycolysis, pentose phosphate pathway, TCA cycle and electron transport system. They are cycles of variable enzymatic activities, designed to shuffle metabolic intermediates where needed and to supplement or replenish nutrients when necessary. With regards to the aromatic hydrocarbon metabolism, ultimate destination of all degradative pathways is the Krebs' cycle as a metabolic intermediate (Fig. 4.3) (Neidhardt, 1990). In 1947, Stainer first conceptualized the principle of enzyme regulation, defining the basis for such regulation as substrate induction of related enzyme synthesis (Stainer, 1947). Metabolic processes seem to converge upon a few compounds, such as catechol, protocatechuate, or benzoate. The principal catabolic intermediates provide alternative routes, or metabolic options, the availability of which is dependent upon the concentration of each principal catabolite and the level of enzymatic activity in pathways that might permit their further degradation. In some *Pseudomonas*, the major metabolic route for degradation of toluene and alkyl-substituted toluene contrasts with that of the beta-ketoadipate pathway. The meta-fission of catechol is different from oxygen into the ring structure via a catechol 2,3 dioxygenase rather than by a catechol 1,2-dioxygenase (Burlage et al., 1989).

4.3.1.2 Degradative pathways of PAHs

Degradation of PAHs containing four or five fused rings by bacteria and fungi has been well studied (Dagely, 1986). The process involving biodegradation is inversely proportional to the ring size of the PAH molecule. Low molecular weight PAHs are degraded more readily while high molecular weight PAHs are difficult to be degraded by normal enzymatic degradation (Ronald, 1987). Microbial catabolism of PAHs mainly depends upon two types of enzymes, mono-oxygenases and dioxygenases. Microbes require molecular oxygen to catalyze initial hydroxylation of substituted PAHs (Gibson, 1982). Biodegradation of naphthalene in soil was first reported in 1927, since then many researchers are still studying the ability of bacteria to utilize naphthalene as the sole source of carbon and energy (Ferrero, 2002). *Pseudomonas* species are successfully applied in degradation of naphthalene by using naphthalene dioxygenase.

Table 4.1 Biodegradation of different compounds in the different environment

Category	Toxic Pollutants	Microorganisms	Environmental compartment	Reference
Aliphatic hydrocarbons	Alkanes/ hydrocarbons/oil	*Acinetobacter, Aeromonas eutrophus, Alcaligenes, Bacillus subtilis, Corynebacterium Micrococcus, Moraxella, Mycobacterium, Psedomonas mendocina, Pseudomonas pickettii, Pseudomonas putida, Pseudomonas stutzeri*	Water Soils	Witholt et al. (1990)
Chlorinated aromatic compounds	Dichloro-, pentachloro and hexachlorobiphenyl	**Bacteria** *Enterobacteriaceae, Ralstonia, Flavobacterium, Gluonbacter, Halobacterium, Hyphomicrobium sp., Micrococcus, Moraxella Nocardia, Xanthomonas, Zoogloea spp, Ralstonia, Nocardia, Psedomonas mendocina, Pseudomonas pickettii, Pseudomonas putida, Pseudomonas stutzeri, Pseudomonas testosterone* **Algae** *Anabaena, Calothrix, Cyanobacteria (Blue green algae), Diatoms, Nostoc.* **Yeast and Fungi** *Aspergillus aculeatus, Aspergillus niger, Coriolus versicolor, Emericella nidulans, Frankia, Ganoderma sp., Geotrichum candidum, Laminaria digitata, Penicillium miczynskii, Phanerochaete chrysosporium, Rhizopus oryzae, Torulopsis bombicola, Trametes versicolor*	Water Soils Sediments	Robert and Robert (2001); Davis et al. (2002); Xiujin et al. (2004)
Aromatic hydrocarbons	Benz(a)anthracene, Benzo (a)pyrene, Chrysene, 7,12 Dimethylbenz(a)anthracene, Dibenz(a,h)anthracene, Fluoranthene, Fluorene Naphthalene, Phenanthrene,	*Xanthomonas, Zoogloea spp, Ralstonia, Nocardia, Psedomonas mendocina, Pseudomonas pickettii, Pseudomonas putida, Pseudomonas stutzeri, Pseudomonas testosterone*	Soil, sediments and water	Canter (1997); Ronald (1987); Autheunisse, et al. (1987); Kappesser et al. (1989); Cafaro et al. (2004)

Table 4.1 Biodegradation of different compounds in the different environment (Cont'd)

Nitrogen compounds	Ammonium, Ammonium carbonate, Nitrate and Cyanide	*Anabaena, Moraxella, Mycobacterium, Nitrosomonas and Nitrobacter SPP., Nostoc, Rhizobium, Bradyrhizobium, Rhodopseudomonas, Spirillum, Thiobacillus*	Effluents, water and soil	Brady (1979); Martin (2003); Stephen (2004)
Metals	Arsenic, Aluminium, Beryllium, Cadmium, Chromium and nickel, Cobalt, Iron, Lead (including differences between inorganic and organic lead compounds), Mercury, Manganese, Vanadium, Zinc and copper	Bacteria, fungi and micro algae are involved in removal of heavy metals through biosorption. *Rhizopus arrhizus, Mycobacterium chlorophenolicium, Bacillus subtilis, Penicillium miczynskii, Emercilla nidulans, aeromonas species, Rhizopus oryzae, Asperigillus niger, Candida lipolytica,*	Effluent and water	Toshiyuki et al. (1998); Zumriye (2005)
Micro pollutants	**Endocrine Disrupting Compounds (EDCs)** 17 β-estradiol, 17α-ethylestradiol estrone, DEHP, Nonylphenol, androsterone, Nonylphenol, NPEOs, steroids (estrogens, progestogens), Alkyl phenols, Bisphenol AOC pesticides PCDDs/DFs, PCBs, PAHs, & Phthalates, Musk xylol and Musk ketone	Anaerobic bacteria are involved in degradation of these micropollutants	Estuary water, river sediment, suspended particles, soil,	Thomas (2004); Petrovic (2004)

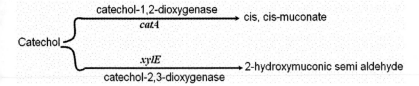

Figure 4.1 Mechanism for degradation pathway of catechol

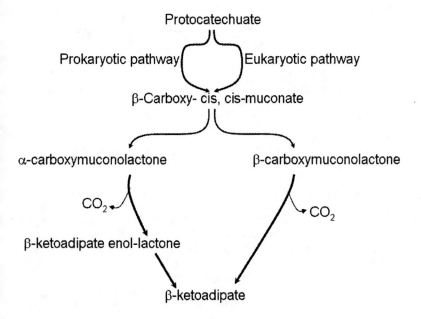

Figure 4.2 Catabolism of protocatechute by prokaryotic and eukaryotic organisms

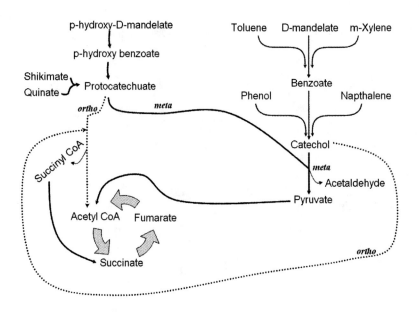

Figure 4.3 Central metabolic pathways for aromatic hydrocarbons

The metabolic pathway incorporates both atoms of molecular oxygen to form cis-naphthalene dihydrodiol. The degradation of naphthalene is accomplished by the following three steps. Firstly, the oxidation of naphthalene by *Pseudomonas* cis-naphthalene dihydrodiol is converted to 1,2-dihydroxynaphthalene catalyzed by cis-naphthalene dihydrodiol dehydrogenase. Secondly, 1,2-dihydroxynaphthalene is cleaved by a dioxygenase to yield cis-2′-hydroxybenzylpyruvate. Finally, further cleavage of cis-2′-hydroxy- benzylpyruvate to pyruvate and slicylaldehyde by an aldolase is followed by subsequent oxidation by dehydrogenase to salicylate (Fig. 4.4) (Cerniglia and Heitkamp, 1989). The oxidation of salicylate by salicylate hydroxylase yields catechol, which undergoes either ortho- or meta-fission depending on the bacterial species, enzyme regulation and its expression whether by inducible or constitutive expression as discussed above. Besides, *Cunninghamela elegans* and a variety of fungi have been found to transform naphthalene to metabolites that are similar to those produced by mammalian enzymes and laboratory animals. Like prokaryotic organisms, fungi do not utilize naphthalene as a sole source of carbon and energy (Skubal et al., 2001) but utilize cytochrome P-450 monooxygenase to form naphthalene 1,2-oxide. The enzyme epoxide hydrolase converts naphthalene 1,2-oxide into trans-naphthalene dihydrodiol.

Blue green algae are also found to degrade naphthalene through oxidation and produce cis-naphthalene dihydrodiol as a minor metabolite. The efficient and complete biodegradation of naphthalene can be achieved using biotechnology process and genetic engineering tools for cloning and genetic improvement of micro-organisms (Thomas, 2004).

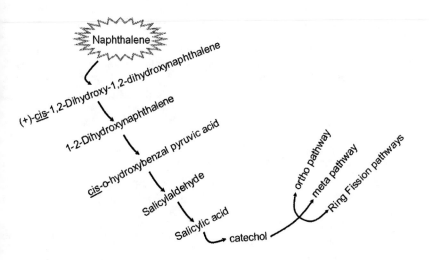

Figure 4.4 Steps involved in bacterial degradation of naphthalene

4.3.1.3 Catabolic pathways of PCBs

Polychlorinated biphenyls (PCBs) are a series of organic compounds with 1 to 10 chlorine atoms attached to biphenyl and a general chemical formula of $C_{12}H_{10-x}Cl_x$. PCBs vary from mobile oily liquids to white crystalline solids and hard non-crystalline resins. PCBs are one of the major toxic substances manufactured by many countries and used in variety of applications such as transformer oils, capacitor dielectrics and heat transfer fluids and are found in large quantities in the environment (D'Silva, 2003).

PCB degrading bacteria are mostly aerobic in nature and found in the PCBs polluted sites. Gram negative bacteria for PCB degradation include *Pseudomonas, Acinetobacter, Achromobacter, Alcaligenes, Moraxella* and *Acetobacter*. Gram positive bacteria include *Arthrobacter* and *cornyebacterium* species, *et al*. These strains utilize PCBs as a sole source of carbon and energy and co-metabolize them into a number of components like chlorobenzoic acids via ring-dioxygenation and meta-cleavge (Kas, 1997). Microbial degradation of PCBs using *Acinetobacter* and *Alcaligenes* species has been discussed. The degradation rate of PCBs decreases with the increase in chlorine substitution. PCBs containing two chlorines in the ortho-position of a single ring and each ring show a striking resistance to degradation. PCBs containing all chlorines on a single ring are generally degraded faster than those containing the same number on both rings. PCBs having two chlorines at the 2,3 position of one ring are susceptible to microbial attack by *Alcaligenes and Acinetobacter* (Xiujin, 2004). The oxidative degradation of PCBs into chloro-benzoic acid by microbes involves four enzymes, biphenyl dioxygenase, dihydrodiol dehydrogenase, 2,3-dihydroxybiphenyl dioxygenase and hydrolase, produced from the bphA, bphB, bphC and bphD genes, respectively in *Alcaligenes and Acinetobacter* (Furukawa et al., 1982). A schematic diagram showing possible mechanism is given in Fig. 4.5. With the development of molecular biology techniques, the clone of bph operon into other soil bacterial species has been established in laboratories in order to understand the evolution of gene and its possible role in ecology.

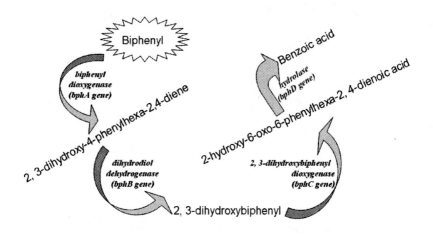

Figure 4.5 Degradative pathways of PCBs

4.3.2 Inorganic pollutants

Removal of inorganic pollutants (metallic and non-metallic) and understanding their roles is very important particularly for the removal of toxic metals. Lead, cadmium, chromium, zinc, arsenic, mercury and copper are major toxic metal pollutants existed in ecosystem (Furukawa et al., 1982). Metal compounds are assimilated by extra cellular compounds and enzymes produced by microorganisms. Most of the algae accumulate these toxic metals on the cell wall. The biological assimilation of inorganic compounds and heavy metals are well studied and defined by using several terms such as bioassimilation, biosorption, bioaccumulation, and biotransformation (Zumriye, 2005). But these removal mechanisms have not been fully studied at molecular level for heavy metals and radioactive isotopes.

4.3.3 Nitrogenous compounds

Nutrient compounds for organisms are mainly nitrogen, phosphorus and sulfur. Nutrients are firstly used by plants and animals, returned to the environment and then re-used. High nutrient concentration causes phytoplankton booms such as, red tides, various yellow and green foams, slimes, and slicks (Toshiyuki, 1998). However, algal and phytoplankton blooms have been considered as an unhealthy ecosystem. The toxicity of these blooms is increasing, which can have a direct effect on the organisms that feed on them. Moreover, phytoplanktons naturally contain dimethyl sulfoxide (DMS), which is released from dead phytoplankton into the atmosphere and can be transformed to sulfuric acid and result in acid rain. Plants can store high levels of nitrate, or translocate it from tissue without deleterious effect. If however, livestock and humans consume plant material rich in nitrates, they may suffer methemoglobinemia. In contrast to nitrate, high levels of ammonium are toxic to both plant and animals (Lea, 1992).These nitrogenous compounds are assimilated and degraded by natural transformation process such as, ammonification, volatilization, nitrification, denitrification, plant and microbial uptake, deposition, and adsorption (Campbell, 1996; Marschner, 1995).

Biochemical pathways and microbial metabolism can also be used for removal of cyanides, a kind of nitrogenous compound (Stephen, 2004). Cyanide plays an important role in the evolution of life on earth (Oro, 1981) and remains provides nitrogen source for microorganisms, fungi and plants. Major sources of cyanide discharge include petrochemical refining, synthesis of organic chemicals and plastics, electroplating, aluminum, metal mining and processing industries. The release of cyanide from these industries has been estimated to be >14 million kg/yr (ATSDR, 1997).

Some microorganisms can biodegrade cyanide compounds. Cyanide degradation takes place by hydrolytic, oxidative, reductive, and substitution transfer reactions. Several reviewers have discussed these pathways by various microorganisms (Raybuck, 1992; Dubey, 1995; Barclay, 2002), and organisms (Adjei et al., 1999; Kwon et al 2002; Yanese et al., 2000). More than one pathway can be utilized for cyanide biodegradation (Ezzi-Mufaddal, 2002) as shown in Figures 4.6. The catalytic pathway is controlled by external factors such as availability of oxygen, pH and cyanide concentration. Besides cyanide bio-availability and solubility in soil-water systems are also important factors for its degradation (Aronstein et al., 1994).

Hydrolytic Reactions

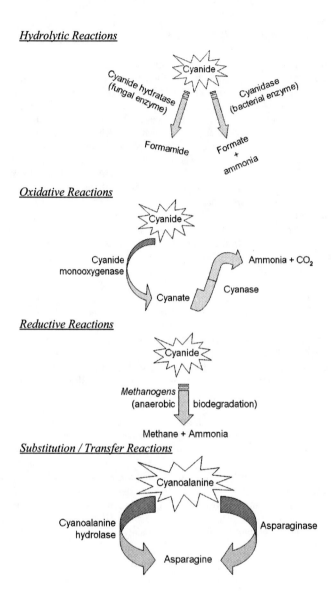

Oxidative Reactions

Reductive Reactions

Substitution / Transfer Reactions

Figure 4.6 Reactions for the degradation of cyanide

4.3.4 Micro pollutants

Micro pollutants such as endocrine disrupting compounds, estrogens and antibiotics which are strictly prohibited in most countries (Williams, 1996) are creating multitudinal impacts on the environment because of their unacceptable risks caused to humans and ecosystems as well (FAO, 2003.). The last decades have seen a growing awareness of the possible adverse effects of micro pollutants. To reduce the risks for humans and ecosystem, reduction of contaminants at source is often necessary. Some of the organic micro pollutants include chiral herbicide mecoprop, chiral pharmaceutical drug ibuprofen as well as the natural compound caffeine (Thomas, 2004). Brominated organic micro pollutants such as tetrabromobisphenol-A, tetrabromophthalic anhydride, polybrominated diphenyl ethers, polybrominated biphenyls, hexabromocyclododecane, are highly toxic and mimic thyroxine hormones. Some of these compounds are more toxic than chlorinated dioxins. These compounds are extracted using solvent extraction, modified silica column chromatography and carbon chromatography; and purification can be done by silica chromatography/alumina chromatography methods (D'silva, 2003; 2004). Chemical analysis of EDCs is generally not sure what in meant here insufficient to assess contaminants present in the environment and to estimate their endocrine potential. Apart from the fact that several new chemicals are synthesized every year and released into the environment with unforeseen consequences, and evidence for their endocrine potential is constantly emerging. However, chemical identification of all compounds responsible for endocrine disrupts is an arduous task. Therefore, detection of EDCs from complex mixtures of environmental contaminants requires specific procedures combining chemical analysis and biological assays, an approach based on the toxicity identification and evaluation (TIE). This procedure was developed by US EPA at the onset of 1990s (Norberg, 1991; Ankley, 1992). However, during the last decade, the TIE approach has become an established and powerful tool for determining acute toxicity, genotoxicity and endocrine disrupting potential from environmental samples.

However, recent studies have revealed that some emerging contaminants are weak endocrine disruptors, such as p-amino musk xylene showed *in vitro* estrogenic activity (Bitsch et al., 2002). Some other reports have shown the presence of synthetic ovulation inhibiting hormones in the aquatic environment (Shore et al., 1993). Several reviewers report levels of various personal care products in the aquatic environment (Daughton et al., 1999). Rimkus et al. (1999) reviewed data on the occurrence of polycyclic musk, musk xylene and musk ketone aminometabolites in water, sediments and suspended particulate matter, sewage sludge and biota. Iodine containing diagnostic agents was reported by Hirsch et al. (1999).

4.4.   Rates of Transformation and Half-lives

Fundamental prerequisites for microbial biodegradation are listed as follows:

(1) A microorganism must exist which has the necessary metabolic capacity to bring about biodegradation ("Principle of Microbial Infallibility").

(2) Biodegrading organism(s) must be present in contaminated environment.
(3) Contaminant must be accessible.
(4) Environmental conditions must be conducive for proliferation of microorganisms.

Before biodegradation of any organic compound, a lag period occurs, called acclimatization period, and induces genes to produce enzymes for biodegradation, growth of initial population of organisms. Concentration of the contaminant affects the rate of degradation. Therefore, a minimum concentration of compound required for the growth of microorganisms and if the compound concentration is lower than the required level, biodegradation may not occur. Typical threshold concentration for organic compounds is in the range between 0.1-10 mg/L H_2O or mg/kg soil (Alexander, 1999). Most of organisms can not tolerate high concentration of contaminants because of their toxicity. Significant degradation does not occur until rare species capable of surviving are able to produce biomass sufficient to catalyze biodegradation as detailed by Chapelle (2001).

4.5   Pollutants and Persistency/Bioavailability

Bioavailability of pollutants is an important factor to determine the quantity of chemical compounds that can be absorbed into systemic circulation by human or ecological receptors is exposed to soil at a particular site. Chemical and physical characteristics of the environments as well as the chemical form of the contaminants can have significant effects on the bioavailability of pollutants. Scientists have been involved in developing and applying site specific oral bioavailability values for inorganic (antimony, arsenic, beryllium, cadmium, chromium, lead, mercury and vanadium) and organic compounds (polycyclic aromatic hydrocarbons) in soil at several sites (D'Silva, 2004).

4.6   Factors Affecting Degradation

4.6.1 Oxygen

Availability of oxygen increases the growth rate/yield and biodegradation is always maximal with oxygen as the electron acceptor. Aerobes have mono- and dioxygenases, which are uniquely effective in oxidation of hydrocarbons (especially aromatics). Presence of O_2 can suppress degradation of halogenated pollutants through inhibition of reductive dechlorination. In case of hydrocarbon degradation, oxygen is rapidly depleted until the development of anaerobic bacteria populations capable of degradation (Donald, 1991).

4.6.2 Inorganic compounds

Availability of inorganic nutrients such as nitrogen/phosphorus may limit the biomass production in presence of excess organic carbon which in turn reduces the rate of degradation for organic pollutant (Johns, 1991). On the other hand, addition of inorganic nutrients such as nitrogen/phosphorus is also used to stimulate degradation of oil spills and soil/subsurface petroleum hydrocarbon degradation (Bitsch, 2002).

4.6.3 Sorption

Sorption of toxic pollutant on solid-phases decreases availability to microbes and hence biodegradation may require growth of organisms with the ability to utilize sorbed compounds. Some examples exhibiting the same mechanism include uptake of selenium by *Chlamydomonas* species and removal of basic blue, acid blue and congo red dyes from aqueous solution by biosorption on *Asperigillus niger* (Fu, 2002)

4.6.4 Solubility

Many organic contaminants like PAHs are highly insoluble and therefore are difficult for microbes to access (Ryan, 1988).

4.6.5 Non-aqueous phase liquids (NAPLs)

NAPLs are the most widespread organic pollutants of the subsurface and cause severe environmental and health hazards as they dissolve into groundwater or soil. However, NAPL is very difficult to be degraded because NAPL system can seriously hinder solubilization /biodegradation of toxic organic contaminants (e.g. chlorinated solvents) (Robert, 2001).

4.6.6 Influences of contaminant structure on bioavailability

Degradation of petroleum hydrocarbons such as branched and linear alkanes, monoaromatics and polyaromatics inhibit the efficiency of enzymatic attack by presence/location of halogens, amine groups, methoxy groups, phenoxy groups from halogenated organics, pesticides, herbicides (Harvey, 1991).

4.6.7 Miscellaneous factors

Microbial amount (concentration of cells), co-metabolism/synergism, plasmid-born degradation, and some additional factors such as pH, suspended solids, soil texture and permeability, soil moisture, free ammonia, biological oxygen demand (BOD), chemical oxygen demand (COD), dissolved oxygen (DO) and total nitrogen, temperature, acidity, alkalinity etc. also affect the rate of biodegradation (Dushenkov, 1997).

4.7   Case Studies

Examples of effective biodegradation of toxic contaminants using microorganisms and plants are presented.

4.7.1 Alkanes and PCBs biodegradation

The biodegradation of aliphatic and aromatic hydrocarbons by natural soil microflora and seven fungi species, including imperfect strains and higher level lignolitic species, was studied in a 90-day laboratory experiment using a natural, not-fertilized soil contaminated with 10% crude oil. The natural microbial soil biofilm isolated from an urban forest area was unable to significantly degrade crude oil, whereas pure fungi cultures effectively reduced the residues by 26-35% in 90 days. Normal alkanes were almost completely degraded in the first 15 days, whereas aromatic compounds (phenanthrene and methylphenanthrenes) exhibited slower degradation. *Aspergillus terreus* and *Fusarium solani*, isolated from oil-polluted areas, showed more efficient degradation of aliphatic and aromatic hydrocarbons, respectively. Overall, imperfect fungi isolated from polluted soils showed a somewhat higher efficiency, but the performance of unadapted, indigenous, lignolitic fungi was comparable, and all three species, *Pleurotus ostreatus, Trametes villosus* and *Coriolopsis rigida*, effectively degraded aliphatic and aromatic components. The simultaneous, multivariate analysis of 22 parameters allowed the elucidation of a clear reactivity trend of the oil components during biodegradation: lower molecular weight n-alkanes > phenanthrene > 3-2-methylphenanthrenes > intermediate chain length n-alkanes > longer chain length n-alkanes > isoprenoids ~9-1-methylphenanthrenes. Irrespective of the individual degrading capacities, all fungi species tested seem to follow this decomposition sequence (Colombo et al., 1996).

Marine pollution by hydrocarbons of crude and refined petroleum products is known to cause serious environmental problems. Several petroleum hydrocarbons are known to be toxic and natural remediation of these substances is known to take months to years leaving high-density hydrocarbons in the environment. These toxic hydrocarbons often enter food chains causing long term effects in marine organisms, fishes, birds, leading to permanent inheritable genetic changes. In view of these environmental issues of pollutant oil spill in marine environment, effective disaster management becomes an important task (Kostal et al., 1998). The rate of biodegradation can be enhanced in such situations by using efficient oil degrading bacteria, slow release nutrients and bioemulsifier for the emulsification and dispersion of pollutant oil.

Degradation of petroleum hydrocarbon using bioemulsifier produced from a consortium of oil degrading bacteria (ODB) isolated from marine environment having ability to degrade and utilize hydrocarbons of high speed diesel (HSD). The bioemulsifier obtained as a dry powder was characterized using different instrumental and analytical techniques (Susan et al., 2004). The bioemulsifier was found to reduce the surface tension of distilled water from 72 mN/m to 36 mN/m at a concentration of 0.25% (w/v). In seawater and synthetic seawater, the material was found to show reduction in surface tension to almost same extent as observed in distilled water. Surface tension reduction is an important property of a bioemulsifier/biosurfactant for the formation of effective stable emulsions of two immiscible phases such as oil in water commonly encountered during marine oil pollution.

Bioemulsifiers/biosurfactants are amphiphilic molecules having affinity both for water and oil. This amphiphilic nature facilitates homogeneous dispersion of oil into the water column as microdroplets thus forming emulsions by reducing surface tension. These surface-active agents help in entry of bacteria into oil layers, increase the bioavailability of hydrocarbons to microorganisms and increase dissolved oxygen for the growth and multiplication of aerobic bacteria.

The bioemulsifier was sprayed along with oil degrading bacteria (ODB) in oil polluted site in west coast in Arabian Sea and resulted in emulsification of oil, which were readily taken up by bacteria as an energy source. Nutrient compounds were also added in hydrophobic phase which helped in the fast removal of oil without any adverse impacts on the environment. Effective biodegradation of floating oil was found to be within a period of 10 days in the area of 25 m^2 in west coast of the Arabian Sea.

Kas et al. (1997) studied the biodegradation of alkanes and PCBs in soil from oil pollution using mixtures of indigenous soil bacteria in the laboratory bioreactos. The oil was biodegraded within three months. The microbiological, biochemical and genetic characteristics of bacterial strain *Pseudomonas* C12B, which degrade petroleum hydrocarbons and alkanes were studied. Bacterial strain was isolated originally for its ability to utilize alkylsulfonates as the sole source of carbon and energy, demonstrating the metabolic ability of bacteria to degrade various chemical pollutants. Biodegradation of PCBs was carried out using two biological approaches, bacterial co-cultures and plant cells cultivated *in vitro*. An industrial mixture of PCBs was used. The bacterial enzymatic degradation showed a decrease in initial PCB concentration to 20% after 45 days (Kas et al., 1997).

Degradation of PCBs using plants cells involved selection of various plant species with differing growth parameters and morphology (amorphous, differentiated shoot forming or "hairy root"), also genetically transformed and/or non-transformed by *Agrobacterium* bacterial strains, were used. Differentiated or hairy root cultures exhibited better degradative abilities than undifferentiated amorphous cultures.
The study confirmed the effectiveness of biodegradation by specific biodegrading bacteria and plants for alkanes and PCBs.

4.7.2 Biotransformation of chlorinated compounds

Skubal et al. (2001) studied natural biotransformation of petroleum hydrocarbons to evaluate biogeochemical characterization in the field and laboratory microcosm. They assessed the biotransformation potential of trichloroethylene (TCE) and toluene in a plume containing petroleum hydrocarbons and chlorinated solvents at the former Wurtsmith Air Force base in Oscoda, MI.

The study investigated the terminal electron accepting process (TEAPs), microbial phylogeny and contaminants composition. Biotransformation was assessed based on the presence of reduced electron acceptors, relevant microbial communities and occurrence of metabolic byproducts like dissolved methane and carbon dioxide. The accumulation of benzene, toluene, ethylbenzene and xylene (BTEX) metabolites and dechlorination products were used to further confirm the biodegradation of BTEX in the methanogenic zone by reductive dechlorination.

Mineralization of TCE and toluene was found to be less in the microcosms during 300 days of incubation, without added electron acceptors and nutrient compounds. Evidence of methanogenesis in microcosms appeared after 8 months and CH_4 and CO_2 were found in the degradation of contaminants other than toluene.

Cis-dichloroethylene was found only in one methanogenic microcosms after 500 days of incubation. The study revealed that dynamic redox potential zonation prevented the reductive dehalogenation of highly chlorinated solvents during the course of year. The predominat TEAP at the highly contaminated water shifted from methanogenesis to iron- and sulfate reduction.

This study recommended that biotransformation should be integrated with other factors such changes in redox zonation, electron donor utilization and co-metabolic substrate transformation to yield a more accurate assessment of bio-attenuation of specific pollutants in aquifers contaminated by undefined waste mixtures.

4.7.3 Biosorption of organic pollutants

Fu and Viraraghavan (2002) investigated the removal of dyes such as basic Blue 9 (cationic), acid Blue 29 (anionic), Congo Red (anionic) and disperse Red 1 (nonionic) dyes from aqueous solution by biosorption process using dead *Asperigillus niger* fungus. They found that *A. niger* removes dyes from aqueous environment utilizing three major functional groups-carboxyl, amino and phosphate. The lipid fraction in the biomass of fungus also played an important role in the biosorption.

Similarly, Aksu (2001) also investigated biosorption of two reactive dyes onto dried activated sludge containing bacteria and protozoa. He inferred that activated sludge has an extensive uptake capacity of organic pollutants due to acidic polysaccharides, lipids, amino acids and other cellular components available on the cell wall of bacteria.

Hong et al. (2000) studied the biosorption of 1,2,3,4 tetrachlorodibenzo-p-dioxin (1,2,3,4-TCDD) and pesticides such as polychlorinated dibenzofurans (PCDFs) by *Bacillus pumilus*. The results indicated that removal of these compounds is more efficient by dead organisms than the live bacterial strains. They suggested that in addition to the attachment to microorganisms itself; extracellular polymorphic substances might also be involved in the biosorption process. Biosorption of pesticide lindane was studied by Ju et al. (1997) under different pH ranging between 2.93 to 6.88 using *E. coli, Z. ramigera, Bacillus megaterium,* and *B. subtilis.* It was proposed that the repulsive electrostatic force for the adsorption of organic halide on the bacterial cell surface decreased when the lower pH generated less negative charges on the cell wall. As the cell and lindane molecules move closer to each other, the van der Waals force was intensified and biosorption was enhanced consequently.

4.7.4 Biodegradation of cyanide and explosives

Stephen (2004) studied the removal of cyanide compounds from the waste products of a number of industrial processes. Biodegradation of cyanide compounds occurred via four reaction modes such as hydrolytic, oxidative, reductive and substitution. He suggested that biodegradation of cyanide using bacteria and plants can be successfully used for the removal of cyanide and cyanide compounds from the waste products.

Contamination of soil and groundwater by explosives wastes and their transformation products is a common problem at explosives manufacturing/ processing, storage and disposal facilities around the world. Of the nearly twenty different energetic compounds used in conventional munitions by the military today, hexahydro-1, 3,5-trinitro-1,3,5-triazine (RDX) and octahydro-1,3,5,7- tetranitro-1,3,5,7-tetrazocine (HMX) are the most powerful and commonly used compound. These compounds are highly toxic in nature and have resulted in severe soil and groundwater contamination (Cataldo, 1990).

Khan et al. (2004) conducted laboratory-scale studies on the removal of RDX and HMX by black grass (Ophiophogon blackei), tomato (Lycopersicon Sps.) and sacred basil (Ocimum sanctum). These plants were grown under simulated and controlled environmental conditions in a greenhouse. The soil used was taken from the Metcalfe House complex, Delhi. Explosive samples of RDX and HMX were obtained from High Energy Materials Research Laboratory (HEMRL), Pune. The purities of the explosives were 99±1%.

Results of this study revealed phyto availability of RDX and HMX and their possibility of removal from the soil plants. Ophiophogon sps. (Black grass) is a versatile plant species that can withstand higher concentrations of RDX and HMX up to 301 mg/kg. Lycopersicon Sps. (tomato) showed a high removal of almost 96 % at a concentration of 25 mg/kg and significant removal rates at other concentrations of 37.5 and 50 mg/kg. Ocimum Sps. was very sensitive to the concentrations of both RDX and HMX. Conjugate removal mechanism operated in the soil by involvement of both biotic and abiotic factors and was a greener and eco-friendly option to tackle such wastes. Phytoremediation was considered as an apt technology for adoption at explosives contaminated sites after extensive geo-morphological study (Thompson et al., 1998; Harvey, 1991).

Stephen et al. (2002) studied the degradation of explosives from pink water generated from military waste water discharge. The hazardous pink-water contained dissolved trinitrotoluene (TNT) and cyclo trimethylene trinitamine (RDX), as well as

TNT & RDX by-products. They studied anaerobic treatment of explosives by fluidized bed reactor (FBR) using granular activated carbon (GAC). They have demonstrated bench scale batch studies using an anaerobic bacterial consortium, which fed ethanol as a sole source of carbon and energy and converted the TNT into triaminotoluene (TAT), which was consequently degraded into undetectable end products.

RDX was also sequentially degraded into nitrioso-, dinitroso-, trinitroso- and hydroxylaminodinitroso-RDX followed by triazine ring cleavage and formed methanol and formaldehyde as major end products. The anaerobic bacterial consortia were isolated from the sludge digesters of municipal waste water treatment plants. The same procedure was studied in field conditions to treat large volume of samples. They concluded that FBR-GAC was cost effective and more efficient method of treatment for explosives compared to the only GAC based system. Thus, this method could be used and was recommended for the treatment of pink water.

Kyung et al. (2002) designed a bench scale anoxic membrane bioreactor (MBR) system for treatment of explosives. It consisted of a bioreactor coupled with ceramic cross flow ultrafiltration module. The system was evaluated to treat synthetic waste water containing alkaline hydrolysis byproducts (hydrolysates) of RDX. The MBR system removed 95% of the carbon sources, 93.3% amount of nitrate and 55% of nitrite. The MBR very efficiently separated the biomass from the effluent.

4.8   Conclusions

Toxic pollutants proliferation and magnification in the environment are going to intensify in the time to come and characteristic alarms of the same are being already felt in the environment. Naturally, physical and chemical processes have been shown as to detoxify and degrade some of these contaminants. However, these process results in the recompartmentalisation of pollutants and transfer of the pollution from one place to another instead of degrading them. At this crux, biological process can treat and handle these pollutants more in the natural context. Further, extensive studies are being carried out to potentially use biological processes as well as ameliorate pre-existing ones in an effort to enhance pollutant removal. Genetics involved in biological degradation are being studied more extensively so as to comprehend molecular changes during various biologically facilitated mechanisms and to arrive at a solution that can offer plenty of advantages. It is evident from this review that many biologically mediated pollutant degradation have been successfully documented for field studies. It would be wise to conclude that, "Dilution is no more a solution to pollution abatement; however, biological option is an eco-friendly and long term potion".

References

Adjei, M.D. and Ohta, Y. (1999) Isolation and characterization of a cyanideutilizing Burkholderia cepacia strain. *World J Microbiol Biotechnol.*, 15:699-704

Aksu, Z. (2001) Biosorption of reactive dyes by dried activated sludge: equilibrium and kinetic modeling. *Biochem Eng* J. 7: 79-84.

Ahlet, R.C., and Peters, R.W. (2001) Treatment of PCB-contaminated soils: I. evalution of in situ Reductive Dechlorination of PCBs. *Environmental Progress*. 108-116.

Ankley, G.T., and Burkhard, L.P. (1992) Environ Toxicol Chem 11:1235.

Aronstein, B.N., Maka, A., Srivastava, V.J. (1994) Chemical and biological removal of cyanides from aqueous and soil-containing systems. *Appl Biochem Microbiol.* 41:700-707.

ATSDR (1997) Toxicological Profile for Cyanide. US Department of Health Human Services, Public Health Service. Atlanta, GA.

Barclay, M., Day, J.C., Thompson, I.P., Knowles, C.J., and Bailey M.J. (2002) Substrate-regulated cyanide hydratase (chy) gene expression in Fusarium solani: the potential of a transcription-based assay for monitoring the biotransformation of cyanide complexes. *Environ Microbiol*, 4:183-189.

Bitsch, N., Dudas, C., Koerner, W., Failing, K., Biselli, S., Rimkus, G., and Brunn H. (2002) *Arch Environ Contam. Toxicol.*, 43:257.

Burlage et al. (1989) The TOL (pWWO) catabolic plasmid. *Appl. Environ. Microbiol.* 55:1323-1328.

Burrows. W.D. (1982) Tertiary treatment of effluent from Holston Army Ammunition Plant, US Army Armament R&D Command, Dover, NJ, Report No. 8207.

Cafaro, V., Izzo, V., Scognamiglio, R., Notomista, E., Capasso, P., Casbarra, A., Pucci, P., and Donato, A.L. (2004) Phenol hydroxylase and toluene/o-xylene monooxygenase from *Pseudomonas stutzeri* OX1: Interplay between two enzymes. *Appl. Env. Microbiol.*, 70(4): 221-2219.

Campbell, W.H. (1996) Nitrate reductase biochemistry comes of age. *Plant Physiol.* 111:355-361.

Cerniglia, C. E., and Heitkamp, M.A. (1989) Microbial biodegradation of polycyclic aromatic hydrocarbons in aquatic environment. In metabolism of *Polycyslic Aromatic Hydrocarbons in the aquatic Environment*. Ed. U. Varanasi. Boca Raton, FL; CRC Press, Inc. 41-68.

Cooper, D.G., and Paddock, D. A. (1984) Production of a biosurfactants from *Torulopsis bombicola. Appl. Env. Microbiol.* 47: 173-176.

Dagely, S. (1986) Biochemistry of aromatic hydrocarbon degradation in *Pseudomonads*. In the Bacteria, X. Ed. J. R. Sokatch, L. N. Ornston. New York: Academic press. 527-555.

Daughton, C.G., Ternes, T.A. (1999) *Environ Health Perspect* 107:907

Davis, J. W., Odom, J.M., DeWeed, K.A., Stahl, D.A., Fishbain, S.S., West, R.J. (2002) Natural attenuation of chlorinated solvents at area 6, Dover Air force base: characterization of microbial community structure . *Jounal of contaminant hydrology*, 57(1-2), 41-59.

Donald, J. and Freeman, P.E. (1991) Application of the powdered activated Carbon/ Activated Sludge (PACT) and wet air oxidation process to the treatment of explosives contaminated waste water, Proceedings of the ninth International Symposium on Compatibility of Plastics and other Materials with Explosives, Propellants, San Diego.

D'Silva, K., Fernandes, A., White, S., Rose, M. (2003) Analysis of polybrominated diphenyl ethers in foodstuffs - extension of an extraction method for PCDD/Fs and PCBs to incorporate PBDEs, *Organohalogen Compounds*, 61: 179-182.

D'Silva, K., Fernandes, A., Rose, M. (2004) Brominated Organic Micropollutants - Igniting the flame retardant issue, *Critical Reviews in Environmental Science and Technology*, 34: 141-207.

Dubey, S.K., and Holmes, D.S. (1995) Biological cyanide destruction mediated by microorganisms. *World J Microbiol Biotechnol*, 11:257.

Dushenkov, S., Vasudev, D., Kapulnik, Y., Gleba,D., Fleisher D., Ting K.C., and Ensley, B. (1997) Removal of Uranium from water using terrestrial plants, *Environ. Sci. Technol.*, 31: 3468-3474.

Eisler, R. (1987) Polycyclic aromatic hydrocarbon hazards to fish, wildlife, and invertebrates: a synoptic review. *Contaminant Hazard Reviews. Biological Report* 85(1.11): 1-55.

Evans, G.M., and Furlong, J.C. (2003) Environmental Biotechnology: theory and applications. West Sussex, UK, John Wiley and Sons.

Ezzi-Mufaddal, I, Lynch, J.M. (2002) Cyanide catabolizing enzymes in *Trichoderma spp. Enzyme Microb Technol*, 31:1042-1047.

FAO, Food and Agriculture Organization of the United Nations, 2003. 'International Code of Conduct on the Distribution and Use of Pesticides', FAO, Rome,

Ferrero, M., Brossa, E.L., Lalucat, J., Valdes, E.G., Mora, R.R., and Bosch, R. (2002) Coexistence of two distinct copies of naphthalene degradation genes in Pseudomonas strains isolated from the western Mediterranean region. *Appl. Environ. Microbiol.* 68(2): 957-962.

Fu, Y., and Viraraghavan, T. (2002) Removal of Congo Red from an aqueous solution by fungus *Asperhillus niger.Advances Environ Res.* 7: 239-247.

Furukawa, K., and Chakrabarty, A.M. (1982) Involvement of plasmids in total degradation of chlorinated biphenyls. *Appl. Environ. Microbiol.* 44: 619-626.

Gibson, D.T. (1982) Microbial degradation of hydrocarbons. Toxicol. *Environ. Chem.* 5: 237-250.

Gibson, D.T. (1988) microbial metabolism of aromatic hydrocarbons and the carbon cycle. In microbial metabolism and the carbon cycle. Ed. S. R. Hagedorn, R. S. Hanson, D. A. kunz. Chur, Switzerland. Harwood Academic Publishers. 33-58

Harayama, S., and Timmis, K.N. (1989) Catabolism of aromatic hydrocarbons by Pseudomonas. In Genetics of bacterial Diversity. Ed. D. A. Hopwood, K. F. Chater, London; Academic press.

Harvey, S.D. (1991) Fate of the explosive hexahydro-1,3,5-trinitro-1,3,5 triazine (RDX) in soil and bioaccumulation in bush bean hydroponic plants, *Environmental Toxicology and Chemistry*,10: 845-855.

Hong, H.B., Hwang, S.H., and Chang, Y.S. (2000) Biosorption of 1,2,3,4-tetrachlorodibenzo-p-dioxin and polychlorinated dibenzofurans by *Bacillus pumilus*. *Water Res.* 34: 349-353.

Johns, H.O., Bragg, J.R., Dash, L.C. and Owens, E.H. (1991) Proceedings of the 1991 International oil spill conference. March 4-7,1991, San Deigo, C.A. pp 167-176.

Ju, Y.H., Chen T. and Liu, J. C. (1997) A study on the biosorption of lindane. *Colloids Surf B*. 9: 187-196.

Kas, J., Burkhard, J., Demnerova, K., Kostal, J., Macek, T., Mackova, M., and Pazlarova, J. (1997) Perspectives in biodegradation of alkanes and PCBs. *Pure &Appl. Chem.*, Vol. 69, No. 11, pp. 2357-2369.

Kassel A.G., et al. (2002) Phytoremediation of trichloroethylene using Hybrid Poplar. *Physiol. Mol. Biol. Plants* 8(1): 3-10.

Khan, S., Brar, S.K., and Misra, K. (2004) Phytoremediation of RDX and HMX by Ophiophogon Blackei and Lycopersicon sps., Chapter 6 in Book Titled` Bio-Technological Applications in Environmental Management' editor; Dr. R.K. Trivedy.

Krämer, U. (2005) Phytoremediation: novel approaches to cleaning up polluted soils, *Current Opinions in Biotechnology*, 16:133-141

Kostal, J., Suchanek, M., Klierova, H., Kralova, B., Demnerova, K., and McBeth, D. (1998) *Pseudomonas* C12-b, an SDS degrading strain , harbours a plasmid coding for degradation of medium chain length n-alkanes. International *Biodeterioration and Biodegradation* . 42:221-228.

Kwon, H.K., Woo, S.H., Park, J.M. (2002) Thiocyanate degradation by Acremonium strictum and inhibition by secondary toxicants. *Biotechnol Lett*. 24:1347-1351.

Kyung, D.Z., and Stenstrom, M.K. (2002) Application of a membrane bioreactor for treating explosives process wastewater. *Water Research*. 36: 1018-1024.

Lajoie, C.A., Zylstra, G.J., DeFlaun, M.F., and Strom, P.F. (1993) Development of field application vectors for bioremediation of soils contaminated with polychlorinated biphenyls. *Appl. Environ. Microbiol*. 59: 1735-1741.

Lasat, M. M. (2002) Phytoextraction of toxic metals: a review of biological mechanisms. *J. Environ. Qual*. 31, 109-20

Lea, P.J., Blackwell, R.D., and Joy Kenneth, W. (1992) Ammonia assimilation in higher plants. *In: "Nitrogen Metabolism of Plants" Proc Phytochem Soc Europe* 33:153-186.

Marschner, H. (1995) *Mineral Nutrition of Higher Plants*. 2nd ed. Academic Press, London.

Martin, B. and Jackson, L.E. (2003) Microbial immobilization of ammonium and nitrate in relation to ammonification and nitrification rates in organic and conventional cropping systems. *Soil biology and biochemistry*. 35: 29-36.

Nandakumar, B. A., Dushenkov, V., Motto, H., Raskinse, I. (1995) Phytoextraction: Use of plants to remove heavy metals from soils, *Environ. Sci. Technol*. 29(5): 1232-1235.

Neidhardt, et al., (1990) Physiology of the bacterial cell. New York: Sinauer.

Norberg King, T.J., Durhan, E.J. and Ankley, G.T. (1991) Environ Toxicol Chem 10:891.

Ornston, L.N., and Yeh, W.K. (1982) Recurring themes and repeated sequences in metabolic evolution., In A. M. Chakrabarty (ed.), Biodegradation and Detoxification of environmental pollutants, CRC press, Miami.

Ornston, L.N., Houghton, J.E., Neidle, E.L., and Gree, L.A. (1990) subtle selection and novel mutation during divergence of the β-ketoadipate pathway. In *Pseudomonas*. Biotransformations, pathogenesis, and evovlving biotechnology. Ed. R. Silver, A. M. Chakrabarty. Ed., ASM publications, Washington D.C., 207-225.

Oro´, J., Lazcano-Araujo, A. (1981) The role of HCN and its derivatives in prebiotic evolution. In Cyanide in Biology. Edited by Vennesland B, Conn EE, Knowles CJ, Westley J, Wissing F., Academic Press, London, 517-541.

Petrovic, M., Eljarrat, E., Lopez de Alda, M.J., Barcelo, D. (2004) Endocrine disrupting compounds and other emerging contaminants in the environment: A survey on new monitoring strategies and occurrence data. *Anal Bioanal Chem* 378 : 549-562.

Providenti, M.A., Lee, H. and Trevors, J. C. (1993) Selected factors limiting the microbial degradation of recalcitrant compounds. *J. Industrial Microbiol.* 12,379-395.

Ratledge, C. (1984) Microbial conversion of alkanes and fatty acids. JAOCS 61,447-453.

Raybuck, S.A. (1992) Microbes and microbial enzymes for cyanide degradation. *Biodegradation*, 3:3-18.

Ryan, A.J., Bell, R.M., Davidson, J.M., and O'Conner, G.A. (1988) Plant uptake of non-ionic organic chemicals from soils, *Chemosphere.* 2299-2323.

Schwartz, O.J., and Jones, L.W. (1997) Chapter 14:Bioaccumulation of xenobiotic organic chemicals by terrestrial plants,In W. Wang, J.W. Gorsuch, and J.S. Hughes (eds.) *Plants for Environmental Studies.* 418-449.

Shore, L.S., Gurevitz, M., and Shemesh, M . (1993) Bull Environ ContamToxicol 51:361

Singh, O.V., and Labana, S. (2003) Phytoremediation: an overview of metallic ion decontamination from soil. *Appli. Microbiol. Biotechnol.* 61, 405-12.

Skubal, K.L., Barcelona, M.J., and Adriaens, P. (2001) An assessment of natural biotransformation of petroleum hydrocarbons and chlorinated solvents at an aquifer plume transect. *Journal of contaminant hydrology.* 49: 151-169.

Stainer, R.Y. (1947) Simultaneous adaptation: A new technique for the study of metabolic pathways. J. Bacterial. 54: 339-348

Stephen, W.M., Adrian, N.R., Hickey, R.F., and Heine, R.L. (2002) Anaerobic treatment of pinkwater in a fluidized bed reactor containing GAC. *Journal of Hazardous Matterials* 92: 77-88.

Stephen, E. (2004) Biological degradation of cyanide compounds. *Current Opinion in biotechnology.* 15: 231-236.

Titus, S., Kumar, P., and Deb, P.C. (2003) Bioemulsifier produced by hydrocarbon utilizing bacteria for the bioremediation of pollutant oil. Paper presented in *World Conference on Disaster Management and Infrastructure Control System*, 10-12 November, Hydrerabad, India

Thomas M.J. (2004) A review of mass-dependent fractionation of selenium isotopes and implications for other heavy stable isotopes. *Chemical geology*, 204: 201-214.

Thompson, P.L., Ramer, L.A., and Schnoor, J.L. (1998) Uptake and transformation of TNT by a poplar hybrid. *Environ. Sci. Tech.* 32(7): 975-80.

Thornton, F.S., Bright, M.I., Lerner, D.N., and Tellam, J.H. (2000) Attenuation of landfill leachate by UK Triassic sandstone aquifer materials. 2. Sorption and degradation of organic pollutants in laboratory columns. *Journal of Contaminant Hydrology* 43: 355-383.

Toshiyuki, M., Ueda, M., Kawaguchi, T., Arai, M., and Tanaka, A. (1998) Assimilation of cellooligosaccharides by a cell surface-engineered yeast expressing β-glucosidase and carboxymethylcellulase from *Aspergillus aculeatus*. *Appl. Env. Microbiol.* 64 (12): 4857-4861.

Williams, F., Henschler, D., Rummel, W., and Starke, K. (1996) 'Allgemeine und spezielle Pharmakologie und Toxikologie', Spektrum Akademischer Verlag, Heidelberg, , p. 981.

Witholt, B., de Smet, M.J., Kingma, J., van Beilen, J.B., Kok, M., Lageveen, R.G. and Eggink, G. 1990. Bioconversion of aliphatic compounds by Pseudomonas putida in multiphase bioreactors; Background and economic potential. *Trends Biotechnol.* 8, 46-52

Xiujin, Q., Melinda, K.S., and Davis, J.W. (2004) Remediation options for chlorinated volatile organics in a partially anaerobic aquifer. *Remediation* Autumn. 39-47.

Yanese H., Sakamoto, A.,Okamoto, K., Kita, K., Sato. Y. (2000) Degradation of the metal-cyano complex tetracyanonickelate (II) by Fusarium oxysporum N-10. *Appl Biochem Microbiol*, 53:328-334.

Zumriye, A.. (2005) Application of biosorption for the removal of organic pollutants: a review. *Process Biochemistry* 40: 997–1026

CHAPTER 5

Photolysis and Photocatalytic Degradation

FENG MAO

5.1   Introduction

Photochemical degradation by sunlight is an important abiotic chemical transformation process in natural environment. Contaminants that are capable of absorbing ultraviolet, visible, or infrared radiation (light) may undergo transformation by direct photodegradation by breaking a specific chemical bond by light. Alternatively, contaminants that do not themselves absorb light can also be degraded in the environment through indirect photodegradation. This process may occur when chromophoric components in an environmental medium absorb photons and subsequently transfer energy to the contaminant of interest, or when highly reactive species such as free radicals are formed photochemically, and subsequently attack and degrade the target contaminant. In many cases, the contribution of photodegradation may be insignificant, but in others it may be the dominant pathway in determining the fate of a contaminant in the natural environment.

5.2   Basic Principles of Photochemistry

5.2.1 Nature of Light

Light is a form of electromagnetic radiation, in which energy is transmitted through space by the interaction of electric and magnetic fields. Light has both wave- and particle-like properties. As a wave, light is characterized in part by its wavelength (λ) and frequency (ν). The wavelengths in extraterrestrial solar radiation range from approximately 100 nm to greater than 3000 nm. Figure 5.1 shows a typical solar irradiance spectrum above atmosphere and at surface. Visible light consists of radiation having wavelength ranging from approximately 400 to 700 nm whereas the ultraviolet (UV) region covers the spectral range of 290 to 400 nm. Wavelengths in the range of 760 to 2000 nm are considered to be infrared (IR) or heat rays. The relationship between wavelength and frequency can be expressed by:

$$v = \frac{c}{\lambda} \qquad \text{(Eq. 5.1)}$$

where c is the speed of light in a vacuum, 3.0×10^8 m/s.

Light can also be described by its particle-oriented nature. Each quantum of light called a photon possesses energy equal to E:

$$E = hv = h\frac{c}{\lambda} \qquad \text{(Eq. 5.2)}$$

where h is Plank's constant, 6.64×10^{-34} J•s. The energy of a photon is dependent on its wavelength with shorter wavelength having higher energy per photon.

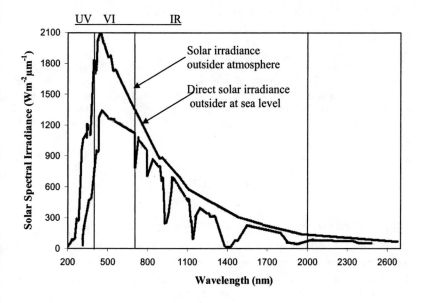

Figure 5.1 Solar irradiance spectrums above atmosphere and at surface (UV – ultravoilet, VI – visible light, IR – infrared)

To understand the phenomenon of photochemical reaction, it is instructive to compare the energies of light at different wavelengths with the energies of the covalent bonds of organic molecules. Table 5.1 shows that the typical energies of common covalent bonds and the approximate wavelength of light corresponding to this energy.

Table 5.1 Typical energies of covalent bonds and the corresponding wavelength of light

Bond	Bond energy, E (kcal/mol) [*]	Wavelength, λ (nm)
C-H (alkanes)	91-99	289-314
C-H (benzene)	103	278
C-C (alkanes)	78-84	340-367
C-O (alcohols)	89-90	318-321
C-N	79	362
C-Cl	78-82	349-367
C-Br	67	427
O-H (alcohols)	100-102	280-286
O-O	35-51	561-867

[*]Source: Harris, J. C. (1982)

5.2.2 Light Absorption by Chemical Substances

5.2.2.1 Beer-Lambert Law

Absorption of light by chemical substances is an essential step in initiating a photochemical reaction. Generally, absorption of light can be described by the Beer-Lambert law, which is an empirical relationship that relates the absorption of light to the properties of the medium (specifically referring to water for this chapter) through which the light penetrates. There are several ways in which the law can be expressed:

$$I_\lambda = I_{0,\lambda}\ 10^{-[\alpha_\lambda + \varepsilon_\lambda C]l} \qquad\qquad \text{(Eq. 5.3)}$$

$$A = log\frac{I_{0,\lambda}}{I_\lambda} = [\alpha_\lambda + \varepsilon_\lambda C]l \qquad\qquad \text{(Eq. 5.4)}$$

where $I_{0,\lambda}$ is the intensity of the incident light at wavelength λ; I_λ is the intensity of light after passing through the water solution; ε_λ is the molar absorptivity or extinction coefficient of the absorbing species of interest in the water; α_λ is the absorption coefficient or attenuation coefficient of the medium, reflecting the absorbance by water or other absorbing species; C is the concentration of absorbing species of interest in the water; l is the distance traveled by light through the water (the path length); and A is the absorbance.

As shown in Eq. 5.3 and Eq. 5.4, light is attenuated not only by the compound of interest absorbing the light, but also by the medium (water) or other species. The impact of the medium and other species on absorbance can be quantified separately, but are often removed by defining $I_{0,\lambda}$ as the intensity of light passing through a sample "blank" or "baseline" or reference sample (the same water solution but with zero concentration of the compound of interest). In this case, the absorbance A becomes:

$$A = \varepsilon_\lambda \, C \, l \qquad\qquad \text{(Eq. 5.5)}$$

Eq. 5.5 states that, the optical absorbance of a compound in a transparent medium varies linearly with both the compound concentration (C) and the optical path length (l). If l and λ are fixed, a plot of absorbance (A) versus concentration (C) would yield a straight line. Measurement of a compound in water or solution by a spectrophotometer is based on this principle.

5.2.2.2 Absorption Spectrum

As indicated by Beer-Lambert law, the absorbance (A) for a particular compound with a known concentration (C) and optical path length (l) is a function of the light wavelength. For a particular compound, the plot of absorbance versus wavelength, called absorption spectrum, can be determined with a spectrophotometer. The normalized absorption spectrum (independent of concentration) is a characteristic of the particular compound and is like the chemical "fingerprint" of the compound. Typically, chemical substances absorb light over a wide wavelength range and their absorption spectrum may exhibit one or more absorption peaks. The wavelengths (λ) of maximum absorptions are of utmost importance since these wavelengths can be used to evaluate or predict whether a compound absorbs ambient light or not. These absorption peaks may also help determine the optimal or suitable wavelength for the photochemical degradation of the compound in an engineering system.

The absorption spectrum of an organic compound strongly depends on its molecular structure. Organic compounds that can significantly absorb UV/visible light in the solar spectrum have one or more *chromophores* (functional groups within the compound that can absorb UV/visible light). The chromophoric groups include alkenes, aromatic and heterocylic compounds, aldehydes and ketones, nitro compounds, and azo dyes (Larson and Weber, 1994; Harris, J. C., 1982). Table 5.2 lists the wavelengths of maximum absorption (λ_{max}) and molar absorptivity (ε) for selected chromophoric groups.

Table 5.2 Chromophoric groups and their typical chromophoric values*

Chromophore	Wavelength, λ_{max} (nm)	Absorptivity, ε
C-C or C-H	<180	1000
C=C	180	10,000
C=O	280	20
C=S	460	Weak
C=C-C=C	220	20,000
C=C-C=O	320	50
N=N	350	50
N=O	300	100
Benzene	260	200
Naphthalene	310	200
Anthracene	350	10,000
Phenol	275	1500

Sources: Larson and Weber (1994); Harris, J. C. (1982)

5.2.2.3 Quantum Yields

Absorption of a photon results in the promotion of a molecule from a lower or ground state (P), to a higher energy level or excited state (P*). The excited species (P*) may then lose their energy and return to their ground state by a variety of physical or chemical pathways. Physically, the energy may be lost as heat in the process of *internal conversion*, or as light (electromagnetic radiation) in the processes of *fluorescence* and *phosphorescence*. Alternatively, the energy may be transferred to another molecule through the process of *photosensitization* and thus the other molecule becomes excited. Chemically, the energy may initiate a chemical reaction within the molecule (primary step) and the products of primary step may further participate in a series of photo, chemical, or biological processes. These reactions are of utmost importance from the perspective of photodegradation since they lead to the transformation and possibly the "removal" of the compound from a given system.

The efficiency at which the initial excited species (P*) undergo a particular pathway, i, is conventionally expressed in terms of its *quantum yield*, ϕ_i, defined as:

$$\phi_i = \frac{Number\ of\ P^*\ undergoing\ pathway\ i}{total\ number\ of\ photons\ absorbed\ by\ P} \qquad \text{(Eq. 5.6)}$$

Typically, the quantum yield will be less than 1 since not all photons are absorbed productively. Quantum yields greater than 1 are possible for photo-induced or radiation-induced chain reactions, in which a single photon may trigger a long chain of transformations.

For a chemical photodegradation process where a molecule falls apart after absorbing a light quantum, the *reaction quantum yield*, ϕ_r is the number of destroyed (transformed) molecules divided by the total number of photons absorbed by the system:

$$\phi_r = \frac{total\ number\ of\ molecules\ transformed}{total\ number\ of\ photons\ absorbed\ by\ P} \qquad \text{(Eq. 5.7)}$$

The reaction quantum yields for the transformation of organic pollutants in natural water are generally less than 0.01 (Harris, 1982), mainly due to low pollutant concentrations and the presence of other species that may inhibit the chain reactions.

5.3   Light Absorption in Natural Water Bodies

5.3.1   Chromophores in Natural Water Bodies

Inorganic chromophores in natural water bodies are rare, as many of the inorganic ions in natural waters, Cl^-, Br^-, SO_4^{2-}, CO_3^{2-}, Na^+, K^+, Mg^{2+}, and Ca^{2+} are transparent to solar radiation. Several exceptions are nitrate, nitrite, and $Fe(OH)^{2+}$, which can absorb solar radiation and produce hydroxyl radicals ($OH^{•}$) (Larson and Weber, 1994). Nitrate photolysis is believed to be the dominant process for $OH^{•}$ formation in natural waters. The reactions and mechanisms involved in the nitrate photolysis will be discussed later.

Simple and known natural organic chromophores are rare, and a few examples are carbonyl compounds, methyl iodide, riboflavin, tryptophane, thiamine, and vitamin B_{12} (Zafiriou et al., 1984). Many of the natural chromophores, however, cannot to be identified and therefore are usually named as unknown photoreactive chromophores (UPC).

The most important chromophore in natural waters is dissolved organic matter (DOM). A major fraction of DOM is humic substances, which may account for up to 90% of the dissolved organic carbon (DOC) content in water (Thurman and Macolm, 1983). When humic substances absorb UV or solar radiation, reactive oxygen intermediates are produced, which play a significant role in various photochemical processes in surface waters.

5.3.2 Mathematical Description of Light Absorption in Natural Water Bodies

The intensity of light at a given depth in a water column depends on three factors: the degree of light transmittance through the atmosphere, the transmittance of light in the air-water interfacial region, and the optical characteristics of the water body. Sunlight can penetrate to relatively great depths in pure water. For example, sunlight can penetrate to depths in excess of 140 m in clear mid-ocean seawater (Larson and Weber, 1994). All natural water, however, contain dissolved and suspended materials that limit the transmittance.

The light absorption process in natural water bodies has been described in detail by Schwarzenbach et al. (1993). For a well-mixed (homogenous) surface water body with a known volume (V) and a surface area (A), the light intensity at a given depth (z) can be approximately described by the Lambert-Beer law:

$$I_{z,\lambda} = I_{0,\lambda}\, 10^{-\alpha_{D,\lambda}\, z} \qquad \text{(Eq. 5.8)}$$

where $I_{z,\lambda}$ is the light intensity for a particular wavelength (λ) at z; I_{λ} is the light intensity at the water surface; and $\alpha_{D,\lambda}$ is commonly referred to as the *attenuation diffusion coefficient*, which is generally determined by in-situ measurement. Light absorption by the water body (A_{λ}) is then estimated by calculating the difference between the light intensity at the water surface and the light intensity at the depth of z:

$$A_{S,\lambda}(\text{based on per unit surface area}) = I_{0,\lambda}[1 - 10^{-\alpha_{D,\lambda} z}] \qquad \text{(Eq. 5.9)}$$

$$A_{V,\lambda} \text{ (based on per unit volume)} = \frac{I_{0,\lambda}[1-10^{-\alpha_{D,\lambda}z}]A}{V} = \frac{I_{0,\lambda}[1-10^{-\alpha_{D,\lambda}z}]}{z} \quad \text{(Eq. 5.10)}$$

The fraction of light absorbed by a particular compound in water bodies is given by:

$$F_{c,\lambda} = \frac{\varepsilon_\lambda C}{\alpha_\lambda + \varepsilon_\lambda C} \quad \text{(Eq. 5.11)}$$

where ε_λ and α_λ have been previously defined. Generally, the pollutant concentration in water bodies is very low and thus the absorption of light by the pollutant is much smaller than the absorption of light by all other chromophores ($\varepsilon_\lambda C \leq \alpha_\lambda$). Eq. 5.11 is simplified to:

$$F_{c,\lambda} \approx \frac{\varepsilon_\lambda}{\alpha_\lambda} C \quad \text{(Eq. 5.12)}$$

Therefore, absorption of light by the particular compound can be calculated by Eq. 5.13 which is a combination of Eq. 5.10 and Eq. 5.12:

$$I_{a,\lambda} = A_{V,\lambda} \times F_{c,\lambda} = \frac{I_{0,\lambda} \varepsilon_\lambda [1-10^{-\alpha_{D,\lambda}Z}]}{z\alpha_\lambda} C = k_{a,\lambda} C \quad \text{(Eq. 5.13)}$$

where $k_{a,\lambda}$ is defined as the specific rate of light absorption of a given compound in a given system.

5.4   Photochemical Reaction in Natural Water Bodies

5.4.1 Direct Photolysis

Direct photolysis is a chemical process where a chemical compound of interest absorbs light and consequently undergoes transformation. Direct photolysis reactions are kinetically simple and easily modeled if the absorption spectrum of the compound and its quantum yield of disappearance are known or can be measured. The reaction rate of direct photolysis at wavelength λ in the medium of interest can be determined experimentally by monitoring the disappearance of reacting substance as well as the appearance of the product. The reaction rate is given by:

$$r_{dp} = -\left(\frac{dC}{dt}\right)_\lambda = k_{p,\lambda} C \quad \text{(Eq. 5.14)}$$

where $k_{p,\lambda}$ is the first-order rate constant of direct photolysis at a wavelength of λ and is given by the slope of a plot of $ln(C/C_0)$ versus time. The rate of a photochemical reaction involving a single light-absorbing species can also be expressed as the product of the quantum yield ($\phi_{r,\lambda}$) and the rate of light absorption ($I_{a,\lambda}$) of the reacting substance at a wavelength of λ :

$$r_{dp} = \phi_{r,\lambda}I_{a,\lambda} = \phi_{r,\lambda}k_{a,\lambda}C \qquad \text{(Eq. 5.15)}$$

Therefore, the relationship between $k_{p,\lambda}$ and $\phi_{r,\lambda}$ can be obtained by combining Eq. 5.14 and Eq. 5.15:

$$\phi_{r,\lambda} = \frac{k_{p,\lambda}}{k_{a,\lambda}} \qquad \text{(Eq. 5.16)}$$

For photolysis test cells with short light path lengths and distilled water as the irradiation matrix in laboratory experiments, the quantum yield of a pollutant can be formulated as (Zepp, 1978):

$$\phi_{r,\lambda} = \frac{k_{p,\lambda}}{2.303I_{0,\lambda}\left(\dfrac{A}{V}\right)\varepsilon_\lambda l_\lambda} \qquad \text{(Eq. 5.17)}$$

where l_λ is the average path length of the photolysis cell that can be determined experimentally at wavelength λ. To calculate the term of $[I_{0,\lambda}l_\lambda A/V]$, a *chemical actinometer*, containing a solution of an appropriate reference compound with a well-defined quantum yield is tested under the same experimental condition. The quantum yield of this pollutant can be determined by comparing the photolysis rates of the pollutant with the reference compound:

$$\phi_{r,\lambda} = \frac{k_{p,\lambda}\varepsilon_{\lambda,ref}}{k_{p,\lambda,ref}\varepsilon_\lambda}\phi_{r,\lambda,ref} \qquad \text{(Eq. 5.18)}$$

where $\varepsilon_{\lambda,ref}$, $k_{p,\lambda,ref}$, and $\phi_{p,\lambda,ref}$ are the molar extinction coefficient, the first-order photolysis rate constant, and the reaction quantum yield for the actinometer at wavelength λ, respectively.

5.4.2 Indirect Photolysis

Chemicals that do not themselves absorb light energy can be transformed through indirect photolysis, i.e., the reaction is initiated through light absorption by chemicals (chromophores) present in the system other than the substrate itself. The chromophore may change irreversibly and undergoes direct photolysis itself, which simultaneously causes the transformation of other substrates present. The transformation of indirect photolysis can be completed through either the transfer of energy from another excited species (*sensitized photolysis*) or the reaction with highly reactive species that are formed in the presence of light.

5.4.2.1 Sensitized Photolysis

Photosensitization is usually defined as the transfer of energy from a photochemcially excited molecule to an acceptor, often oxygen, to form a reactive, transient form of oxygen, singlet oxygen ($^1O_2$). Under environmental conditions, about 1-2% of the UV-absorbing chromophores give rise to long-lived triplet states of high enough energy to interact with dissolved oxygen to form singlet oxygen. The singlet oxygen in water is quenched back to the ground state oxygen by water and has a half-life of 3 μs (Zafiriou et al., 1984). At typical dissolved organic matter (DOM) concentrations of most surface waters (dissolved organic carbon (DOC) < 20 mg/L), quenching of $^1O_2$ by DOM can be neglected (Haag and Hoigne, 1986).

5.4.2.2 Radical-Producing Photochemical Reactions

In addition to energy transfer to molecular oxygen, environmental photoreactions often produce free radicals. Typical free radicals produced in natural water include organic peroxy radicals ($ROO^•$), hydroxyl radicals ($OH^•$), and carbonate radicals ($^•CO_3^-$).

Organic peroxy radicals ($ROO^•$) are formed by the reaction of excited DOM chromophores with $^3O_2$ (triplet oxygen). $ROO^•$ may not be scavenged significantly by DOM (at least for DOC concentrations < 5 mg/L), and they react predominately with easily oxidizable compounds including alkyl phenols, aromatic amines, thiophenols, and imines. The most likely reaction mechanism is the abstraction of a hydrogen atom $H^•$ from the compound as peroxy radicals typically act as electrophilic species.

There are several possible mechanisms by which hydroxyl radicals ($OH^•$) may be formed in surface water. These include the photolysis of nitrate and nitrite, photolysis of $Fe(OH)^{2+}$, reactions of excited humic materials, and reaction of H_2O_2 with iron(II) (Fenton's reaction). In fresh waters, the nitrate photolysis may be the dominant process for hydroxyl radicals ($OH^•$) formation (Zepp et al., 1987; Haag and Hoigne, 1985):

$$NO_3^- \xrightarrow{\;h\nu\;} NO_3^{-*} \longrightarrow NO_2^- + O(^3P) \qquad \text{(Eq. 5.19)}$$

$$\rightarrow NO_2 + {}^•O^- \qquad \text{(Eq. 5.20)}$$

$$^•O^- + H_2O \longrightarrow OH^• + OH^- \qquad \text{(Eq. 5.21)}$$

The quantum yield for $OH^•$ production at 313 nm in lake water was about 0.015 (Zepp et al., 1987). The radicals of $OH^•$ derived from nitrate photolysis readily participate in organic reactions. Many organic compounds rapidly react with $OH^\bullet$ and the reaction rate constants are generally diffusion controlled (Atkinson, 1986).

In natural water, carbonate is always abundant. Carbonate radicals ($^\bullet CO_3^-$) can be generated from the reaction of $HO^\bullet$ with either carbonate or bicarbonate ions:

$$OH^\bullet + CO_3^{2-} \rightarrow \ ^\bullet CO_3^- + OH^- \qquad\qquad \text{(Eq. 5.22)}$$

$$OH^\bullet + HCO_3^- \rightarrow \ ^\bullet CO_3^- + H_2O \qquad\qquad \text{(Eq. 5.23)}$$

Carbonate radicals are more selective than $OH^\bullet$ in organic reactions and therefore has a higher concentration in natural water relative to $OH^\bullet$ in some cases.

Table 5.3 summarizes the concentrations and rate constants of some reactive transients in natural waters.

Table 5.3 Concentrations and rate constants of reactive transients in natural waters*

Reactive Transients	Concentration (M)	Rate Constant (1/M·s)
$OH^\bullet$	10^{-8}-10^{-14}	10^7-10^{10}
$ROO^\bullet$	10^{-9}	10^3
$^\bullet CO_3^-$	10^{-15}-10^{-13}	10^6-10^7
$^1O_2$	10^{-14}	10^7

Adapted from Lam et al. (2003)

5.4.2.2 Kinetics

The kinetics of indirect photolysis of organic compounds has been described in detail by Schwarzenbach et al. (1993). The rate of formation of a given photooxidant (Ox) by radiation over a range of wavelengths can be described by

$$r_{f,ox} = \left(\frac{d[Ox]}{dt} \right) = \int k_{a,A}(\lambda)\phi_{r,A}(\lambda)[A]d\lambda \cong \left(\sum k_{a,A}(\lambda)\phi_{r,A}(\lambda) \right)[A] \qquad \text{(Eq. 5.24)}$$

where $k_{a,A}(\lambda)$ is the specific light absorption rate of the chromophores for the production of Ox, [A] is the bulk concentration of the chromophores; $\phi_{r,A}(\lambda)$ is the overall quantum efficiency for the production of Ox.

Photooxidants are also consumed by various processes including physical quenching by the water itself or by chemical reactions with various water constituents. The rate of consumption of a given Ox can be described by:

$$r_{c,ox} = -\frac{d[Ox]}{dt} = \sum_i (k_{ox,i})[Ox] \qquad\qquad \text{(Eq. 5.25)}$$

where $k_{ox,i}$ is a pseudo-first order rate constant for the Ox-consuming process "i". The Ox-consuming processes are chemical processes and therefore, $r_{c,ox}$ is light-independent. When steady state is reached, the formation rate of Ox is equal to its consumption rate, i.e.:

$$r_{f,ox} = r_{c,ox} \qquad \text{(Eq. 5.26)}$$

Therefore, the concentration of photooxidant at steady state $[Ox]_{ss}$ will be:

$$[Ox]_{ss} = \frac{\sum k_{a,A}(\lambda)\phi_{r,A}(\lambda)[A]}{\sum_i k_{ox,i}} \qquad \text{(Eq. 5.27)}$$

Assuming $k_{ox} \ll \sum k_{ox,i}$ (i.e., $[Ox]_{ss}$ is a constant), the degradation rate of a given pollutant is given by

$$r_d = -\frac{dC}{dt} = k'_{ox}[Ox]_{ss} C = k^0_{ox} C \qquad \text{(Eq. 5.28)}$$

where k'_{ox} and k^0_{ox} are the second-order and pseudo-first-order reaction rate constants, respectively.

5.4.3 Photocatalytic Degradation

Different definitions have been suggested for photocatalysis (Kisch and Hennig, 1983; Carassiti, 1984; Mirbach, 1984). Here, photocatalytic degradation is denoted as cyclic photoprocesses in which the target substrate undergoes the photogradation with spontaneous regeneration of catalysts and regeneration continues indefinitely until the substrate is destroyed. This process strongly depends on the ability of the catalyst to create electron-hole pairs which generate free radicals or strong oxidant species to further secondary reactions.

In natural water, one of the most important catalysts for photodegradation is metal oxide such as ZnO, MnO_2, Fe_2O_3 (hematite), and TiO_2 in suspended particles. Semiconductor oxides emit an electron to the conduction band and leave a positively charged site in the valence band when irradiated with UV/visible light having greater energy than their band gap energy:

$$\text{Metal oxide} \xrightarrow{\ hv\ } h_{vb}^{+} + e_{cb}^{-} \qquad \text{(Eq. 5.29)}$$

If the rate of recombination of the electron-hole pair is slow, the conduction-band electrons may be transferred to O_2, causing the formation of superoxide anions:

$$O_2 + e_{cb}^{-} \rightarrow O_2^{\bullet -} \qquad \text{(Eq. 5.30)}$$

The valence-band holes may oxidize suitable substrates by accepting an electron from them, or by reaction of the holes with surface hydroxyl groups to produce hydroxyl radicals:

$$h_{vb}^+ + S - OH_2 \rightarrow S - OH^\bullet + H^+ \qquad \text{(Eq. 5.31)}$$

Theoretically, hydrogen peroxide could also be formed either by reduction of oxygen by conduction-band electrons or oxidation of water by valence-bands holes (Hoffman et al., 1995):

$$O_2 + 2e_{cb}^- + H^+ \rightarrow H_2O_2 \qquad \text{(Eq. 5.32)}$$

$$2h_{vb}^+ + 2H_2O \rightarrow H_2O_2 + 2H^+ \qquad \text{(Eq. 5.33)}$$

H_2O_2 may contribute to the degradation of organic electron donors by acting as direct electron acceptor or as a direct source of hydroxyl radicals due to hemolytic scission.

It should be noted that the equations above are very schematic and are not able to describe many complicated heterogeneous photoreactions that occur on solid surfaces. For further information on metal oxides-mediated photodegradation, the reader is referred to Sulzberger and Hug (1994) and Hoffman et al. (1995).

5.5   Photoreactions of Organic Compounds

5.5.1   Polycyclic Aromatic Hydrocarbons

Polycyclic aromatic hydrocarbons (PAHs) are common toxic and carcinogenic micropollutants found in the aquatic environment. Surface waters are contaminated by PAHs in different ways such as atmospheric deposition, discharge of industrial and municipal wastewaters, and watershed runoff. While PAHs resist biodegradation (especially PAHs with four or more rings), they can undergo fairly rapid transformations in aqueous solutions when exposed to UV light, implying photodegradation is an important process for PAHs removal from aquatic environment.

PAHs may be degraded through either direct or sensitized photochemical reactions (Zepp and Schlotzhauer, 1979). While sensitized photodegradation does not appear to be significant for most natural waters, most PAHs can absorb surface solar radiation, allowing for the possibility of direct photodegradation. Table 5.4 summarizes the quantum yields and half-lives for the direct photolysis of common PAHs. As shown in Table 5.4, the quantum yields vary from 10^{-5} to 10^{-2} and the half-lives for most of PAHs are less than 10 hours.

Three mechanisms have been proposed for the direct photodegradation of PAHs (Fasnacht and Blough, 2003; Miller and Olejnik, 2001; Sigman et al., 1998; Zepp and Schlotzhauer, 1979). They are:

Photoionization. Photodegradation can be initiated by the photoionization process, which leads to the formation of PAH radical cation and a hydrated electron

$$PAH \xrightarrow{\hphantom{xx}hv\hphantom{xx}} PAH^{\bullet+} + e_{aq}^{-} \qquad \text{(Eq. 5.34)}$$

In aerated solutions, the hydrated electron produced reacts rapidly with O_2 to form superoxide ($O_2^{\bullet-}$). PAH radicals can react with water (or hydroxide ion) to form secondary (radical) intermediates which can react further to form stable products.

Electron transfer oxidation. An alternative, mechanism for electron transfer oxidation is through the excitation of PAH-oxygen contact charge transfer pairs:

$$PAH + O_2 \rightarrow [PAH\text{-}O_2] \xrightarrow{\hphantom{xx}hv\hphantom{xx}} [PAH^{\bullet+}\text{-}O_2^{\bullet-}] \qquad \text{(Eq. 5.35)}$$

The resultant charge transfer complex may undergo further solvent separation to form PAH radicals and superoxide as in photoionization, undergo charge recombination to regenerate the ground-state PAH and O_2, or react within the collision complex to form products. Figure 5.2 depicts an electron transfer mechanism proposed for the oxidative photolysis of pyrene in water.

Direct reaction of the excited triplet state of PAH with O_2. Photodegradation can also be initiated by the direct reaction of O_2 with the excited triplet state of PAH [$^3PAH^*$] to produce a [$^3PAH^*$-$^3O_2$] complex

$$PAH \rightarrow PAH^* \rightarrow\, ^3PAH^* + O_2 \rightarrow [^3PAH^*\text{-}^3O_2] \qquad \text{(Eq. 5.36)}$$

Within the collision complex produced, a [PAH-$^1O_2$] complex may be further formed by energy transfer from $^3PAH^*$ to $^3O_2$. Reactions within the two types of collision complex may form stable products.

It remains unknown which mechanism predominately occurs in the direct photodegradation of PAHs under natural sunlight conditions. For some PAH compounds, the presence or absence of dissolved O_2 has insignificant effect on their photolysis rates (Zepp and Schlotzhauer, 1979; Miller and Olejinik, 2001). This fact implies that reactive oxygen species are unimportant in their photolysis and the photoionization process may be the dominant step in the initiation of the photodecomposition. However, Fasnacht and Blough (2003) argued that photoioinzation is unimportant based on the following observations in their study: the photodegradation quantum yields of the PAHs examined did not correlate with ionization or oxidation potentials; all PAHs examined exhibited wavelength-independent photodegradation quantum yields; photodegradation rate constants were

Table 5.4 Quantum yields and half-lives of PAHs photodegradation under sunlight or simulated Sunlight

PAH	Quantum yields $\phi \times 10^3$	Half-life (hr)	References
Acenaphthene	9	6.4[a]	Fasnacht and Blough (2002)
Anthracene	4.2	0.4[a]	Fasnacht and Blough (2002)
	3	0.8	Zepp and Schlotzhauer (1979)
Benzo[a]anthracene	4.4	0.4[a]	Fasnacht and Blough (2002)
	3.3	3.3	Mill et al. (1981)
Benzo[a]pyrene	5.4	0.09[a]	Fasnacht and Blough (2002)
	0.89	1	Mill et al. (1981)
	10.8~12.1	-	Miller and Olejnik (2001)
Benzo[b]fluoranthene	0.13	6.4[a]	Fasnacht and Blough (2002)
Benzo[k]fluoranthene	0.14	6.4[a]	Fasnacht and Blough (2002)
Chrysene	2.5	2.1[a]	Fasnacht and Blough (2002)
	3	4.4	Zepp and Schlotzhauer (1979)
	2.5~6.0	-	Miller and Olejnik (2001)
Fluoranthene	0.032	38[a]	Fasnacht and Blough (2002)
	0.2	21	Zepp and Schlotzhauer (1979)
Fluorene	3.3	214[a]	Fasnacht and Blough (2002)
	3.8		Miller and Olejnik (2001)
Perylene	0.54	0.4[a]	Fasnacht and Blough (2002)
Phenanthrene	3.5	21[a]	Fasnacht and Blough (2002)
	10	8.4	Zepp and Schlotzhauer (1979)
Pyrene	-	0.5[a]	Fasnacht and Blough (2002)
	2	0.7	Zepp and Schlotzhauer (1979)
	2.8	-	Sigman et al. (1998)

[a]*Half-lives are calculated based on the given first-order rate constants*

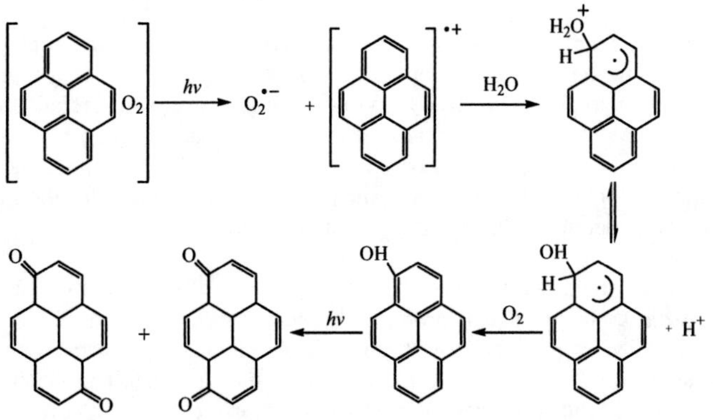

Figure 5.2   Electron transfer mechanism proposed for the oxidative photolysis of pyrene in water (adapted from Sigman et al., 1998)

not enhanced in the presence of N_2O, which will react with hydrated electron to form hydroxyl radical and thus accelerate the degradation; and photodegradation quantum yields for all PAHs increased with increasing O_2 concentration (which appeared contradictory to the findings of the two previous studies). They concluded that PAHs photodegrade mainly via reactions of O_2 with the excited states of these compounds.

A few studies have attempted to identify the final products of the photodegradation of PAHs. Mill and co-workers (1981) identified 7,12-benz[a] anthracenequinone as a product from the photolysis of aqueous solutions of benz[a] anthracene. A study of anthracene photolysis has shown that in the presence of oxygen the primary products were anthraquinone and the endoperoxide 17A, whereas in argon-purged solution the dimmer endoperoxide 17B dominates (Sigman et al., 1991). A later study of pyrene photolysis in water revealed that the stable products were 1,6- and 1,8-pyrenequinones (Sigman et al., 1998).

Fewer efforts have been made to evaluate the relative toxicity and biodegradability of the photoproducts and parent compounds. Photolysis of benzo[a]pyrene in methanol followed by incubation in soil or sewage sludge demonstrated a noticeable increase in biodegradability and a decrease in mutagenicity (Miller et al., 1988). Thus, it would appear that, at least in some cases, more biodegradable products were formed from the photolysis of PAH.

5.5.2 Pesticides

Current agricultural practices rely heavily on pesticides for crop production as illustrated by the fact that million of tons of pesticides are applied to soil and foliage each year. Considering the adverse effects of many pesticides on human health and the equilibrium of ecological system, it is important to evaluate their transformation and fate in the environment.

Photodegradation is one of the most important abiotic degradation pathways of pesticide in the environment. Photodegradation of pesticide can occur via direct photodegradation, photosensitized degradation or photocatalytic degradation (Burrows et al., 2002). Many studies are available for the photocatalytic degradation with TiO_2, but such reactions is expected to be, in general, of only limited importance in aquatic environment. Figure 5.3 depicts the primary reactions proposed for the initial step in the photodegradation of pesticides under natural sunlight.

Direct irradiation of pesticides or the reaction of pesticides with excited states of sensitized compounds will promote the pesticides to their excited singlet states, which may further produce triplet states. Such excited states can then undergo: (i) homolysis process, during which the excited states dissociate and generate two free radicals; (ii) heterolysis process, during which the excited states cleave and produce a cation and an anion; and (iii) photoionization, during which a pesticide radical cation and a hydrated electron are formed.

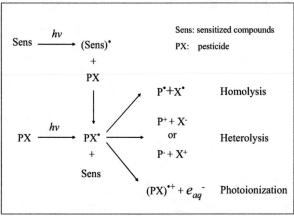

Figure 5.3 Primary reactions that initiate the photodegradation of pesticides under natural sunlight (adapted from Burrows et al., 2002)

Table 5.5 summarizes the half-lives of photodegradation of common pesticides under natural sunlight or simulated sunlight. The half-lives for most of pesticides varies from several hours to around 70 days. It is not surprising that the reported half-lives for a certain compound vary with different studies since the photodegradation reaction rate is strongly dependent on the environmental conditions, including geographic location, season, daylight exposure time, and water source. As a detailed discussion of photodegradation for each pesticide is beyond the scope of this chapter, atrazine and metolachlor will be used as two examples to discuss possible reaction pathways and final products associated with photodegradation process.

Atrazine. Atrazine is currently one of the most widely used agricultural pesticides in the world and is commonly detected throughout the entire hydrological cycle. Atrazine can be degraded through either direct or indirect photolysis (Pape and Zabik, 1970; Pape and Zabik, 1972; Ruzo and Zabik, 1973; Rejto et al., 1983; Pelizzetti et al., 1990; Torrents et al., 1997; Konstantinou et al., 2001; Evgendou and Fytianos, 2002).

Direct photolysis of atrazine proceeds via excitation of atrazine followed by dechlorination and hydroxylation (step 1 in Figure 5.4) while dealkylation (step 2 in Figure 5.2) is the dominant photodegradation pathway in the indirect photolysis of atrazine (Torrents et al., 1997; Evgendou and Fytianos, 2002). The intermediate products generated in step 1 and step 2 may undergo further degradation, resulting in complex and interconnected pathways. The final products may be ammelide (aminodihydroxy-s-troazine) and chlorodiamino-s-triazine (Torrents et al., 1997). A Microtox® toxicity test revealed that the toxicity of atrazine in natural waters was reduced, mainly due to photodegradation (Lin et al., 1999).

Table 5.5 Half-lives of photodegradation of common pesticides under sunlight or simulated sunlight

Groups	Compounds	Conditions	Half-life	References
Anilide	Alachlor	Simulated sunlight, SFW[a]	9 hr	Wilson & Mabury (2000)
	Butachlor	Natural sunlight, RW[b]	58 hr	Lin et al. (2000)
		Simulated sunlight, SFW[a]	8 hr	Wilson & Mabury (2000)
	Metolachlor	Simulated sunlight, SFW[a]/RW[b]	8/12 hr	Wilson & Mabury (2000)
		Natural sunlight, RW[b]/LW[c]	578/694hr	Dimou et al. (2005)
		Natural sunlight, LW[c]	22 day	Kochany & Maguire (1994)
	Propachlor	Natural sunlight, RW[b]/LW[c]	70/71 day	Konstantinou et al. (2001)
	Propanil	Natural sunlight, RW[b]/LW[c]	55/60 day	Konstantinou et al. (2001)
Carbamate	Albendazole	Natural sunlight, DW[d]	1.8-4.7 hr	Weerasinghe et al. (1992)
	Carbaryl	Simulated sunlight, DW[d]	21 day	Das (1990)
	Molinate	Natural sunlight, RW[b]/LW[c]	61/62 day	Konstantinou et al. (2001)
Organophosp-horus	Parathion	Simulated sunlight, DW[d]	65 h	Mansour et al. (1983)
Oxadiazole	Oxadiazone	Natural sunlight, RW[b]	29 h	Lin et al. (2000)
Phenol-based	Lampricide	Simulated sunlight, DW[d]	22-92 h	Ellis & Mabury (2000)
	MCPA	Natural sunlight, RW[b]	168 h	Stangroom et al. (1998)
Thiocarbama-te	Thiobencarb	Natural sunlight, DW[d]	3 day	Cheng & Hwang (1996)
Triazine	Atrazine	Natural sunlight, RW[b]/LW[c]	43/53 day	Konstantinou et al. (2001)
		Natural sunlight, DW[d]	6-18 day	Comber (1999)
	Propazine	Natural sunlight, RW[b]/LW[c]	53/65 day	Konstantinou et al. (2001)
	Simazine	Natural sunlight, DW[d]	6-37 day	Comber (1999)
	Prometryne	Natural sunlight, RW[b]/LW[c]	52/51 day	Konstantinou et al. (2001)

SFW[a]: synthetic field water; RW[b]: river water; LW[c]: lake water; DW[d]: DI water.

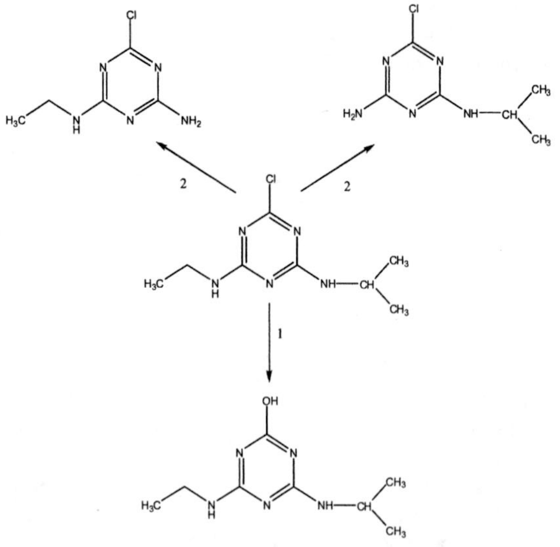

Figure 5.4   Initial products of atrazine photodegradation in natural water (Adapted from Torrents et al., 1997; Evgendou and Fytianos, 2002)

Metolachlor. As a germination inhibitor used mainly for weed control of grasses, metolachlor is one of the most heavily used agricultural pesticides in the United States and Canada. etolachlor is among one of those herbicides most frequently detected in surface water (Jaynes et al., 1999) and the concentrations of metolachlor have been reported for the period 1979-1985 in the range of 0.3-4.4 µg/L for various locations along the Mississippi River (Chester et al., 1989). Since metolachlor is non-volatile and has a long hydrolysis half-life (Chiron et al., 1995), photolysis is an important abiotic degradation pathway for metolachlor under natural aquatic environment (Kochany and Maguire, 1994; Mathew and Khan, 1996; Wilson and Mabury, 2000; Dimou et al., 2005).

Metolachlor can be degraded through either direct or indirect photolysis, involving a series of reactions such as hydroxylation, dehalogenation, oxoquinoline formation, and demethylation (Mathew and Khan, 1996). In Kochany and Maguire's study (1994), four dechlorinated photoproducts (Figure 5.5) (species I, II, III, and IV) were identified in lake water after 40 days of sunlight irradiation. Wilson and Mabury (2000) focused on the extent of monochloroacetic acid (MCA) formation under simulated solar irradiation (Figure 5.5) (compound V), and found that direct photolysis resulted in 5.2% conversion of metolachlor to MCA while indirect photolysis in river water resulted in 12.6% conversion. In a recent study (Dimou et al., 2005), different aqueous media were used under solar and simulated solar irradiation and nine photoproducts were identified, among which 4-(2-ethyl-6-methyl-ohenyl)-5-methyl-3-morpholinone (Figure 5.5) (compound III) was found as the main product in natural waters. This study further evaluated the toxicity of photoproducts and found that metolachlor photolysis produced more toxic compounds, which conflicted with the results from another study (Lin et al., 1999).

5.5.3 Chelating Agents

The environmental fate of chelating agents has received considerable attention, mainly due to their influence on metal mobility and bioavailability and their persistence in the environment (Nowack, 2002). Fe(III) and Cu(II) complexes of aminopolycarboxylates such as EDTA, NTA, DTPA, and EDDS are rapidly photodegraded in natural water (Trott et al., 1972; Langford et al., 1973; Lockhart and Blakeley, 1975a, 1975b; Kari et al., 1995; Metsärinne et al., 2001) while most other complexes are only slightly photodegradable or resist to photodegradation. Two common chelating agents (NTA and EDTA) and their photodegradation pathways and products in natural waters will be discussed.

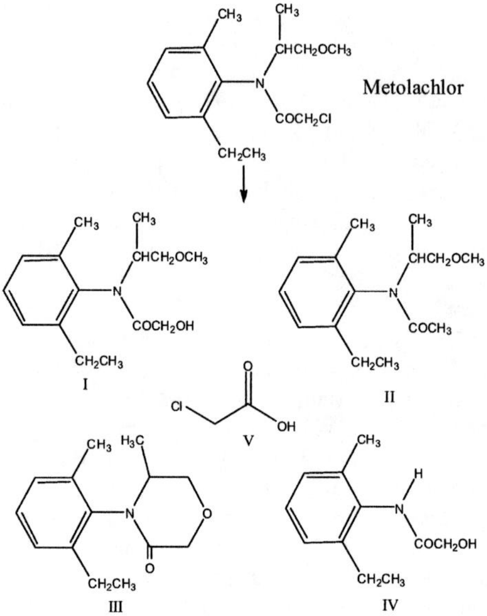

Figure 5.5 Products of metolachlor photodegradation in natural water

NTA. Nitrilotriacetate (NTA) is the first chemical synthesis of aminopolycarboxylic acids (APCAs). NTA is easily biodegradable and the biodegradation process is typically complete, as indicted by the compound iminodiacetate (IDA) does not accumulate in nature (Anderson et al., 1985). IDA is formed intracellularly during the microbial breakdown of NTA. Besides biodegradation, photodegradation has been demonstrated as another important pathway in limiting the accumulation of NTA (Trott, 1972; Langford et al., 1975). Photodegradation of Fe(III)NTA is rapid with a half-life of 1.5 hr under sunlight irradiation (Stolzberg and Hume, 1975), whereas that for CuNTA is much slower and the half-life is more than 100 times higher. The photodegradation of Fe(III)NTA is initiated by the excitation of the metal complex, during which an electron transfer from ligand to metal, resulting in a reduced metal ion and the formation of a ligand radical. The ligand radical then undergoes sequential decarboxylation, producing CO_2, formaldehyde, and iminodiacetate (IDA) (Trott et al., 1972; Lanford et al., 1973):

$$2Fe(III)NTA \rightarrow Fe(II) + Fe(II)NTA + IDA + CH_2O + CO_2 \quad \text{(Eq. 5. 37)}$$

Oxygen from air reoxidizes Fe(II) and Fe(III) is replenished to continue the reaction. Fe(III)IDA is further photodegraded forming glycine. The extent of NTA photodegradation is strongly determined by its speciation. Fe(III)NTA and Cu(II)NTA are readily photodegradable in sunlight while photodegradation for other metal complexes (such as Cd(II)NTA, Pb(II)NTA, Mg(II)NTA and Cr(III)NTA) are not likely to occur due to the poor overlap of their absorption spectrum with the solar emission spectrum (Stolzberg and Hume, 1975).

EDTA. Ethylenediaminetetraacetic acid (EDTA) is widely used to remove metal ions that interfere with industrial processes. As a recalcitrant compound, EDTA resists biological degradation in sewage treatment plants, resulting in significant release of EDTA into surface waters. In European rivers and lakes, EDTA concentrations of 10-60 μg/L are regularly found (Frimmel et al., 1989). Transport and fate of EDTA in surface waters has been widely studied. Among various physical, chemical and biological reactions, photodegradation may be a dominant pathway for the removal of EDTA from surface waters (Frank and Rau, 1990).

Generally, direct photochemical transformation of EDTA (in the form of Na_2-EDTA) is not possible in sunlight since Na_2-EDTA do not absorb light within the natural UV radiation range (Metsärinne et al., 2001). However, EDTA in natural waters are predominately in the form of metal complexes. For example, EDTA has a tendency to complex dissolved Fe(III) and adsorb at the surface of Fe(III) (hydr)oxide, forming Fe(III)-EDTA complexes (Chang et al., 1983). Kari et al. (1995) and Kari and Giger (1995) reported that Fe(III)-EDTA complex is the only EDTA species that undergo the direct photolysis in natural waters. All other EDTA-species (such as Ca-EDTA, Zn-EDTA, Cu-EDTA, Ni-EDTA and Mn-EDTA) present in the river water are stable in the presence of light (Kari and Giner, 1995) because of their low concentrations in surface waters and low quantum yields for their photochemical degradation. Excited chromophores present in natural waters may generate photooxidants (such as hydroxyl radicals and singlet oxygen) but the impact of indirect photolysis on EDTA photodegradation is generally minor due to the extremely low concentrations of these photooxidants in natural waters (Kari et al., 1995). The key parameters to evaluate the photodegrdation of EDTA in natural waters are the fraction of EDTA species present as Fe(III)-EDTA and the direct photolysis rate of Fe(III)-EDTA under sunlight conditions (Kari et al., 1995).

Fe(III)-EDTA complexes are known to photodegrade with relatively high quantum yields (an order of 10^{-2}) (Lockhard and Blakeley, 1975; Kari et., 1995). Rapid photodegradation of Fe(III)EDTA results in a mean half-life of a few hours in river water during summer and several days in winter (Kari et al., 1995). In aerobic conditions, Fe(III)EDTA is degraded in a stepwise process under successive decarboxylation   (Lockhart and Blakeley, 1975a) and the major products of photodegradation are CO_2,   formaldehyde, N-carboxymethyl-N, N'-ethylenediglycine (ED3A), N,N'-ethylenediglycine (EDDA-N,N'), iminodiacetic acid (IDMA), N-carboxymethyl-N-aminoethyleneglycine   (EDDA-N,N),   N-amino-ethyleneglycine (EDMA), and glycine (Lockhart and Blakeley, 1975b):

$$Fe(III)EDTA \rightarrow Fe(II)ED3A + CO_2 + HCOH$$
$$Fe(III)ED3A \rightarrow Fe(II)EDDA + CO_2 + HCOH \qquad \text{(Eq. 5.38)}$$

$$Fe(III)EDDA \rightarrow Fe(II)EDMA + CO_2 + HCOH$$

The formed Fe(II)-complexes are rapidly oxidized to Fe(III)-complexes in the presence of oxygen. The final product EDMA is stable in the presence of Fe(III) and light (Lockhard and Blakeley, 1975b). The photoproducts of Fe(III)EDTA are readily biodegradable compared to the recalcitrant parent compound (Nowack and Baumann, 1998) .

5.5.4 Pharmaceutical Compounds

Pharmaceuticals and personal care products (PPCPs) are an emerging class of aquatic contaminants that have been increasingly detected in natural waters globally (Halling-Sorensen et al., 1998; Ternes, 1998; Kolpin et al., 2002; Calamari et al., 2003). Potential concerns from the environmental presence of these compounds include abnormal physiological processes and reproductive impairment, increased incidences of cancer, the development of antibiotic-resistant bacteria, and the potential increased toxicity of chemical mixtures (Kolpin et al., 2002). Despite the fact that many PPCPs are produced and used in significant amounts and maybe pose an environmental threat, relatively little is known about their occurrence and fate in natural waters.

There are several indications that photochemical degradation may be one of the potentially significant removal mechanisms for PPCPs in aquatic environment. Many of PPCPs have functional groups such as aromatic rings, heteroatoms, phenol, nitro, and napthoxyl, which can either absorb solar radiation, or react with free radicals and other reactive oxygen species generated by photosensitizers (Boreen et al., 2003). Moreover, many PPCPs are not completely removed or degraded in biological process of wastewater treatment plants (Ternes, 1998; Buser et al., 1999), which implies that photodegradation may be more important than biodegradation in determining the fate of PPCPs in sunlit waters.

The importance of photodegradation of PPCPs in aquatic environment has been demonstrated by several lab or field studies. Table 5.6 summarizes the quantum yields and half-lives of direct photolysis for common PPCPs under sunlight or simulated sunlight. The reported photodegradation rates vary widely, with half-lives ranging from less than 1 hour to more than 100 days. Some PPCPs (such as diclofenace, naproxen, and several estrogens) are very susceptible to photodegradation, whereas some PPCPs (such as carbamazepine and clofibric acid) degrade much slower. Some PPCPs that are resistant to direct photolysis, such as cimetidine (H_2 blocker), have been shown to react with singlet oxygen ($^1O_2$) and hydroxyl radical ($OH^\bullet$) and undergo indirect photolysis in natural waters (Latch et al., 2003b). Such indirect photolysis reactions are expected to dominate the photochemical fate of some PPCPs but so far this area remains poorly studied (Boreen et al., 2003).

Several PPCPs and their photodegradation pathways and products in natural waters will be discussed. These compounds are commonly detected in aquatic environment and have been extensively studied in the laboratory or field.

Naproxen.   Naproxen may be the most studied pharmaceutical compound in the area of environmental photochemistry (Moore and Chappuis, 1988; Bosca et al., 1990; Vargas et al., 1991; Condorelli et al., 1993; Jimenez et al., 1997; Albini and Fasani, 1998; Martinez and Scaiano, 1998; Bosca et al., 2001; Packer et al., 2003). Naproxen is very susceptible to photodegradation under sunlight and the half-life is estimated to be around 1 hour (Packer et al., 2003; Lin and Reinhard, 2005). The high photolability of naproxen is due to the fact that absorption spectrum of naproxen exhibits a significant overlap with the solar spectrum (Packer et al., 2003).

The photodegradation pathways of naproxen (Figure 5.6) have been proposed by several studies (Moore and Chappuis, 1988; Bosca et al., 1990; Jimenez et al., 1997). The first step is conversion of the carboxylate ($RC(=O)O-$) group to a carboxyl radical ($RC(=O)O^{\bullet}$) by photoionization. Decarboxylation then occurs, yielding carbon dioxide and a benzylic radical. The benzylic radical further abstracts a hydrogen atom from a suitable donor resulting in a product with an ethyl side chain, or reacts with molecular oxygen, eventually leading to an alcohol or ketone moiety in place of the carboxylate group. Overall, irradiation of naproxen leads to photodecarboxylation products with ethyl, 1-hydroxyethyl and acetyl side chains. These photoproducts, however, may be more toxic than the parent drug (DellaGreca et al., 2004).

Diclofenac. Diclofenac is an excellent example to illustrate the importance of photochemistry on the fate of some PPCPs in aquatic environment. Diclofenace resists the chemical and biological degradation under natural conditions but it photodegrades rapidly under sunlight (Buser et al., 1998). It is not surprising, since diclofenac has an absorbance maximum at 273 nm that tails well over 300 nm, which indicates significant amount of light absorbance in the solar spectrum. The decomposition quantum yield determined ranges from 0.038 to 0.22 while the half-lives ranged from less than 1 hour to several hours (Moore et al., 1990; Buser et al., 1998; Andreozzi et al., 2003; Packer et al., 2003).   Photodegradation has been demonstrated as the dominant pathway for diclofenac elimination in a lake (Buser et al., 1998; Poiger et al., 2001).

The primary process of photodegradation of diclofenace (Figure 5.7, step 1) is the loss of HCl and yield 8-chlorocarbazole-1-acetic acid. This intermediate is then dechlorinated to carbazole-1-acetic acid in the presence of H-source (Figure 5.7, step 2) or form 8-hydroxycarbazole-1-acetic acid via photosubstitution in the absence of H-source (Figure 5.7, step 3). 8-hydroxycarbazole-1-acetic acid is not stable under natural conditions and may further photodegrade rapidly. These photoproducts were identified in the laboratory experiments but were not detected under the natural conditions in the lake (Buser et al., 1998), possibly due to the extremely low concentrations of these compounds (less than the current detection limits) in the lake (Poiger et al., 2001).

Table 5.6 Quantum yields and half-lives of photodegradation for PPCPs under sunlight or simulated sunlight

Chemical	Quantum yields ϕ	Half- life	References
Carbamazepine	0.000048	~100 day	Andreozzi et al. (2003)
	0.00013	115 hr	Lam and Mabury (2005)
Ciprofloxacin	0.001		Albini and Monti (2003)
Clofibric acid	0.0055	~30 day	Andreozzi et al. (2003)
	0.002	50 hr	Packer et al. (2003)
Diclofenac	0.22		Moore et al. (1990)
	0.13	< 1hr	Buser et al. (1998)
	0.038	~ 6 hr	Andreozzi et al. (2003)
	0.094	< 1hr	Packer et al. (2003)
Enoxacin	0.13		Albini and Monti (2003)
Estriol	0.0048	2.9 hr	Lin and Reinhard (2005)
Estrone [E1]	0.0296	2.3 hr	Lin and Reinhard (2005)
17β-estradiol [E2]	0.0048	2.0 hr	Lin and Reinhard (2005)
17α-ethinylestradiol[EE$_2$]	0.0048	2.3 hr	Lin and Reinhard (2005)
Furaltadone	-	0.086-0.36 hr	Edhlund et al. (2006)
Furazolidone	-	0.24-0.90 hr	Edhlund et al. (2006)
Ibuprofen	-	14.8 hr	Lin and Reinhard (2005)
	-	2.6 hr	Packer et al. (2003)
Levofloxacin	0.05	-	Lam and Mabury (2005)
Lomefloxacin	0.55	-	Albini and Monti (2003)
Naproxen	0.012	-	Moore and Chappuis (1988)
	0.036	<1 hr	Packer et al. (2003)
	0.026	1.4 hr	Lin and Reinhard (2005)
Nitrofurantoin	-	0.44-1.7 hr	Edhlund et al. (2006)
Norfloxacin	0.06	-	Albini and Monti (2003)
Ofloxcin	0.000078	~ 2 day	Andreozzi et al. (2003)
Propranolol	0.002	~ 2 days	Andreozzi et al. (2003)
	0.0052	1.1 hr	Lin and Reinhard (2005)
Ranitidine	0.0053	< 1 h	Latch et al. (2003a)
Sulfamethoxazole	0.004	~ 10 hr	Andreozzi et al. (2003)
	0.02	-	Lam and Mabury (2005)
Triclosan	0.03	-	Latch et al. (2003b)

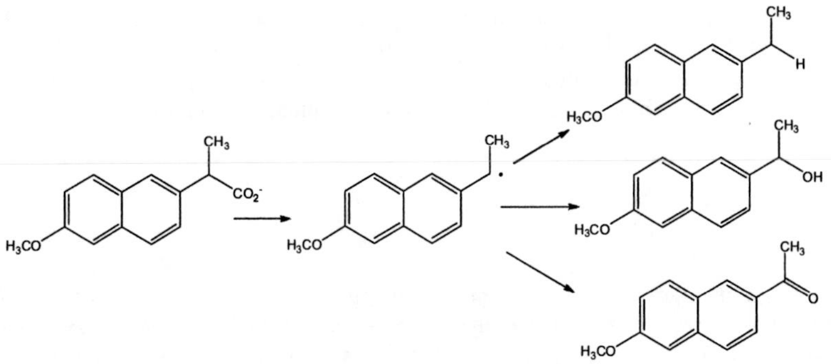

Figure 5.6   Products of naproxen photodegradation in natural water
(adapted from Jimenez et al., 1997)

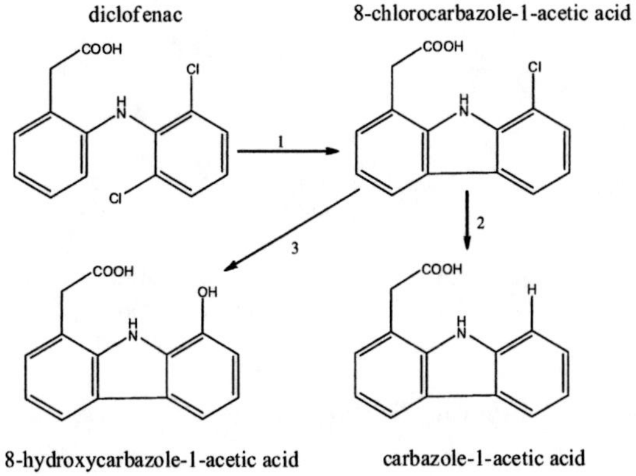

Figure 5.7   Products of diclofenac photodegradation in natural water
(adapted from Poiger et al., 2001)

Carbamazepine. Carbamazepine serves as an example demonstrating that some PPCPs that commonly undergo slow photodegradation may photodegrade rapidly under a specific environmental condition. In the distilled water irradiated by sunlight, the photodegrdation of carbamazepine is very slow, with a quantum yield of 4.8×10^{-5} and a half-life of 100 days (Andreozzi et al., 2003). However, in the presence of chloride, photodegradation of carbamazepine was found to be substantially enhanced in estuarine waters (Chiron et al., 2006). The chloride-enhanced carbamazepine photodegradation may result from the interaction between Fe(III) colloids and chloride ions under sunlight irradiation, yielding $Cl_2^{\bullet-}$:

$$= Fe^{III} - OH + Cl^{-} \xrightarrow{hv} Fe^{2+} + OH + Cl^{\bullet} \qquad \text{(Eq. 5.39)}$$
$$Cl^{\bullet} + Cl^{-} \Leftrightarrow Cl^{\bullet-} \qquad \text{(Eq. 5.40)}$$

Various intermediates of carbamazepine photodegradation have been detected under both direct and indirect photolysis conditions (Chiron et al., 2006). A major photoproduct of direct photolysis is acridine (Figure 5.8), which is toxic, mutagenic, carcinogenic and more harmful than parent compound. In the chloride-enhanced photodegradation process, various hydroxylated/oxidized intermediates as well as chloroderivatives are formed. The health and environmental impact of these photoproducts remain unknown.

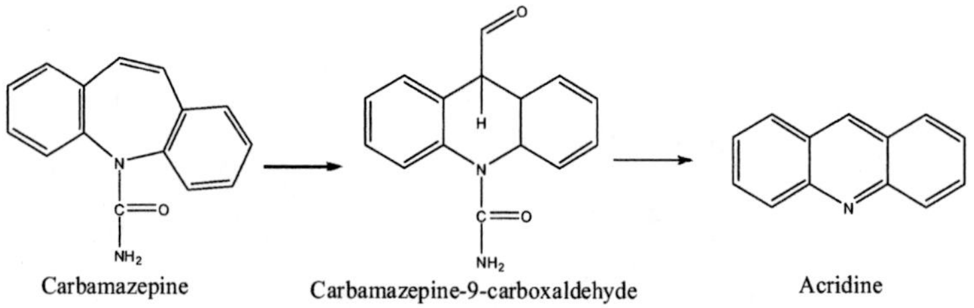

Carbamazepine   Carbamazepine-9-carboxaldehyde   Acridine

Figure 5.8   Formation of acridine during photodegradation of carbamazepine (adapted from Chiron et al., 2006)

5.6   Factors Influencing Photolysis

5.6.1 Suspended Particles

In the environment, varying amounts of suspended particles are present in the water and the particles (sediments) may change the rate or mechanism of photolysis of a contaminant in several ways. Suspended particles may reduce the photolysis rate by either shielding the organic from the available light or by quenching the excited states of the organic molecules before they react to form products. It is also possible for suspended particles to enhance the rate of organic photolysis if sediment absorption of light produces excited state or free radicals that can then react with the organics. In this way it is possible for organic materials that do not absorb sunlight to be photolyzed indirectly.

This type of indirect photolysis does occur with semiconductors such as TiO_2, which are common constituents of clays, sediments and soils. A question is that, whether the mechanism analogous to photosensitized semiconductor reactions is import in natural waters? Oliver et al. (1979) evaluated the role of suspended sediments and clays on the photolysis of pollutants in water and concluded that suspended particles do not appear to enhance the photodegradation of organic pollutants, even though suspended particles can contain TiO_2 and other semiconductor in the 5-10% range. In contrast, suspended particles were found to reduce the rates of photolysis by shielding the pollutant from the available light.

A more complicated issue is the photolytic transformation of sorbed compounds. Although the role of semiconductors may be insignificant in natural water, the indirect photolysis of sorbed compounds may be important. For example, the rate of photolytic transformation of a series of alkylated anilines was found to be significantly accelerated by algae (Zepp and Wolfe, 1987). The accelerating effect increased with increasing hydrophobicity of the compound, indicating that sorption of the compound was rate determining.

5.6.2 Dissolved Organic Matter

It is well known that dissolved organic matter (DOM) plays an important part in sunlight-induced photochemical processes in surface waters. DOM can act as inner filter, radical scavenger or precursor of reactive species:

Inner filter. DOM can absorb solar radiation in a broad range of wavelengths and thus reduce the available energy for the target compound in the solution.

Radical scavenger. DOM is one of the most important radical scavengers in photochemical reactions.

Precursor of reactive species. After being activated by solar UV photons, DOM can be promoted to a transient excited state (triplet state), in which they may react with oxygen forming reactive species (Haag and Hoigne, 1986), or react directly with other organic species, thus promoting their phototransformation (Zeep et al., 1985).

The overall effect of DOM on the phototransformation rate of an aquatic contaminant will therefore depend on a balance of the above three contributions. When DOM acts mainly as an inner filter, the addition of DOM will result in a decrease of photolysis rate. On the other hand, if DOM mainly acts as a precursor of reactive species, the presence of DOM will accelerate the photolysis process. In the hydroxyl radical ($OH^•$) -mediated photochemical reactions, DOM will scavenge $OH^•$ and thus inhibit the photodegradation of contaminants.

Table 5.7 summarizes the effects of DOM on the photodegradation of some organic compounds. Generally, DOM reduces the photolysis rate for most of the compounds that photodegrade rapidly via the direct photolysis. It is not surprising, since DOM–induced indirect or sensitized photolysis plays insignificant role in their photodegradation and DOM effectively compete with the target compounds for the solar radiation. The photodegradation of some compounds could be enhanced in natural waters through reactions with intermediates produced photochemically from chromophoric DOM. However, the degree of enhancing effects on photodegradation depends on the concentration of DOM in water. It has been found that the degradation rate of a target compound increased rapidly with the amount of DOM added but the rate reached a plateau at high levels of DOM (Doll and Frimmel, 2003; Miller and Chin, 2005). In this case, DOM may also act as a sink for reactive species produced from DOM-induced reactions and the enhancing effects compromised due to scavenging of these reactive species.

Table 5.7 Effects of DOM on photodegradation of some compounds

Chemical	Concentration of [a] DOC/DOM (mg/L)	[b] Effect	Reference
12 PAH compounds	Unknown	o	Fasnacht and Blough (2002)
3 nitrofuran antibiotics	5.9 (DOC)	o	Edhlund et al. (2006)
5 sulfa drugs	5.9 (DOC)	+	Boreen et al. (2005)
Acetochlor	2.3-7.4 (DOC)	-	Brezonik ans Brekken (1998)
Alachlor	2-14 (DOC)	+	Miller and Chin (2005)
Atrazine	5.3 (DOC)	-	Torrents et al. (1997)
	10.2-12.8 (DOM)	-	Konstantinou et al. (2001)
Carbamazepine	5 (DOM)	-	Andreozzi et al. (2003)
	1-7 (DOC)	+	Doll and Frimmel (2003)
Carbofuran	5-30 (DOC)	-	Bachman and Patterson (1999)
Cimetidine	16 (DOC)	+	Latch et al. (2003)
Clofibric acid	5 (DOM)	+	Andreozzi et al. (2003)
	16 (DOC)	+	Packer et al. (2003)
Dehydroabietic acid	25.5 (DOC)	+	Corin et al. (2000)
Diclofenac	5 (DOM)	-	Andreozzi et al. (2003)
	16 (DOC)	o	Packer et al. (2003)
Fipronil	2.5-10 (DOM)	-	Walse et al. (2004)
Ibuprofen	Unknown	+	Lin and Reinhard (2005)
	16 (DOC)	+	Packer et al. (2003)
Metolachlor	4-24 (DOM)	-	Dimou et al. (2005)
	0.5 (DOC)	-	Wilson and Mabury (2000)
Ofloxacin	5 (DOM)	+	Andreozzi et al. (2003)

[a] DOC: dissolved organic carbon; DOM: dissolved organic matter

[b] o: no effect; +: enhance photolysis rate; -: reduce photolysis rate

The effects of DOM on nitrate-induced, OH$^{•}$ mediated degradation of aquatic contaminants have been extensively studied (Miller and Chin, 2005; Lam et al., 2003; Brezonik and Fulkerson-Brekken, 1998; Torrents, 1997). As mentioned before, nitrate is the primary source of OH$^{•}$, one of the principal intermediates in natural water photochemistry. DOM is a significant natural OH$^{•}$ scavenger and the importance of DOM as a OH$^{•}$ sink can be directly estimated from the dissolved organic carbon (DOC) concentration of a water (Brezonik and Brekken, 1998). Research by Lam et al. (2003) showed that the effects of DOM on the photodegradation of target organic compounds strongly depended on the concentration of nitrate in water. At low concentrations of nitrate, OH$^{•}$-mediated degradation was unimportant and increasing the concentration of DOM resulted in an increase in the photolysis rate, mainly due to the DOM-induced reactions. However, at high concentrations of nitrate, the opposite effect was observed; an increase in DOM concentration significantly reduced the photolysis rates because of the inhibition of $OH^{\bullet}$-mediated degradation caused by scavenging of $OH^{\bullet}$ by DOM.

Another particular case is the binding of organic compound to DOM. For example, the binding of some PAHs to dissolved organic matter (DOM) in natural waters is known to cause fluorescence quenching (Gauthier et al., 1986; Backhus et al., 1990; Schlautman et al., 1993), which could inhibit or enhance PAH photodegradation depending on the quenching mechanism. Inhibition would occur if the binding to DOM increased the rate of relaxation of the PAH excited singlet state to the ground state without the formation of a reactive intermediate, whereas enhanced degradation could occur if the quenching produced a reactive PAH intermediate.

5.6.3 pH

Changes in acidity and alkalinity of the water medium may influence the photolysis rate of a contaminant in several ways. First of all, pH plays a significant role in determining the speciation of a contaminant in water solution. As different species may have different photoreactive characteristics, the overall photodegradation of the compound may be strongly dependent upon pH. For example, photodegradation of Fe (III)-EDTA was found to be much faster at pH 4.5 than at pH 8.5 (Lockhart and Blakeley, 1975), mainly due to the formation of non-photoreactive dimmers of Fe(III)-EDTA at alkaline pH (Carey and Langford, 1973). Another example was the effect of pH on the photodegradation of pentachlorophenol (PCP) (Cui and Huang, 2004). The photolysis rate of PCP in the alkaline conditions was found to be higher than in the acidic condition because ionized PCP is easier to be excited and degraded than nonionized PCP.

pH may also affect the formation of photocatalysts or reactive species in a photochemical reaction. Chiron et al. (2006) investigated the photodegradation of carbamazepine in the presence of Fe(III) and found that the degradation was faster in acidic conditions. This was due to the fact that acidic conditions favor the formation of $FeOH^{2+}$, which dominates Fe(III) speciation between pH 3 to 5 and is the most photoactive Fe(III) species (Waite, 2005).

Finally, pH can indirectly influence the free radical-mediated photochemical reaction via the impacts on radical scavengers. In an $OH^{\bullet}$-mediated degradation system with the presence of bicarbonate/carbonate, the photodegradation rate of model compounds decreased with increasing of pH (Lam et al., 2003). One possible reason was that, at higher pH the dominant carbon species in water is carbonate which has an $OH^{\bullet}$ scavenging ability that is 28 times more than bicarbonate ions (Brezonik and Fulkeron-Brekken, 1998).

5.6.4 Ionic Strength

Increasing ionic strength in the water medium could inhibit or enhance the photodegradation of a contaminant, depending on the degradation mechanism involved. The presence of halide ions (Cl⁻, Br⁻, and I⁻) may slow down the photolysis rate by acting as quenchers of fluorescence of aromatic molecules. Such effect has been demonstrated by the photodegradation of a fungicide, fenarimol, in natural waters (Mateus et al., 2000), in which the addition of NaCl and NaBr reduced both photodegradation rates and quantum yields. As the most effective $^1O_2$ quencher, halides were also found to inhibit the photolysis of chrysene absorbed to a smectite clay (Kong and Ferry, 2003).

Some organic compounds may strongly bind to metal ions, resulting in a significant increase of photolysis rates. An example, photodegradation of EDTA in the presence of Fe(III), has been previously presented in 5.5.3. Werner et al. (2006) investigated the effects of water hardness on the photolysis of tetracycline, and found that at different Mg^{2+} and Ca^{2+} concentrations the photodegradation rate constants were enhanced by up to an order of magnitude.

5.7   Case Studies

5.7.1 Photodegradation of Methylmercury in Lakes

Methylmercury (CH_3Hg^+) is the major form of organic mercury that is most easily bioaccumulated in organisms. It has been long believed that the biological demethylation dominates methylmercury removal in natural fresh waters. However, the long-accepted view was challenged by the discovery of the importance of methylmercury photodegradation (Sellers et al., 1996).

In this study, a significant loss of methylmercury in lake water was found under sunlight but not under dark conditions. To avoid the biotic effects, lake water was filtered with 0.45-μm filters to remove all photosynthetic organisms and most bacteria. Filtering the water did not inhibit the photodegradation of methylmercury, implying that the photodegradation process was abiotic. To further confirm the abiotic nature of the reaction, unfiltered but sterilized water was used and the sterilization was found to have an insignificant effect on the photodegradation process. The light-induced reaction was finally demonstrated by moving samples from dark to light and loss of methylmercury was only detected when the samples were exposed to sunlight.

The finding that photodegrdadation is an important pathway for methylmercury degradation may be useful in engineering design for the mitigation of methylmercury problems. For effluent water containing high concentrations of methylmercury, for example, the wastewater could be retained in shallow ponds before discharge. The finding also fundamentally changes the existing understanding of mercury cycle in aquatic environments. Sellers et al. calculated an annual mass-balance budget for methylmercury in a lake without and with the inclusion of photodegradation in the methylmercury cycle. If photodegradation is not included in the cycle, there would be a net flux of methylmercury from water to sediments; however, an opposite direction of net flux (i.e., from sediments to water) would be occur when photodegradation process is taken into consideration. Based on the mass-balance analysis, Sellers et al. concluded that the in-lake production of methylmercury from sediments must be an important source of methylmercury in lake water.

5.7.2 Photodegradation of Agricultural Pollutants in Wetland Waters

Wetlands have several features that favor the occurrence of photochemical reactions. Wetlands generally have large surface areas and shallow depths that would allow significant sunlight exposure and penetration throughout the water column. Wetland water is also abundant in dissolved organic matter (DOC), the dominant photosenstizer in natural waters. For example, wetlands across Canadian prairies typically have DOC concentrations of >10 mg/L and sometimes >100 mg/L (Curtis and Adams 1995). The high levels of DOC lead to a greater near-surface photoreactivity in wetland water in comparison with other natural waters (Valentine and Zepp, 1993). For wetlands that receive drainage from agricultural runoff, their surface water is also characterized by a relatively high concentration of nitrate, one of the primary precursors in the photoproduction of the extremely reactive oxidant - hydroxyl radical ($OH^{\bullet}$) in nature water photochemistry. Miller and Chin (2002, 2005) investigated the indirect photolysis of two agricultural pollutants (Carbaryl and Alachlor) in natural and engineering wetland waters, and evaluated the influence of water constituents on the photodegradation of the two compounds.

In these studies, water samples were taken from a natural coastal wetland and four engineering wetlands in Ohio in the United States. All of the water samples had a relatively high pH of around 8. DOC varied greatly among the samples, ranging from 2 to 36 mg C/L. The concentration of nitrate was found to fluctuate with seasons, and the highest concentration (~1 mM) (62 mg/L) was detected in the natural wetland in June, most likely resulted from the runoff from fertilizer application.

The photodegradation of carbaryl and alachlor strongly depended on pH. At natural pH ~ 8, the light-induced reactions for carbaryl degradation was negligible while for alachlor the photodegradation was important only at high nitrate levels (~1mM). In pH adjusted natural water samples (~ 4), significant photodegradation was detected for both compounds. The light-induced reactions accounted for 87-98% of the overall degradation of carbaryl while the photolysis rate constants for alachlor increased 3-18 times of that determined at the natural pH. The authors attributed the promotion of the photochemical reactions at acidic pH to DOM or Fe-DOM complexes associated photolysis pathways, although the relationship between pH and the photochemical behavior of DOM remains unclear.

Nitrate played a significant role in the photodegradation of the two compounds. Nitrate photolysis was found to be the dominant mechanism for the degradation of alachlor at natural pH. At an adjusted pH ~ 4, the photodegradation of carbaryl exhibited seasonal and spatial dependence, which was strongly correlated with the concentrations of nitrate in water samples.

DOM was also an important photosensitizer for the photodegradation of carbaryl and alachlor in wetland waters. In low-nitrate-containing water, DOM-induced reactions accounted for up to 73% photodegradation of carbaryl. DOM was also a key component in the promotion of alachlor degradation at acidic pH, as demonstrated by enhanced degradation in water samples with high levels of DOM but low levels of nitrate. DOM may also act as a scavenger of free radicals, but there was no evidence that DOM had a negative effect on the degradation of the two compounds.

It appeared that nitrate and DOM were two principal constituents responsible for the observed indirect photolysis processes. The quenching experiments using methanol further provided evidence that hydroxyl radical (OH$^\bullet$) was the primary reactive transient in these wetland water systems.

5.7.3 Photodegradation of Petroleum in Marine Environment

Released crude oil and refined products from spills and shipwrecks are subject to physical, chemical, and biological processes that change their composition, chemical and physical properties (NRC, 2002). Over the past 20 years, it has been recognized that photochemical process may play a significant role in petroleum weathering in the marine environment, particularly under the tropical sea conditions because of the high solar radiation and the inhibition of biological process due to the lack of nutrients (Ehrhardt et al., 1992). The photochemical transformation of petroleum in marine environment has been reviewed by Payne and Phillips (1985) and Nicodem et al. (1997).

Photochemical process may significantly change the composition, physical properties and toxicity of the exposed parent oil. Such changes can occur at rates comparable to evaporation and at rates greater than biodegradation processes (Payne and Phillips, 1985).

Change of Composition. The composition of crude petroleum is extremely complex, as it consists alkanes, olefins, aromatic hydrocarbons, nitrogen and sulfur heterocyclics, and oxygen-containing components (NRC, 2002). Typically polycyclic aromatic hydrocarbons (PAH) constitute about 0.2 percent to more than 7 of crude petroleum (NRC, 2002). Petroleum samples showed an increased oxygenation after exposure to solar irradiation, and the primary photooxidation products identified included aliphatic and aromatic acids, carbonyl compounds, phenols, alcohols, ethers, hydroperoxides, and sulfoxides (Nicodem et al., 1997; Payne and Phillips, 1985). The photochemical process resulted in a decrease in the aromatic fraction and an increase in the polar fraction in petroleum (Daling et al., 1990). Aliphatic components were also degraded by photooxidation and the branched compounds were found to be decomposed more readily than the straight chain compounds (Hansen, 1975). Moreover, aliphatic sulfides and aromatic thiophenes were readily photooxidized to their corresponding oxides such as sulfoxides and sulfones (Burwood and Speers, 1974; Patel et al., 1979).

Change of Physical Properties.   Photochemical reactions may alter the petroleum viscosity and water-in-oil emulsification tendencies. Solar irradiation has been found to increase the viscosity of spilled oil and promote the formation of water-in-oil emulsification, mainly due to the increase of asphaltene fraction upon irradiation (Nicodem et al., 1997). Photooxidation can also increase the "oil solubility" in water via the formation of polar derivatives including ketones, alcohols, hydroperoxides, sulfoxides, phenols and carboxylic acids (Nicodem et al., 1997).

Change in Toxicity. Photochemical process is known to affect the toxicity of oil and the water-soluble fraction of oil. Upon solar irradiation, surface oil films become less toxic due to the loss of polycyclic aromatic hydrocarbons (PAH) but the water-soluble fraction becomes more toxic due to the formation and dissolution of photooxidation products (Larson and Berenbaum, 1988; Sydnes et al., 1985). For example, hydroperoxides were found to be responsible for much of the phototoxicity of the water-soluble fraction of No.2 fuel oil (Larson et al., 1977). It is also possible that the soluble organic compounds at low concentrations may support the growth of microorganism and hence promote biodegradation. It has been demonstrated that sunlight irradiation may increase the bioavailability of crude oil samples (Maki et al., 2001).

The mechanisms involving the photooxidation of petroleum products are generally complicated. First of all, singlet oxygen ($^1O_2$) are generated by energy transfer from electronic excited states of photosensitizers (such as polycyclic aromatic hydrocarbons) to molecular oxygen. Quantum yields for $^1O_2$ formation were found to range from 0.5 to 0.8 for a number of crude oils (Lichtenthaler et al., 1989). The importance of $^1O_2$ was demonstrated by the inhibition of the phototoxidation of a No.2 fuel oil in the presence of β-carotene, an excellent quencher of $^1O_2$ (Larson and Hunt, 1978). Another mechanism may involve the reaction of singlet oxygen ($^1O_2$) with a reactive acceptor such as substituted olefins and polyunsaturated fatty acids to form hydroperoxides (Larson et al., 1979).

The photolysis rates of petroleum can be inhibited or accelerated in the presence of several organic compounds. Common to many crude oils, thiocyclanes have been demonstrated to effectively restrict the formation of radical species because of their propensity to produce thiocyclane oxides (Larson et al., 1979). β-carotene was also found to rapidly quench the singlet-oxygen mediated photooxidation at a diffusion-controlled rate (Larson and Hunt, 1978). Humic substances in marine water can also inhibit or accelerate the photodegradation of petroleum compounds by acting as an inner filter, or a radical scavenger, or a photosensitizer (Mill et al., 1981; Zepp et al., 1981).

5.8   Conclusion and Implications

Photodegradation plays a significant role in environmental degradation of anthropogenic chemicals, such as petroleum, pesticides, pharmaceutical products, and metal-organic complexes. The mechanisms of photodegradation include direct photolysis, indirect photolysis via energy transfer or free radical reactions, and heterogeneous photocatalysis that mainly mediated by suspended natural particles. Estimates of photolysis rates and quantum yields can be affected by natural variations in solar radiation, pH, ionic strength, suspended particles, dissolved organic materials and the presence of inhibitors.

Although numerous contaminants have been demonstrated to be photodegradable in laboratory or field studies, the engineering experience in applying the natural photodegradation using natural sunlight to remove pollutants is rare. The effectiveness of photodegradation under natural conditions is seasonal, spatial, and water composition dependent, making it difficult to extrapolate conclusions from one study to another. Moreover, it remains a tremendous challenge to evaluate the environmental effects of natural photodegradation process, mainly due to the fact that this process generally cannot lead to the mineralization of organic contaminants. Identification and toxicity evaluation of photoproducts become critical issues since the products may potentially have an environmental impact that is severe or more severe than the parent contaminant. In most of photodegradation cases, however, products identification can be one of the most difficult tasks, particularly under the field conditions. Coupling photochemical studies with a biological assay may provide a compromised approach to address the photoproducts' biological activity without identifying them.

Under field conditions, the effects of photodegradation on the chemical transformation and fate of compounds generally cannot be differentiated from those of other natural attenuation processes. Unfortunately, the interaction between photodegradation and other processes and their quantification is less well understood. There is intriguing evidence that photodegradation may increase the bioavailability and accelerate the biodegradation process. Therefore, it could be that photodegradation followed by biological metabolism provides a significant pathway for the removal of organic contaminants under natural conditions, but in the absence of supporting studies this possibility cannot be properly evaluated. More work in this area is definitely required.

References

Albini, A. and Fasani, E. (1998) Photochemistry of Drugs: An Overview and Practical Problems. *Special Publication – Royal Society of Chemistry*. 225: 1–73.
Albini, A. and Monti, S. (2003) Photophysics and Photochemistry of Fluoroquinolones. *Chemical Society Reviews*. 32: 238–250.

Anderson, R.L., Bishop, E.B. and Campbell, R.L. (1985) A review of the Environmental and Mammalian Toxicology of Nitrilotriacetic Acid. *Crit. Rev. Toxicol.* 15: 1–102.

Andreozzi, R., Raffaele, M. and Nicklas, P. (2003) Pharmaceuticals in STP Effluents and Their Solar Photodegradation in Aquatic Environment. *Chemosphere.* 50: 1319–1330.

Atkinson, R. (1986) Kinetics and Mechanism of the Gas Phase Reactions of the Hydroxyl Radical with Organic Compounds under Atmospheric Conditions. *Chem. Rev.* 86: 69-201.

Bachman, J. and Patterson, H. H. (1999) Photodecomposition of the Carbamate Pesticide Carbofuran: Kinetics and the Influence of Dissolved Organic Matter. *Environ. Sci. Technol.* 33:874-881

Benkelberg, H.-J. and Warneck, P. (1995) Photodecomposition of Iron(III) Hydroxo and Sulfato Complexes in Aqueous Solution: Wavelength Dependence of OH and SO_4^- Quantum Yields. *J. Phys.Chem.* 99: 5214–5221.

Boreen, A. L., Arnold, W. A. and McNeill, K. (2003) Photodegradation of Pharmaceuticals in the Aquatic Environment: A review. *Aquat. Sci.* 65: 320–341.

Boreen, L., Arnold, W. A. and McNeill, K. (2005) Triplet-sensitized Photodegradation of Sulfa Drugs Containing Six-membered Heterocyclic Groups: Identification of an SO_2 Extrusion Photoproduct. *Environ. Sci. Technol.* 39: 3630–3638

Bosca, F., Marin, M. L. and Miranda, M. A. (2001) Photoreactivity of the Nonsteroidal Anti-inflammatory 2-arylpropionic Acids with Photosensitizing Side Effects. *Photochemistry and Photobiology.* 74: 637–655.

Bosca, F., Miranda, M. A., Vano, L. and Vargas, F. (1990) New Photodegradation Pathways for Naproxen, a Phototoxic Nonsteroidal Anti-inflammatory Drug. *Journal of Photochemistry and Photobiology A: Chemistry.* 54: 131–134.

Brezonik, P.L. and Fulkerson-Breken, J. (1998) Nitrate-induced Photolysis in Natural Waters: Controls on Concentrations of Hydroxyl Radical Photo-intermediates by Natural Scavenging Agents. *Environ. Sci. Technol.* 32: 3004–3010.

Burrows, H.D., Canle, L.M., Santaballa, J.A. and Steenken, S. (2002) Reaction Pathways and Mechanisms of Photodegradation of Pesticides. *J Photochem Photobiol. B: Biol* 67: 71-108.

Burwood, R. and Speers, G.C. (1974) Photo-oxidation as a Factor in the Environmental Dispersal of Crude Oil. *Estuarine Coastal and Mar. Sci.* 2: 117-135.

Buser, H.R., Poiger, T. and Mueller, M. D. (1999) Occurrence and Environmental Behavior of the Chiral Pharmaceutical Drug Ibuprofen in Surface Waters and in Wastewater. *Environ. Sci. Technol.* 33: 2529–2535.

Calamari, D., Zuccato, E., Castiglioni, S., Bagnati, R. and Fanelli, R. (2003) Strategic Survey of Therapeutic Drugs in the Rivers Po and Lambro in Northern Italy. *Environ Sci Technol.* 37: 1241-1248.

Carassiti V. (1984) What Means Photocatalysis? Another Reply. *EPA Newslett.* 21: 12-16.

Carey, J.H. and Langford, C.H. (1973) Photodecomposition of Fe(III) Aminopolycarboxylates. *Can. J. Chem.* 51:3665-70

Cheng, H.M. and Hwang, D.F. (1996) Photodegradation of Benthiocarb. *Chem. Ecol.* 12: 91-101.

Chesters, G., Simsiman, G.V., Levy, J., Alhajjar, B.J., Fathulla, R.N., and Harkin, J.M. (1989) Environmental Fate of Alachlor and Metolachlor. *Rev. Environ. Contam. Toxicol.* 110:1–74.

Chiron S., Minero, C. and Vione, D. (2006) Photodegradation Processes of the Antiepileptic Drug Carbamazepine, Relevant To Estuarine Waters. *Environ. Sci. Technol.* 40:5977 –5983.

Chiron, S., Abian, J., Ferrer M., Sanchez-Baeza, F., Messeguer, A., and Barcelo, D. (1995) Comparative Photodegradation Rates of Alachlor and Bentazone in Natural Water and Determination of Breakdown Products. *Environ. Toxicol. Chem.* 14:1287–1298.

Comber, S.D.W. (1999) Abiotic Persistence of Atrazine and Simazine in Water. *Pestic. Sci.* 55: 696–702.

Condorelli, G., Guidi, G. De, Giuffrida, S. and Costanzo, L. L. (1993) Photosensitizing Action of Nonsteroidal Antiinflammatory Drugs on Cell Membranes and Design of Protective Systems. *Coordination Chemistry Reviews.* 125:115–127.

Corin, N.S., Backlund, P.H., and Kulovaara, M.A.M. (2000) Photolysis of the Resin Acid Dehydroabietic Acid in Water. *Environ Sci Technol.* 34:2231–2236.

Cui, J. and Huang, G.L. (2004) Photodegradation of Pentachlorophenol by Sunlight in Aquatic Surface Microlayers. *Journal of Environmental Science and Health, Part B: Pesticides, Food Contaminants, and Agricultural Wastes.* 39: 65-73.

Curtis, P.J. and Adams, H.E. (1995) Dissolved Organic Matter Quantity and Quality from Freshwater and Saltwater Lakes in Alberta. *Biogeochemistry.* 30: 59-76.

Daling, P.S., Brandvik, P.J., Mackay, D. and Johansen, O. (1990) Characterization of Crude Oils for Environmental Purposes. *Oil and Chem. Pollut.* 7: 199–224.

Das, Y.T. (1990) Photodegradation of [1-naphthyl-14C]carbaryl in Aqueous Solution Buffered at pH 5 under Artificial Sunlight, Vol.169-208 #87094, Department of Pesticide Regulation, Sacramento, CA.

Della Greca, M., Fiorentino, A., Iesce, M.R., Isidori, M., Nardelli, A. and Previtera, L. (2003) Identification of Phototransformation Products of Prednisone by Sunlight. Toxicity of the Drug and Its Derivatives on Aquatic Organisms. *Environ Toxicol Chem* 22: 534–539.

Dimou, A.D., Sakkas, V.A. and Albanis, A.T. (2005) Metolachlor Photodegradation Study in Aqueous Media under Natural and Simulated Solar Irradiation. *J. Agric. Food Chem.* 53: 694 –701.

Doll, T.E. and Frimmel, F.H. (2003) Fate of Pharmaceuticals—Photodegradation by Simulated Solar UV-light. *Chemosphere.* 52: 1757–1769.

Edhlund, B.L., Arnold, W. A., and McNeill, K. (2006) Aquatic Photochemistry of Nitrofuran Antibiotics. *Environ. Sci. Technol.* 40: 5422 –5427

Ehrhardt, M.G., Burns, K.A., and Bicego, M.C. (1992) Sunlight-Induced Compositional Alterations in the Seawater-Soluble Fraction of a Crude Oil. *Mar. Chem.* 37: 53–64.

Ellis, D. and Mabury, S. (2000) The Aqueous Photolysis of TFM and Related Trifluoromethylphenols. An Alternate Source of Trifluoroacetic Acid in the Environment. *Environ. Sci. Technol.* 34: 632–637.

Evgenidou, E. and Fytianos, K. (2002) Photodegradation of Triazine Herbicides in Aqueous Solutions and Natural waters. *J. Agric. Food Chem.* 50: 6423–6427.

Fasnacht, M.P. and Blough, N.V. (2002) Aqueous Photodegradation of Polycyclic Aromatic Hydrocarbons. *Environ. Sci. Technol.* 36: 4364–4369.

Fasnacht, M.P. and Blough, N.V. (2003) Mechanisms of the Aqueous Photodegradation of Polycyclic Aromatic Hydrocarbons. *Environ. Sci. Technol.* 37:5767–5772

Frank, R. and Rau, H. (1990) Photochemical Transformation in Aqueous Solution and Possible Environmental Fate of Ethylenediaminetetraacetatic acid (EDTA). *Ecotox. Environ. Saf.* 19: 55–63.

Frimmel, F.H., Grenz, R., Kordik,E., and Dietz, F. (1989) Nitrilotriacetat (NTA) and ethylendinitrilotetraacetat (EDTA) in Fliessgewasser der bundes republik deutchland. *Vom Wasser.* 72:175-184

Haag, W.R. and Hoigné, J. (1985) Photo-sensitized Oxidation in Natural Water via OH Radicals. *Chemosphere.* 14: 1659–1671.

Haag, W.R. and Hoigné, J. (1986) Singlet Oxygen in Surface Waters. 3. Photochemical Formation and Steady-state Concentrations in Various Types of Waters. *Environ. Sci. Technol.* 20: 341–348.

Halling-Sørensen, B., Nors Nielsen, S., Lanzky, P.F., Ingerslev, F., Holten Lützhøft, H.C. and Jorgensen, S.E. (1998) Occurrence, Fate and Effects of Pharmaceutical Substances in the Environment—A review. *Chemosphere.* 36: 357–393.

Hansen, H.P. (1975) Photochemical Degradation of Petroleum Hydrocarbon Surface Films on Seawater. *Mar. Chem.* 3:183–195.

Harris, J. C. (1982) Rate of Aqueous Photolysis. Chapter 8 *in Handbook of Chemical Property Estimation Methods.* Lyman, W. J., Reehl, W. F., and Rosenblatt, D. H. Eds. McGraw-Hill., NY.

Hoffmann, M. R., Martin, S. T., Choi, W., and Bahnemann, D. W. (1995) Environmental Applications of Semiconductor Photocatalysis. *Chem. Rev.* 95:69-96.

Jaynes, D.B., Hatfield, J.L., and Meek, D.W. (1999) Water Auality in Walnut Creek Watershed: Herbicides and Nitrate in Surface Waters. *J. Environ. Qual.* 28:45–59.

Jimenez, M. C., Miranda, M. A. and Tormos, R. (1997) Photochemistry of Naproxen in the Presence of b-Cyclodextrin. *Journal of Photochemistry and Photobiology A: Chemistry.* 104:119–121.

Kari, F.G. and Giger, W. (1995) Modelling the Photochemical Degradation of Ethylene Diaminetetraacetate in River Glatt. *Environ. Sci. Technol.* 29: 2814–2827.

Kari, F.G., Hilger, S. and Canonica, S. (1995) Determination of the Reaction Quantum Yield for the Photochemical Degradation of Fe(III)–EDTA: Implication for the Environmental Fate of EDTA in Surface Waters. *Environ. Sci. Technol.* 29: 1008–1017.

Kisch, H. and Hennig, H. (1983) What Means 'Photocatalysis'? A Proposal to Initiate Further Discussions. *EPA Newslett.* 19: 23–26.

Kochany, J., and Maguire, R.J. (1994) Sunlight Photodegradation of Metolachlor in Water. *J. Agric. Food Chem.* 42: 406–412.

Kolpin, D.W, Furlong, E.T., Meyer, M.T., Thurman, E.M., Zaugg, S.T., Barber, L.B. and Buxton, H.T. (2002) Pharmaceuticals, Hormones and Other Organic Wastewater Contaminants in US Streams, 1999–2000: a National Reconnaissance. *Environ. Sci. Technol.* 36:1202–1211

Kong, L. and Ferry, J.L. (2003) Effect of Salinity on the Photolysis of Chrysene Adsorbed to a Smectite Clay. *Environ. Sci. Technol.* 37: 4894 – 4900.

Konstantinou, I.K., Zarkadis, A.K., and Albanis, T.A. (2001) Photodegradation of Selected Herbicides in Various Natural Waters and Soils under Environmental Conditions. *J. Environ. Qual.* 30:121–130.

Lam, M.W., Tantuco, K., and Mabury, S.A. (2003) PhotoFate: A New Approach in Accounting for the Contribution of Indirect Photolysis of Pesticides and Pharmaceuticals in Surface Waters. *Environ. Sci. Technol.* 37:899–907.

Lam, M.W, and Mabury, S.A. (2005) Photodegradation of the Pharmaceuticals Atorvastatin, Carbamazepine, Levofloxacin, and Sulfamethoxazole in Natural Waters. *Aquat. Sci.* 67:177–188

Langford, C.H., Wingham, M. and Sastri, V.S. (1973) Ligand Photooxidation in Copper(II) Complexes of Nitrilotriacetic Acid. *Environ. Sci. Technol.* 7: 820–822.

Larson, R.A. and Berenbaum, M.R. (1988) Environmental Phototoxicity. *Environ. Sci. Technol.* 22:354–360.

Larson, R.A., Boil, T.L., Hunt, L.L. and Rogenmuser, K. (1979) Photooxidation Products of a Fuel Oil and Their Antimicrobial Activity. *Environ. Sci. Technol.* 13: 965–969.

Larson, R.A., Hunt, L.L. and Blankenship, D.W. (1977) Formation of Toxic Products from a #2 Fuel Oil by Photooxidation. *Environ. Sci. Technol.* 11: 492–496.

Larson, R.A. and Hunt, L.L. (1978) Photooxidation of a Refined Petroleum Oil: Inhibition by Carotene and Role of Singlet Oxygen. *Photochem. and Photobiol.* 28: 553–555.

Larson, R. A. and Weber, E. J. (1994) *Reaction Mechanisms in Environmental Organic Chemistry.* Lewis Publishers, Boca Raton, FL.

Latch, D. E., Packer, J. L., Arnold, W. A. and McNeil, K. (2003a) Photochemical Conversion of Triclosan to 2,8-Dichlorodibenzo-Pdioxin in Aqueous Solution. *Journal of Photochemistry and Photobiology, A: Chemistry.* 158: 63–66.

Latch, D. E., Stender, B. L., Packer, J. L., Arnold, W. A. and McNeill, K. (2003b) Photochemical Fate of Pharmaceuticals in the Environment: Cimetidine and Ranitidine. *Environ. Sci. Technol.* 37: 3342–3350.

Lichtenthaler, R.G., Haag, W.R. and Mill, T. (1989) Photooxidation of Probe Compounds Sensitized by Crude Oils in Toluene and as an Oil Film on Water. *Environ. Sci. Technol.* 23: 39–45

Lin, A.Y. and Reinhard, M. (2005) Photodegradation of Common Environmental Pharmaceuticals and Estrogens in River Water. *Environ Toxicol Chem.* 24:1303-1309.

Lin,Y.J., Karuppiah, M., Shaw, A. and Gupta, G. (1999) Effect of Simulated Sunlight on Atrazine and Metolachlor Toxicity of Surface Waters. *Ecotoxicol Environ Safe.* 43:35-37.

Lin, Y.J., Lin, C., Yeh, K.J. and Lee, A. (2000) Photodegradation of the Herbicides Butachlor and Ronstar Using Natural Sunlight and Diethylamine. *Bull. Environ. Contam. Toxicol.* 64: 780–785.

Lockhart, H.B. and Blakeley, R.V. (1975a) Aerobic Photodegradation of X(N) Chelates of Ethylenedinitrilotetraacetate (EDTA): Implications for Natural Waters. *Environ Lett.* 9:19–31.

Lockhart, H.B. and Blakeley, R.V. (1975b) Aerobic Photodegradation of Fe(III)–(Ethylenedinitrilo)tetraacetate (Ferric EDTA). *Environ. Sci. Technol.* 9:1035–1038.

Maki, H., Sasaki, T., and Harayama, S. (2001) Photo-Oxidation of Biodegraded Crude Oil and Toxicity of the Photo-Oxidized Products. *Chemosphere.* 44: 1145–1151.

Mansour, M., Thailer, S. and Korte, F. (1983). Action of Sunlight on Parathion. *Bull. Environ. Contam. Toxicol.* 30:358–364.

Martinez, L. J. and Scaiano, J. C. (1998) Characterization of the Transient Intermediates Generated from the Photoexcitation of Nabumetone: a Comparison with Naproxen. *Photochemistry and Photobiology.* 68: 646–651.

Mateus, M.C.D.A., Da Silva, A.M. and Burrows, H.D. (2000) Kinetics of Photodegradation of the Fungicide Fenarimol in Natural Waters and in Various Salt Solutions: Salinity Effects and Mechanistic Considerations. *Water Res.* 34: 1119–1126.

Mathew, R. and Khan, S.U. (1996) Photodegradation of Metolachlor in Water in the Presence of Soil Mineral and Organic Constituents. *J. Agric. Food Chem.* 44: 3996-4000.

Metsarinne, S., Tuhkanen, T. and Aksela, R. (2001) Photodegradation of Ethylenediaminetetraacetic Acid (EDTA) and Ethylenediamine Disuccinic Acid (EDDS) within Natural UV Radiation Range. *Chemosphere.* 45: 949–955.

Mill, T., Hendry, D.G. and Richardson, H. (1980) Free Radical Oxidants in Natural Waters. *Science.* 207: 886-887.

Mill, T., Mabey, W.R., Lan, B.Y. and Baraze, A. (1981) Photolysis of Polycyclic Aromatic Hydrocarbons in Water. *Chemosphere.* 10:1281–1290.

Miller, P. L. and Chin, Y. P. (2005) Indirect Photolysis Promoted by Natural and Engineered Wetland Water Constituents: Processes Leading to Alachlor Degradation. *Environ. Sci. Technol.* 39: 4454–4462.

Miller, P. L. and Chin, Y. P. (2002) Photoinduced Degradation of Carbaryl in a Wetland Surface Water. *J. Agric. Food Chem.* 50: 6758-6765.

Miller, J.S. and Olejnik, D. (2001) Photolysis of Polycyclic Aromatic Hydrocarbons in Water. *Water Res.* 35: 233-243.

Miller, R.M., Singer, G.M., Rosen, J.D. and Bartha, R. (1988) Photolysis Primes Biodegradation of Benzo[a]pyrene. *Appl. Environ. Microbiol.* 54:1724–1730.

Mirbach, M.J. (1984) What Means 'Photocatalysis'? A reply: On the Classification of Photocatalysis. *EPA Newslett.* 20:16–19.

Moore, D. E. and Chappuis, P. P. (1988) A Comparative Study of the Photochemistry of the Non-steroidal Anti-inflammatory Drugs, Naproxen, Benoxaprofen and Indomethacin. *Photochemistry and Photobiology.* 47: 173–180.

Moore, D. E., Roberts-Thomson, S., Zhen, D. and Duke, C. C. (1990) Photochemical Studies on the Anti-inflammatory Drug Diclofenac. *Photochemistry and Photobiology A: Chemistry.* 52: 685–690.

Nicodem, D.E., Conceicao, M., Fernandes, Z., Guedes, C. and Correa, R.J. (1997) Photochemical Processes and the Environmental Impact of Petroleum spills. *Biogeochem.* 39: 121–138.

Nowack, B. (2002) Environmental Chemistry of Aminopolycarboxylate Chelating Agents. *Environ. Sci. Technol.* 36: 4009–4016.

Nowack, B. and Baumann, U. (1998) Biologischer Abbau der Photolyseprodukte von Fe^{II}IEDTA. *Acta Hydrochim Hydrobiol.* 26:1–8.

National Research Council. (2002) *Oil in the Sea III: Inputs, Fates, and Effects.* National Academy Press, Washington, D.C.

Oliver, G. J., Cosgrove, E. G., and Carey, J. H. (1979) Effect of Suspended Sediments on the Photolysis of Organics in Water. *Environ. Sci. Technol.* 13: 1075-1077.

Packer, J. L., Werner, J. J., Latch, D. E., McNeill, K. and Arnold, W. A. (2003) Photochemical Fate of Pharmaceuticals in the Environment: Naproxen, Diclofenac, Clofibric Acid, and Ibuprofen. *Aquat. Sci.* 65: 342–351.

Pape, B.E. and Zabik, M.J. (1972) Photochemistry of Bioactive Compounds. Solution-Phase Photochemistry of Symmetrical Triazines. *J. Agr. Food Chem.* 20:316-320.

Pape, B.E. and Zabik, M. (1970) Photochemistry of Bioactive Compounds. Solution-Phase Photochemistry of Symmetrical Triazines. *J. Agric. Food Chem.* 18:202–210.

Patel, J.R., Overton, E.B. and Laster, J.L. (1979) Environmental Photooxidation of Dibenzothiophenes Following the Amoco Cadiz oil spill. *Chemosphere.* 8: 557–561.

Payne, J.R. and Phillips, C.R. (1985) Photochemistry of Petroleum in Water. *Environ. Sci. Technol.* 19: 569–579

Pelizzetti, E., Maurine, V., Minero, C., Pramauro, E., Zerbinati, O., and Tosato, M.L. (1990) Photocatalytic Degradation of Atrazine and other *s*-Triazine Herbicides. *Environ. Sci.Technol.* 24:1559–1565.

Poiger, T., Buser, H.R. and Müller, M.D. (2001) Photodegradation of the Pharmaceutical Drug Diclofenac in a Lake: Pathway, Field Measurements and Mathematical Modelling. *Environ. Toxicol. Chem.* 20: 253–256.

Rejto, M., Saltzmann, S., Acher, A.J., and Muszkat, L. (1983) Identification of Sensitized Photooxidation Products of s-Triazine Herbicides in Water. *J. Agric. Food Chem.* 31:138–142.

Ruzo L.O., Zabik, M.J., and Schuetz, R.D. (1973) Photochemistry of Bioactive Compounds: Kinetics of Selected *s*-Triazines in Solution. *J. Agric. Food Chem.* 21:1047–1049.

Schlautman, M. and Morgan, J.J. (1993) Effects of Aqueous Chemistry on the Binding of Polycyclic Aromatic Hydrocarbons by Dissolved Humic materials. *Environ Sci Technol.* 27: 961–969.

Schwarzenbach, R.P., Gschwend, P.M., and Imboden, D.M. (1993) *Environmental Organic Chemistry*. John Wiley and Sons, NY.

Schwarzenbach, R.P., Gschwend, P.M., and Imboden, D.M. (2003) *Environmental Organic Chemistry*. John Wiley and Sons, NY.

Sellers, P., Kelly, C.A., Rudd, J.W.M. and MacHutchon, A.R. (1996) Photodegradation of Methylmercury in Lakes. *Nature*. 380: 694–697.

Sigman, M.E., Schuler, P.F., Ghosh, M.M., and Dabestani, R.T. (1998) Mechanism of Pyrene Photochemical Oxidation in Aqueous and Surfactant Solutions. *Environ Sci Technol.* 32:3980–3985.

Sigman, M.E., Zingg, S.P., Pagni, R.M. and Burns, J.H. (1991) Photochemistry of Anthracene in Water. *Tetrahedron Lett.* 32: 3737–3740

Stangroom, S.J., MacLeod, C.L. and Lester, J.N. (1998) Photosensitized Transformation of the Herbicide 4-chloro-2-methylphenoxy Acetic Acid (MCPA) in water. *Water Res.* 32: 623–632.

Stolzberg, R.J. and Hume, D.N. (1975) Rapid Formation of Iminodiacetate from Photochemical Degradation of Fe(III) Nitrilotriacetate Solutions. *Environ. Sci. Technol.* 9: 654–656.

Sulzberger and Hug. (1994) Light-Induced Process in the Aquatic Environment. Chapter 8 *in Chemistry of Aquatic Systems: Local and Global Perspectives*. Bidoglio, G. and Stumm, W. Eds. Kluwer Academic Publishers, Dordrecht, Netherlands.

Sydnes, L.K., Hemmingsen, T.H., Skare, S. and Hansen, S.H. (1985) Seasonal Variations in Weathering and Toxicity of Crude Oil on Seawater under Arctic Conditions. *Environ. Sci. Technol.* 19:1076–1081.

Ternes, T. (1998) Occurrence of Drugs in German Sewage Treatment Plants and Rivers. *Water Res.* 32: 3245–3260.

Thurman, E. M. and Malcolm, R. L. (1983) Structure Study of Humic Substances: New Approaches and Methods. Chapter 1 *in Aquatic and Terrestrial Humic Materials*. Christman, R. F. and Gjessing, E. T. Eds. Ann Arbor Science, Ann Arbor, MI.

Torrents, A., Anderson, B.G., Bilboulian S., Johnson, W.E. and Hapeman, C.J. (1997) Atrazine photolysis: Mechanistic Investigations of Direct and Nitrate-mediated Hydroxy Radical Processes and the Influence of Dissolved Organic Carbon from the Chesapeake Bay. *Environ. Sci. Technol.* 31:1476–1482.

Trott, T., Henwood, R. W. and Langford, C. H. (1972) Sunlight Photochemistry of Ferric Nitrilotriacetate Complexes. *Envir. Sci. Technol.* 6: 367–368.

Valentine, R. L. and Zepp, R. G. (1993) Formation of CO from the Photodegradation of Terrestrial DOM in Natural Waters. *Environ. Sci. Technol.* 27: 409-412.

Vargas, F., Rivas, C., Miranda, M. A. and Bosca, F. (1991) Photochemistry of the Non-Steroidal Anti-inflammatory Drugs, Propionic Acid-Derived. *Pharmazie.* 46: 767–771.

Walse, S.S., Morgan, S.L., Kong, L. and Ferry, J.L. (2004) Role of Dissolved Organic Matter, Nitrate, and Bicarbonate in the Photolysis of Aqueous Fipronil. *Environ. Sci. Technol.* 38: 3908 – 3915

Weerasinghe, C. A., Lewis, D. O., Mathews, J. M., Jeffcoat, A. R., Troxler, P. M. and Wang, R. Y. (1992) Aquatic Photodegradation of Albendazole and Its Major

Metabolites. 1. Photolysis Rate and Half-Life for Reactions in a Tube. *J. Agric. Food Chem.* 40: 1413-1418.

Werner, J. J., Arnold, W.A. and McNeill, K. (2006) Water Hardness as a Photochemical Parameter: Tetracycline Photolysis as a Function of Calcium Concentration, Magnesium Concentration, and pH. *Environ. Sci. Technol.* 40: 7236 – 7241.

Wilson, R.I. and Mabury, S.A. (2000) The Photodegradation of Metalochlor: Isolation, Identification and Quantification of MCA. *J. Agric. Food Chem.* 48: 944–950.

Zafiriou, O.C., Joussot-Dubien, J., Zepp, R.G., and Zika, R.G. (1984) Photochemistry of Natural Waters. *Environ. Sci. Technol.* 18: 358–371.

Zepp, R.G. (1978) Quantum Yields for Reaction of Pollutants in Dilute Aqueous Solution. *Environ. Sci. Technol.* 12: 327–329.

Zepp, R.G., Baughman, G.L. and Schlotzhauer, P.F. (1981) Comparison of Photochemical Behavior of Various Humic Substances in Water: Sunlight Induced Reactions of Aquatic Pollutants Photosensitized by Humic Substances. *Chemosphere.* 10: 109–117

Zepp, R. G. and Cline, D. M. (1977) Rates of Direct Photolysis in Aquatic Environment. *Environ. Sci. Technol.* 11(4): 359–366.

Zepp, R.G., Hoigné, J. and Bader, H. (1987) Nitrate-Induced Photooxidation of Trace Organic Chemicals in Water. *Environ. Sci. Technol.* 21: 443–450.

Zepp, R. G. and Schlotzhauer, P. F. (1983). Influence of Algae on Photolysis Rates of Chemicals in Water. *Environ. Sci. Technol.* 17: 462-468.

Zepp, R.G. and Schlotzhauer, P.F. (1979) Photoreactivity of Selected Polycyclic Aromatic Hydrocarbons in Water. In *Polynuclear Aromatic Hydrocarbons.* Jones, P.W. and Leber, P. Eds. Ann Arbor Science, Ann Arbor, MI.

Zepp, R. G. and Wolfe, N. L. (1987) Abiotic Transformation of Organic Chemicals at the Particle Water Interface. *in Aquatic Surface Chemistry Chemical Processes at the Particle Water Interface.* Stumm, W. Ed. Wiley Interscience, NY

Zepp, R.G., Schlotzhauer, P.F., and Sink, R.M. (1985) Photosensitized Transformations Involving Electronic Energy Transfer in Natural Waters: Role of Humic Substances. *Environ. Sci. Technol.* 19:74–81

CHAPTER 6

Phytoprocesses

SHANKHA K. BANERJI, RAO Y. SURAMPALLI, PASCALE CHAMPAGNE,
R.D. TYAGI, BALA SUBRAMANIAN, AND S. YAN

6.1   Introduction

Soils contaminated with hazardous wastes can be remediated either by in-situ or ex-situ processes. In-situ processes have the advantage of less exposure of workers to hazardous substances during clean up operations and often cost less with minimum disruption of activities at the site. Most phytoprocesses for remediation of contaminated soils are in-situ processes that have been successfully applied to both organic and inorganic (metallic) contaminants. However, ex-situ use of this technology to clean up contaminated groundwater may also be possible. This chapter describes the various types of phytoprocesses that have been used for the remediation of soil and groundwater. A somewhat similar process using wetland for remediation of hazardous wastes will not be discussed here but in Chapter 7.

6.2   Definitions

Phytoprocesses use green plants, including grasses, forbs and woody species to remove, detoxify or contain hazardous contaminants such as organic compounds, heavy metals and radioactive compounds in soil or groundwater. There are six different types of phytoprocesses that have been used for remediation of contaminated soils and groundwater: phytodegradation, rhizodegradation, phytostabilization, phytovolatilization, phytoextraction and rhizofiltration. Figure 6.1 schematically illustrates the various phytoprocesses.

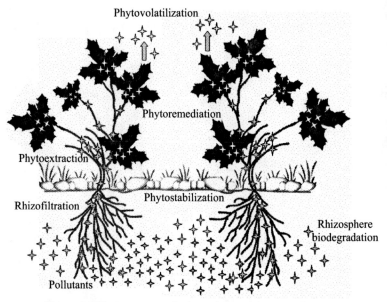

Figure 6.1 Phytoprocesses for hazardous waste removal

6.2.1 Phytodegradation

Phytodegradation or plant-mediated biodegradation is a process by which the plants take up hazardous compounds and biochemically transform constituents through metabolic processes within the plant or external to the plant through the effects of plan-based enzymes to innocuous products within the plants. This process is particularly applicable for the transformation of organic compounds and is also referred to as phytoreduction, phytooxidation, phytolignification and phytotransformation. Figure 6.2 illustrates the phytodegradation process. Contaminants are transformed into products that are subsequently used as nutrients and incorporated into plant tissue or by-products that can be broken down further by microorganisms. Plant transformation pathways vary depending on the plant species, tissue type and nature of the contaminant. These pathways are generally categorized as reduction, oxidation, conjugation, and sequestration (Champagne and Bhandari, 2007). Plants synthesize a large number of enzymes during primary and secondary metabolisms which are useful for phytodegradation. Contaminant degradation by these plant-based enzymes can occur in an environment free of microorganisms and even in environments that are not ideal for biodegradation (Suthersan, 2002).

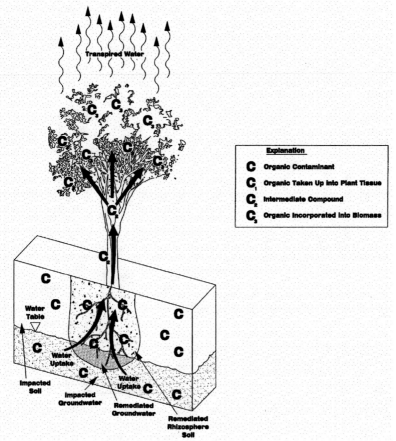

Figure 6.2 Phytodegradation of organic contaminants

6.2.2 Rhizodegradation

Rhidegradation is a process that involves the exudation of organic substances (amino acids, carbohydrates, nucleic acid derivatives, growth factors, carboxylic acids and enzymes) and oxygen by plants to the rhizosphere (root zone) which can stimulate microorganisms (fungi and bacteria), and consequently enhance the biodegradation of hazardous organic compounds within the root zone. As much as 60% of the oxygen transported to the root zone of the plants can be transferred to the rhizosphere generating aerobic conditions for the thriving microbial community associated with the root zone. Plant exudates, root necrosis and other processes provide the organic carbon and nutrients necessary to spur soil bacteria growth to as much as 20 times that is normally found in non-rhizosphere soil. The conditions also favor enzyme induction and cometabolic enzyme degradation by mycorrhizal fungi and other microorganisms in the rhizosphere. Plants can also enhance biodegradation by physical rooting mechanisms which loosen the soil in the root zone and facilitate oxygen and water transport (Champagne and Bhandari, 2007). Rhizodegradation (Figure 6.3) is also called phytostimulation, plant-assisted bioremediation, rhizosphere biodegradation or enhanced rhizosphere biodegradation. This process is particularly useful in the remediation of hazardous organic compounds.

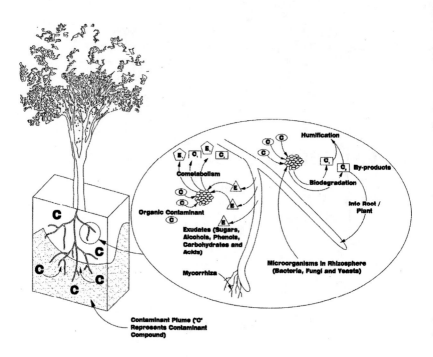

Figure 6.3 Rhizodegradation of organic contaminants

6.2.3 Phytostabilization

Phytostabilization is the use of plants to prevent the migration of contaminants in the soil and groundwater through erosion prevention, absorption and accumulation by roots, adsorption onto roots, or precipitation within the rhizosphere. This process is illustrated in Figure 6.4. It is also known as biomineralization, phytosequestration and lignification and can be applied to organic and inorganic constituents (Champagne and Bhandari, 2007). Plant exudates released to the root zone can increase the local soil pH and the soil oxygen content producing a significant effect on the redox conditions of the soil, promoting oxidation, and causing speciation and adsorption to form stable mineral deposits (Carman and Crossman, 2001). Humification, lignification and irreversible binding of some organic compounds can also occur as a result of plant functions (McCutcheon and Schnoor, 2003).

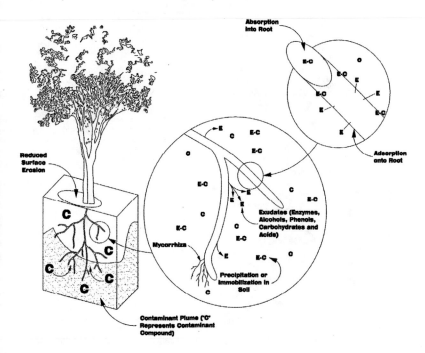

Figure 6.4 Phytostabilization of organic and inorganic contaminants

6.2.4 *Phytovolatilization*

In this process, plants take up water and contaminants which are transported and/or transformed and then transpired to the atmosphere through the leaves as shown in Figure 6.5. This mechanism is also referred to as biovolatilization and phytoevaporation and typically applies to soluble organic compounds and some inorganic constituents suc has selenium, tritium, arsenic, mercury (Champagne and Bhandari, 2007). In some cases, organic compounds that are extracted, are subsequently either degraded or transformed within plant to form catabolic intermediates or end-products that can be volatilized from the plant tissue (Suthersan, 2002). Inorganic contaminants that are extracted and accumulated can be methylated and subsequently volatilized from the plant tissue (Carman and Crossman, 2001).

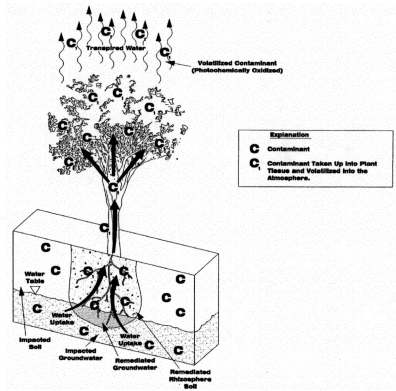

Figure 6.5 Phytovolatilization of organic and inorganic contaminants

6.2.5 *Phytoextraction*

Phytoextraction, also called phytoaccumulation, hyperaccumulation, phytoconcentration, phytomining or phytotransfer is the use of plants to uptake, translocate and accumulate constituents from the plant root zone into above-ground plant tissue. In this process, plant roots extract metals and organic contaminants from the subsurface as shown in Figure 6.6 (Champagne and Bhandari, 2007). Contaminants are not transformed but are accumulated in plant shoots and leaves. This process is particularly applicable to heavy metal contaminants and organic compounds that are not tightly bound to soils. These constituents are extracted from the soil and pore water by cation pumps, absorption and other mechanisms (McCutcheon and Schnoor, 2003). Hyperaccumulator plants can accumulate and tolerate very high levels (2 to 5%) of metals in their biomass (Carman and Crossman, 2001). Hyperaccumulation is a special case of phytoaccumulation that results in greater than 100 times the normal plant accumulation of a particular element. The shoots and leaves can later be harvested for disposal or metal recovery.

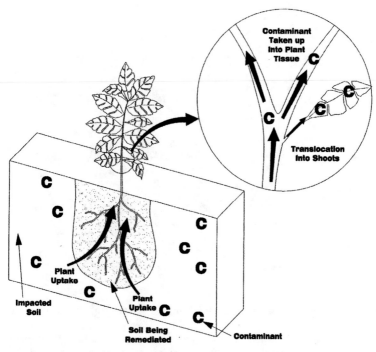

Figure 6.6 Phytoextraction of inorganic contaminants

6.2.6 Rhizofiltration

Rhizofiltration: This process is similar to that of phytoextraction, and involves the uptake, absorption, adsorption, or precipitation of constituents in the soil solution around the root system onto or within the plant root, young shoots, fungi, algae and bacteria (Carman and Crossman, 2001). Contaminants are not tranformed but are stored in plant tissues. The process of rhizofiltration first involves the containment of the constituent via immobilization or accumulation on or within the plant or microorganism. Plant or microbial uptake, accumulation and translocation may follow depending on the contaminant. The process is also referred to as phytofiltration, blastofiltration, phytosorption, biosorption, biocurtain and biofilter (Champagne and Bhandari, 2007). Plant root exudates may also lead to precipitation of some metals (Suthersan, 2002). It can be used for ex-situ treatment of contaminated ground water as illustrated in Figure 6.7. The groundwater can be pumped to the surface to irrigate plants that are grown on artificial soil media (e.g. sand mixed with perlite). Upon saturation with contaminants, roots can be harvested and disposed of safely.

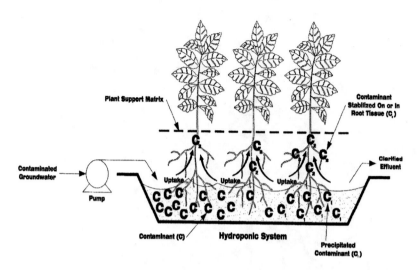

Figure 6.7 Rhizofiltration of inorganic contaminants

6.3   Influence of Environmental Factors Affecting Phytoprocesses

Environmental factors play an important role in the effectiveness of phytoprocesses. More specifically, the soil environment; water, oxygen and nutrient availability; temperature; and contaminant transformations are of particular concern.

6.3.1 Soil Structure, texture and organic content

Soil consists of minerals, organic matter, water and air. Minerals are derived from the weathering of different rocks over long time periods. The composition of soil minerals varies spatially as well as temporally. Soil structure refers to the way soil particles are held together in clumps or larger masses. Structure provides pore space between soil particles that contain and convey air or water. The structure of the soil is developed as a result of wetting and drying, compaction, the burrowing of animals and the growth of plant roots. Texture, on the other hand, refers to size of the individual particles of soil, such as sand, silt and clay. If the soil structure has very small pores (< 100 nm) and contaminants are sorbed at these locations, microorganisms may not have access to these sites (Alexander, 1997). Texture also affects the bioavailability of contaminants. Clayey soils can bind some molecules more readily than silty or sandy soils. Carmichael and Pfaender (1997) found that soils with larger particles contributed to a greater mineralization of polyaromatic hydrocarbons (PAHs) than soils with smaller particles, possibly due to the bioavailability of the compounds in the sandy soils.

Organic matter in soil can bind lipophilic compounds making them less available for biodegradation or plant uptake. A soil containing high organic matter (>5%) may sorb some organic contaminants rendering them less biologically available (Otten et al; 1997). Soil type may also have some affect on the quality and quantity of root exudates affecting phytoremediation processes (Bachmann and Kinzel, 1992). However, studies by Siciliano and Germida (1997) showed that soil type had no effect on the phytoremediation of 2-cholorobenzoic acid.

6.3.2   Water and oxygen availability

Plants and microbes in the root zone are affected markedly by the presence of water and oxygen. Water is a major component of living cells. It serves to transport nutrients to the cell and carry away waste products. Low moisture contents can reduce microbial activity and cause plant dehydration. Excessive moisture limits gas exchange and can lead to the development of anoxic zones where biodegradation by anaerobic microorganisms may be favoured. In some plants, oxygen may be a significant plant exudate. Herbaceous wetland plants have the ability to transport oxygen from the leaves to the root zone, resulting in localized aerobic conditions despite a generally saturated zone in that area (Colmer, 2003). Woody, non-wetland plants do not typically possess this oxygen transport capacity towards the root zone.

6.3.3   Temperature

Temperature has a significant impact on the efficiency of a number of phytoprocesses in a remediation scheme. It is well known that the rate of microbial biodegradation doubles for every $10^{\circ}C$ increase in temperature. Several studies have shown better phytoremediation and biodegradation of hydrocarbons in soils during the summer months compared to the winter months (Wright et al; 1997; Simonich and Hite, 1994).

6.3.4   Nutrients

Plants and microorganisms in the rhizosphere need adequate nutrients to grow. The presence of xenobiotics like petroleum hydrocarbons may stress microorganisms, which in turn make nutrients less available in soil environment (Xu and Johnson, 1997). The biodegradation of petroleum hydrocarbons by soil microorganisms consume the available nutrients causing nutrient deficiency in the soil. Petroleum hydrocarbons can also make nutrients less available to plants and microorganisms by limiting the availability of water in which nutrients are dissolved (Schwendinger, 1968). These nutrient deficiencies can be overcome by adding fertilizers or green manure to the soil.

6.3.5   Degradation

Contaminants in the soil environment degrade with time. The degradation process of contaminants may include volatilization, evapotranspiration, photomodification, hydrolysis, and biotransformation. Degradation results in the reduction of easily degradable compounds, leaving recalcitrant compounds relatively unchanged. The contaminants remaining are generally non-volatile or semi-volatile and can get adsorbed onto soil organic matter or soil fine particles, which limits their bioavailability and biodegradability (Carmichael and Pfaender, 1997).

6.4   Processes Responsible for Contaminant Removal/Detoxification

The processes that are responsible for the removal or detoxification of contaminants in soil/subsurface water environments involve complex interactions between plants and soil microbes. This is a symbiotic relationship which provides the plants with nutrients, protection and enhanced water uptake capacity. In turn, the soil microbes obtain the substrate and nutrients requied for their growth from the plants (Tsao, 2003). Several processes that occur in the soil environment and in the plant itself help to remove or detoxify the existing contaminants.

Microbes present in the soil environment include bacteria, fungi, actinomycetes, molds, algae, protozoa and viruses. Typical numbers of organisms in the bulk soil (upper horizon) vary from 10^5 to 10^8 CFU/g. These numbers decrease to 10^3 to 10^6 in lower horizon soils. Conversely, in the rhizosphere (plant root zone) these numbers can be 1 to 2 orders of magnitude higher than the bulk soil (Tsao, 2003). The numbers and distribution of the various species change depending on the types of contaminants present. These organisms may metabolize the organic contaminants under aerobic or anaerobic conditions present in the soil As previously stated, some plants can provide an oxygen rich environment for the soil microorganism in the rhizosphere. Rates of aerobic degradation are typically much higher than under anaerobic conditions. In some cases the microorganisms may not be able to metabolize the contaminants but can co-metabolize them in presence of other organic compounds. Example of such co-metabolism is the biodegradation of trichloroethylene(TCE) by methanotrophic bacteria in the presence of oxygen and methane.

By degrading certain toxic contaminants in the root zone, soil microorganisms provide protection to the plants by restricting their direct interaction contact with these consituents. Some mycorrhizal fungi also protect plants by restricting the uptake of non-essential inorganic metals such as cadmium, nickel and lead. The fungi also enhance the growth of the plants by seeking nutrients beyond the reach of plant roots through their hyphae, and exchanging them with plant exuded carbohydrates.

Plant processes include the uptake of water and dissolved inorganic nutrients through the roots, the solar radiation mediated photosynthesis of simple hydrocarbons from atmospheric carbon dioxide, the release of oxygen and transpiration of water through leaves and exudation of nutrients, oxygen and simple hydrocarbons to the rhizosphere. Plants may uptake non-essential inorganic elements such as arsenic, cadmium, selenium and lead, which could be toxic to the plant at high concentrations. Water uptake by plants and its eventual transpiration control a number of phytoprocesses by controlling the movement of contaminants from the soil, through the root zone and plant, their transformation or accumulation in various tissues of the plant system, and their potential volatilization from the leaves for certain contaminants. The photosynthetic process fixes atmospheric carbon dioxide to simple sugars which are then used as a precursors for other organic constituents such as cellulose, lignin, starch, sugars, amino acids, proteins and other cellular components. These compounds are distributed throughout the plant including the root system. It has been estimated that as much as 50% of the total compounds derived from photosyntheis could be exuded to the rhizosphere. These exudates are quite varied in chemical structure. Some are used as a mode of defense from potential soil pathogens and against other plants competing for nutrients in the vicinity. Exuded constituents could include terpenoids, flavonoids, alkaloids, anthraquinones and polyphenolics (Tsao, 2003). In addition, woody plants often slough off a portion of the root system on an annual basis, generally prior to winter dormancy. This deposits a lot of organic matter in the root zone which can maintain rhizosphere microorganism activity.

6.5   Mechanisms Responsible for Phytoprocesses

6.5.1 Heavy metal contaminants

Plants that can take up large amounts of heavy metals must have capabilities to tolerate these toxic substances. These plants have mechanisms capable of isolating, sequestering or stabilizing these inorganic compounds such that they do not migrate to more sensitive plant tissues (USEPA, 1999). One of the mechanisms for sequestering these metals is by binding them with exuded chemicals that can form a precipitate or result in the retention of the metal on the soil matrix (Figure 6.7). In some cases, the toxicity of the inorganic compound is reduced by forming an organometallic constituent with the exudates. The metal that enters the specific cell will require detoxification and any cell damage caused by the presence of the constituent will need to be repaired. Cell metal tolerance can occur as a result of a number of possible strategies, e.g., metal binding to cell wall, metal tolerance of the cell membrane, reduced metal transport through the membrane, active egress of the metals from the cell, evolution of metal-tolerant enzymes, accumulation of the metals in compartments (vacuoles), chelation of the metal by ligands (phytochelates), and precipitation of metals with low solubility. There is evidence of heavy metal resistant enzymes such as cell wall acid phosphatases which have been shown to play a role in transformation and detoxification of metals in plants (Thurman, 1981). In some instances, it has been shown that metal tolerant genes are involved in some plants which can store metals in vacuoles, e.g., cadmium tolerant yeast cells (Oritz et al., 1995). The storage of certain metals in vacuoles or leaves may be an effective strategy for the plant to avoid the toxic effects of these compounds in other parts of the plant (Vazquez et al., 1994). It has been shown that accumulation of zinc in vacuoles is associated with chelation by organic acids (Figure 6.6) (Mathys, 1977) or precipitation as Zn-phytate (Van Steveninck et al. 1987). On the other hand, copper may be bound to thiol-rich peptides (phytochelatins), as it accumulates in vacuoles (Salt et al; 1989). Verkleij et al. (1991)

proposed that the translocation of metals to leaves and their subsequent fall may be a possible mechanism of the detoxification process (Figure 6.4).

6.5.2 Organic contaminants

Most of the knowledge regarding plant tolerance to toxic organic compounds has emerged from studies on pesticide resistance. There are three mechanisms that plants and microorganisms use to remediate organic contaminants and they include: degradation (Figures 6.2 and 6.3), containment (Figures 6.4, 6.6 and 6.7) and transfer from soil to the atmosphere by phytovolatilization (Figure 6.5).

Organic contaminants, unlike inorganic compounds, can be altered by biological systems in plants and/or microorganisms. The uptake of organic compounds by the roots depends on three factors: physicochemical properties of the compound, environmental conditions and plant characteristics. Root uptake of organic compounds has been found to be directly related to the n-octanol/water partition coefficient (K_{ow}) of the compound. Lipophilic compounds partition better into roots than lipophobic compounds. The translocation of these compounds to shoots and leaves are generally higher for compounds with intermediate polarity (log K_{ow} ~ 1.8) (Cunningham et al., 1997). The bioavailability of the organic contaminants in the root zone depends on soil environmental factors such as pH, moisture content and organic matter content. Certain plant species can modify the soil environment to some extent which may allow better uptake of some organic compounds. Root surface area can affect the amount of organic contaminant uptake. Selection of plants with increased root surface areas can improve their uptake of selected contaminants.

The fate of organic compounds in the plants, particularly pesticides, have been studied extensively. Plants have many metabolic capabilities to alter the characteristics of the organic compounds that have been taken up (Figure 6.2). These compounds can be incorporated in the cell biomass, may be transpired or metabolized to nontoxic forms. In one study with hybrid poplar trees it was reported that TCE from groundwater was absorbed, metabolized and incorporated into cells (Gordon et al., 1997).

In addition to rhizosphere biodegradation processes (Figure 6.3), plants also have the ability to produce enzymes and cofactors to degrade a variety of toxic compounds. Certain plant-derived dehalogenases have been reported to be capable of removing halogens from chlorinated solvents (McCutcheon, 1996). The rhizosphere biota changes as plants grow, age and certain tissues die. Initially when the plant is growing large amounts of organic matter is exuded through the roots providing substrate for the microorganisms to metabolize or co-metabolize organic contaminants. In aged root zones, carbon may be limiting causing higher forms such as nematodes or protozoa to graze on the bacteria communities. Mycorrhizal fungi may colonize on older root zones that may change the types of exudates therefore giving rise to another set of microorganisms to grow. As plant roots die yet another group of microorganisms may thrive on the carbon released from these dead plant tissues.

6.6   Common Compounds Remediated by Phytoprocesses

6.6.1 Heavy metal and inorganic contaminants

Heavy metals in the environment come from natural geological processes as well as from human activities. The weathering of minerals and rock containing heavy metals, atmospheric deposition from volcanic activities and displacement of metal ions through groundwater movement are some of the natural sources of these compounds. The more common sources from human activities include: industrial, solid and liquid effluents, application of wastewater sludge to land, deposition of air-borne industrial discharges, military activities, mining operations, leachates from un-lined landfills, applications of agricultural chemicals, automobile exhausts and energy production effluents. The level of heavy metals in soils may vary from a low level of 1 mg/kg to in excess of 100,000 mg/kg. In many locations, the levels are high enough to harm humans, animals and the ecosystem. More common metal/metalloids found in the environment are: arsenic, cadmium, chromium, copper, lead, mercury, nickel, selenium, silver and zinc (USEPA, 1997). The processes that can remediate heavy metal contaminated site are: phytoextraction (Figure 6.6), phytostabilization (Figure 6.4) and rhizofiltration (Figure 6.7).

Remediation technologies commonly used for the remediation of heavy metal in contaminated soils include:
- Contaminated soil is physically removed by excavation, transported and disposed off site in secure locations (landfills).
- The contaminants from soil are removed by soil washing and subsequently precipitated or solidified.
- Contaminants are mobilized using electrokinetic process.
- Contaminants are contained in place with a low permeability cap (clay, asphalt and geomembrane).
- Contaminants are vitrified where the soil/metal matrix is melted and converted to a glassy monolith.
- Conversion of the contaminants to a more stable and less mobile form by chemical reduction/oxidation process.

These remediation processes are costly, energy intensive, disrupt the site, may have logistical problems, and generally have less public acceptance. The phytoprocess options to remediate these sites have the following advantages (Sogorka et al., 1998):
- Operational and maintenance costs are low
- Treatment can be applied to large area in a cost effective manner
- The contaminated residue mass after treatment can be considerably reduced
- Public acceptance for the process is high
- Metals concentrated in plant tissues can be recovered and recycled

- Treatment is applicable to a wide variety of metals and radionuclides
- Soil excavation and the subsequent exposure of contaminants to workers are minimized.

There are certain limitations for the phytoremediation processes that must be recognized;

- The process is slow requiring several years to establish itself
Soil properties, salt content and the presence of other chemicals can affect the growth of the desired plants and, hence, metal remediation
- The depth of clean up depends on the root penetration, so deeper contaminants may not be accessible to the roots of the plants and may require alternative treatment strategies.
- Some forms of metals in soils are not biologically available to plants even with amendment additions. In such a case, the sites may not be remediated to desired concentrations even after several years of operation.

6.6.1.1 Phytoextraction

The success of the phytoextraction (Figure 6.6) process is dependent on the site conditions and the type of pollutants. The metal species present in the soil is also important. Metals in soil can exist in several different forms. These could be present in soluble form, as exchangeable ions sorbed onto inorganic phases, as metal complexes with soluble or insoluble organic compounds, as non-exchangeable ions or inorganic metal compounds. Normally only a small fraction of the total metals in soil may be available to the plants. For metal uptake in higher plants, soil conditions often need to be changed. Lowering the pH can solublize metal residues and increase plant uptake but the lowered pH must not affect plant growth. The addition of chelates (e.g., EDTA) to increase metal solubility can be effective in increasing plant metal uptake (Huang et. al., 1997). However, these chelators can render the metals more mobile and may cause the migration of these contaminants downwards towards the groundwater table.

Plant species that have the capacity to tolerate and accumulate metals in their tissues at high levels are called hyperaccumulators. They can accumulate 0.1% of Lead, Cobalt, Chromium and about 1% of Manganese, Nickel, or Zinc in their plant shoots (Baker and Brooks, 1989). These hyperaccumulators (*Thalspi caerulescens*, *Alyssum murale*, etc.), despite their capacity to accumulate large amounts of metals in their tissues, have a slow growth rate and a low biomass production, which can limit their use for phytoextraction. However, with suitable soil amendments it is possible to use some selected species to remediate specific sites (Blaylock and Huang, 2000). Another concern associated with these plants is that accumulated metals from the root zone do not translocate well to the shoot or leaves. Above ground plant tissues are much easier to harvest than roots. However, chelation of the metal with organic compounds (EDTA) can increase the translocation to shoot zone (Blaylock and Huang, 2000). Thus, addition of amendments with chelators can improve the remediation process. Use of non-hyperaccumulating plants which have higher growth rates has been suggested for remediating sites with metal pollution. Indian mustard plants in the presence of chelating agents have been shown to accumulate high amounts of lead from contaminated soil (Blaylock et al., 1997).

However, to date, this process has not been validated through extensive field studies. More research and field evaluations are necessary before phytoextraction can become a commercially acceptable remediation strategy for inorganic contaminants (McCutcheon and Schnoor, 2003).

6.6.1.2 Phytostabilization

In phytostabilization (Figure 6.4), amendments are used to interact with soluble inorganic contaminants to convert them to insoluble form, which are less likely to leach through the soil matrix. Plants are also employed to cover the soil surface to reduce erosion and water percolation. The process does not remove the contaminants but converts them to constituents that are less hazardous to humans and the environment. The process begins with soil being plowed before the addition of lime, fertilizer and amendments to inactivate the soluble metal species. Plant seed or seedlings are introduced and the site is irrigated as necessary. The quantities of amendments added vary according to the type of soil and the metal species present. The most effective soil amendments to decrease the availability of metals are phosphate fertilizers, bio-solids, iron or manganese oxy-hydroxides, clay minerals, or mixture of these chemicals. These amendments involve precipitation, humification, sorption and redox processes. For example, phosphate salts can precipitate lead in the soil to minerals such as lead pyromorphites (Berti and Cunningham, 2000). In this process, plants play a secondary role where they physically stabilize the soil via the root system, preventing erosion, human contact with the soil and minimizing precipitation and runoff effects due to the dense canopy. The plants chosen should not accumulate metal in their tissues, but they should be tolerant of the anticipated soil metal concentrations and other soil properties such as salinity, soil pH, soil structure and water content. The plants must grow quickly with dense rooting systems and canopies. Several grass seed mixtures used for road side stabilization could be used for this purpose provided they can tolerate the high concentrations of metals in the soil.

6.6.1.3 Rhizofiltration

In rhizofiltration processes, (Figure 6.7) plants are grown hydroponically with an extensive root system that can remove heavy metals from the solution by absorption, concentration or precipitation. The process can remediate contaminated groundwater pumped through beds of appropriate plants. The mechanisms suggested for the removal of heavy metals by plant roots may include extracellular precipitation, cell wall precipitation and adsorption, intracellular uptake and storage in vacuoles. Phytochelatins which are metal binding polypeptides may help in accumulating and detoxifying metals in cells. Several phytochelatins may be present which can interact with different metals. The plants used for rhizofiltration should be capable of removing large quantities of specific metals, easy to grow with dense root system and with low maintenance and operations costs. Dushenkov and Kapulnik (2000) reported that sunflower plants in a rhizofiltration application had roots up to 50 g dry weight/m^2 per day. The plants were cultivated in a constructed rhizofiltration unit and nutrients fed through a feeder artificial soil layer at the top of the hydroponic system. Dushenkov et al. (1995) conducted a series of rhizofiltration experiments and found that lead accumulation in the roots was highest in sunflower plants followed by Indian mustard, tobacco, rye, spinach and corn. The lead concentration in roots was 5.6 to 16.9% on a dry weight basis.

At the present time these phytoprocesses are under commercial considerations for the remediation of soils contaminated with nickel, cobalt, selenium and lead. Field scale studies have been conducted for mercury, zinc, cesium and uranium contaminated soils, while laboratory studies have been completed for cadmium and strontium contaminated soils (McIntyre, 2003).

DeSouza et al. (2000) reported that the removal of selenium by phytoremediation process is dependent on the chemical species present. Generally, oxidized species such as selenate or selenite which are highly soluble are more easily removed compared to the insoluble selenide or elemental selenium. Removal of selenium by plants substantially occurs through accumulation in plant tissues and volatilization. The plants belonging to the *Brassicaceae* (Indian mustard) group were excellent volatilizers of selenium and sulfur. It is thought that selenium and sulfur are taken up and volatilized via the same pathway.

Microorganisms play an important role in the uptake of selenium by plants. For selenate about 77-87% of the uptake by Indian mustard roots was aided by the bacteria, while only 16-21% of selenite uptake was bacteria induced. Indian mustard plants can volatilize selenate and selenite species at 0.32 and 0.95 g/g dry weight/d, respectively.

Decontamination of perchlorate-contaminated water by plants was reported by Nzegung et al. (1999). Willow trees removed perchlorate in aqueous solution concentrations of 10 to 100 mg/L to less than 2 μg/L. The processes responsible for the removal included uptake and plant tissue degradation and rhizosphere degradation. High nitrate concentrations interfered with the perchlorate rhizosphere degradation, which indicates that perchlorate phytoremediation may be affected by the presence of competing electron acceptors.

Schnoor (2000) reported on the use of hybrid poplar trees for the phytostabilization of metal and low level radionuclide contaminated sites, where the contamination was in the upper 2-3 m depths. The risks associated with contaminant release from these sites include groundwater contamination and windblown dusts. The poplar trees can take up large amounts of moisture and prevent vertical migration of contaminants. Amendments (nutrients, lime and phosphate fertilizers) can be added to bind some of the heavy metals in the soil and further prevent their migration.

6.6.2 Organic Compounds

Organic contaminants unlike inorganic constituents ones can be degraded by biochemical processes to non-toxic or even to mineralized state. Conventional in-situ methods for treating soils with organic matter include air stripping for volatile organic compounds (VOCs), steam stripping, surfactant flushing followed by ex-situ treatment of the recovered washings and bioremediation. These processes are expensive, may cause site disruptions, have the possibility of harm to site workers, may cause off-site migration of contaminants if proper the system is not desgined adequately and may require subsequent treatment of residues. Phytoremediation and rhizosphere biodegradation processes may be alternatives that can overcome some of the difficulties associated with conventional processes for clean up of these sites at a lower cost. Some of advantages and limitations noted for heavy metal remediation by phytoprocesses are also applicable to clean up of site with organic contaminants. Soil organic contaminants that have been phytoremediated are polyaromatic hydrocarbons, pesticides (atrazine), volatile organic compounds(VOCs), trinitrotoluene, methyl *tert*-butyl ether (MTBE), phenols, DDT, polychlorinated biphenyls (PCB) and petroleum hydrocarbons (McCutcheon and Schnoor, 2003).

McCutcheon and Schnoor (2003) found that the kinetics of rhizosphere biodegradation and phytoremediation were comparable or better than microbial transformations and much faster than neutral and alkaline hydrolysis processes. Table 6.1 shows the transformation kinetics of various xenobiotics due to hydrolysis, microbial transformation (Figure 6.3) and phytodegradation (Figure 6.2).

Table 6.1 Transformation Kinetics of selected Compounds by Hydrolysis, Microbial and Phytoremediation (McCutcheon and Schnoor, 2003)

Chemical	Half life, days		
	Hydrolysis[a]	Microbial Transformation[b]	Phytoremediation[c]
Carbon Tetrachloride	>10 years	39 days	38 hr
Hexachloroethane	>10 years	>30 days	8 hr
Tetrachloroethylene	>10 years	35 days	5 days
Trinitrotoluene	NA	3 days	20 min
Hexahydro-1,3,5-Trinitro-1,3,4-Tianzine (RDX)	NA	>20 days	21 hr
Benzonitrile	0.3 years	30 days	11 hr[d]

[a] For pH 7, including both neutral and alkaline hydrolysis (Jeffers et al., 1989)
[b] Based on bacterial densities of 10 organisms/L and a cell mass of 2 x 10 g (Bowen, 1966, Uhlmann, 1979, Gaudy and Gaudy 1980).
[c] *Spirogyra* spp. at a density of 200 g /L (Lee Wolfe et al., 1995, unpublished data, USEPA, Athens , GA).
[d] Laura Carriera (1995, unpublished data USEPA, Athens, GA)
NA: Not applicable.

6.6.2.1 Phytodegradation

Newman et al., (1999) reported a field study where hybrid poplar trees (*Populous trichocarpa* x *P. deltoides*) were used to clean up groundwater contaminated with TCE (concentration range 0.038 to 0.76 mM) in an outdoor lined test cell. TCE removal by this system during the growing season was over 99%. A very small fraction (~9%) of the TCE was transpired (phytovolatilization) through the leaves. TCE degradation was indicated to be by plant metabolism rather than by rhizosphere biodegradation. Chloride levels in the soil increased in proportion to the TCE loss, indicating dechlorination of the TCE by the plant tissues, and the release of chlorides to the soil by the roots.

In laboratory and field studies to remediate and hydraulically contain MTBE in a groundwater plume, Hong et al. (2001) used hybrid poplar deep-rooted trees and reported 36.5 to 67.0 % MTBE removals within 10 days. The plant roots were also reported to contain the groundwater plume from spreading. The roots extended to 9-10 ft below the ground surface near the capillary fringe zone of the aquifer.

Wang et al. (2004) studied the fate of carbon tetrachloride (CT) during phytodegradation by hybrid poplar trees (*Populous trichocarpa* x *P. deltoides*) in a field test bed. The CT concentration fed varied from 12 to 15 mg/L. Over a period of about 5 months, the CT removal was greater than 99 % but no CT was reported to be transpired through the leaves. Degradation of CT in the rhizosphere was found to be negligible. It was surmised that CT was being metabolized to mineral state by the plant tissues, possibly the root cells. No accumulation of CT or its metabolites in plant tissues were observed.

Laboratory results showed that munition wastewater containing trinitrotoluene (TNT) could be treated using a continuous flow phytoreactor containing the plant parrotfeather (*Myriophyllum aquaticum*) (Medina et al., 2002). The removal efficiency varied from 85% to 100% depending on the loading rate. The influent TNT levels varied from 1 to 15 mg/L. Analysis of the plant tissues indicated no accumulation of TNT at lower loading rates (up to 0.967 g/m^3) , but at higher loading rates (2.488 g/m^3) some TNT was found in the tissues. A breakdown product of TNT, aminodinitrotoluene (ADNT), was found in plant tissues and in the aqueous phase. In a similar hydroponic bench scale laboratory study, Thompson and Schnoor (1997) reported that hybrid poplar trees (*Populus deltoids X nigra* DN 34) were able to reduce the concentration of TNT in the solution phase. Examination of plant roots and stem tissues indicated that TNT was being transformed to several metabolites such as 4-amino-2,6 dinitrotoluene, 2-amino-4,6-dinitrotoluene and 2,4-diamino-6 nitrotoluene. No significant mineralization of TNT to CO_2 was observed. Transpiration data indicated that TNT at 5 mg/L was not toxic to the plants but prolonged exposure at this concentration affected its survival.

A three-year field study on the phytoremediation of a site contaminated with polyaromatic hydrocarbon (PAH) was reported by Qui et al. (1997). Prairie buffalo grass (*Buchloe dactyloides* var. *'Prairie'*) accelerated the naphthalene reduction in the site containing clayey soil. The naphthalene concentration in surface soil with the grass cover decreased from about 800 mg/kg of soil to less than 20 mg/kg of soil over 1000 days. The data for the removal of other PAHs were not convincing due to analytical variability. Another grass, Kleingrass (*Panicum coloratum* var *'Verde'*), had a greater capacity to remove PAHs from rhizosphere soils than other grasses tested. Grass tissue analysis indicated no concentration of PAHs, which would minimize the issue of food chain effects.

In a study examining the remediation of sites contaminated with crude oil or diesel, Reynolds and Wolf (1999) used two cold-hardy plants, arctared red fescue and annual rye grass. Initially the soil had total petroleum hydrocarbons (TPH) levels of 6,200 mg/kg soil and diesel concentration was about 8,350 mg/kg soil. After 640 days, crude oil concentrations in the soil planted with both species was 1,400 mg TPH/kg (77% reduction), while the unplanted soil contained about 2,500 mg TPH/kg (60% reduction). Similarly, the planted soil reduced the diesel concentrations to about 700 mg/kg (92% reduction) compared to 2,200 mg/kg (74% reduction) in the unplanted area.

6.6.2.2 Rhizosphere biodegradation

The removal of aircraft deicing chemical (ethylene glycol and additives) by soils taken from the rhizospheres of several plants was investigated by Rice et al. (1997). The degradation of the ethylene glycol by soil microorganisms was tested at three different temperatures (-10° C, 0° C and 20° C). After 28 days at 0° C, 60.4 %, 49.6 % and 24.4 % of the ethylene glycol was mineralized to CO_2 by alfalfa (*Medicago sativa*), Kentucky bluegrass (*Poa pratensis*) and non-vegetated soils, respectively. Higher temperatures were found to increase the mineralization of the compound.

6.7 Case Studies

It can be seen from the references cited that both metallic contaminants and organic contaminants can be remediated by phytoprocesses. Two case studies are presented to provide some more insight into these processes. The first one is on metal remediation and the second one deals with volatile organic compounds.

6.7.1 Case Study 1: Inorganic compounds removal

This study involves a site contaminated with arsenic, cadmium, zinc and lead in the surface soils, and mine tailings and groundwater in Whitewood Creek, South Dakota as a result of past gold mining activities (Schnoor, 2000). The metal contamination of the site, in the form of arsenic, cadmium, zinc and lead, was mostly located in the upper 6 ft (1.82 m) of soil and mine tailings. Arsenic concentrations in soil varied from 1059 mg/kg at the surface to 934 mg/kg at a depth of 3.5 ft (1.07 m). Concentrations of cadmium, lead and zinc at the surface were 0.22, 4.64 and 754.47 mg/kg respectively, while at a depth of 3.5 ft (1.07 m), the levels were 0.21, 3.82 and 106.6 mg/kg, respectively. In order to apply a phytostabilization remediation strategy, approximately 3,100 hybrid poplar trees were initially planted on 0.4 hectare of the site, with the root zone extending to a depth of 1.6 m, but later more trees were planted. The area was fertilized with a commercial fertilizer at recommended N/P/K rates. The established root zone prevented leachate generation and the downwards percolation of contaminated water to groundwater. The trees grew to a height of about 4 ft (1.22 m) and exhibited signs of phytotoxicity caused by the elevated metal concentrations in the tailings. The uptake and translocation of arsenic or cadmium to the leaves was quite small, averaging 1.1 and 27 mg/kg of dry weight, respectively. These concentrations were below the limits set by USEPA for biosolids field applications. A weak relationship was established between arsenic in the soil and arsenic in stems. In addition, there was a direct relationship between the uptake of zinc and cadmium in poplar leaves, however zinc concentrations were found to increase, while the cadmium uptake leveled off. There appeared to be a competitive relationship between zinc and cadmium uptake by poplar trees. Cadmium concentrations in leaves varied from 1 to 3 mg/kg but zinc concentrations ranged from 100 to 600 mg/kg. After 4 years, only 150 poplar trees were still alive out of the total 6,000 trees planted at the site. This could be attributed to a number of factors including; toxicity effects due to the high metal concentrations in the soil, difficult climatic conditions and animal grazing. Severe ice storms were responsible for the death of a number of trees. Animal grazing was heavy at the site but fencing could assist in the reduction of tree foliage losses. At an abandoned smelter site in Kansas, the addition of aged manure and straw were found to enhance the survival of these trees (Schnoor, 2000).

Results showed that many of the metals, particularly lead, were either adsorbed on the inorganic and organic complexes causing immobilization and stabilization. The immobilization was dependent on pH, where lower pH increased the mobilization of the metals. The addition of lime and limestone could be necesary at sites where soil pH levels are less than 5.5.

6.7.2 Case Study 2: Organic compounds removal

This study was conducted at Aberdeen Proving Grounds, Maryland where the groundwater was contaminated with volatile organic compounds (VOCs) such as 1,1,2,2-tetrachloroethane (TCA) and trichloroethylene (TCE) (Hirsh et al., 2003). In the spring of 1996, a pilot-scale phytoremediation study was started. The goals of the study involved the hydraulic control of a contaminated plume and the treatment potential of phytovolatilization, phytodegradation or rhizosphere degradation for these organic contaminants. The studies were conducted over 5 years. The J-field was a site for weapons testing and their disposal, which resulted in the release of many chemicals to soil and groundwater. This site is located near a marsh area and is a suitable habitat for wetland plants such as reed grass, which could promote the natural attenuation of VOCs and also their biodegradation.

The source of VOCs was the Toxic Burning Pits which were used for the disposal of chlorinated solvents. The water drained through the pits and surrounding surface recharge area for the surficial aquifer and the freshwater marsh received the recharged groundwater, which was contaminated. The chemicals in the groundwater of the Toxic Burning Pits included TCA (~ 4400 mg/L), TCE (~240 mg/L), cis-1, 2-dichloroethylene (cDCE) (~ 280 mg/L) and smaller concentrations of several other chemicals such as tetrachloroethylene (PCE), 1,1,2-trichloroethane (TCA), trans-1,2-dichloroethylene (t-DCE) and vinyl chloride. A thin upper layer of the groundwater was aerobic overlaying an anaerobic layer where sulfate-reducing and methanogenic conditions prevailed. A dense nonaqueous phase liquid (DNAPL) phase was noted below the Toxic Burning Pits which acted as a reservoir for the soluble VOCs.

The soil overlying the site was suitable for planting hybrid poplar trees. An area of approximately 1 acre over the contaminated groundwater area and surrounding the Toxic Burning Pits was selected for the planting of 183 hybrid poplar trees which were planted at 3 m intervals. The treatment strategy was that the poplar tree roots would intercept the VOC plume and prevent its entry to the marsh. Within two years (by 1998), many of the poplar trees had died due to drought. Replacement trees consisting of hybrid poplars, as well as the planting of two native species, tulip and silver maple. Monitoring of the site consisted of the following measurements: tree transpiration rates, groundwater hydraulic and constituent concentrations to assess the fate of VOCs, rhizosphere contaminant concentrations (including VOCs, organic acids, chloroacetic acids, TOC, chloride, dissolved gases and root exudates), tree tissue, transpiration gases, air emissions and soil organisms.

The results from the 5 year study of the site indicated that hybrid poplar trees were successful in remediating the dissolved-phase VOC plume. These trees could seasonally transpire enough water to contain the plume. The VOCs in the transpired water were found in the tree tissues and in the transpiration gases at barely detectable concentrations. The trees were biotransforming TCE to trichloroacetic acids in their tissues with the possibility of some photodegradation of the transpired compounds in the atmosphere. The successful reduction of TCE in the groundwater by poplar trees included the following processes: adsorption to root and tree tissues, rhizosphere degradation, biotransformation and volatilization through plant tissues and leaves.

6.8   Field Applications of Phytoprocesses

Phytoprocesses can be used for hydraulic containment and for contaminant treatment. In most instances, it is difficult to separate these two processes as water uptake by the plants move the contaminants through the biologically active rhizosphere where degradation of the contaminant may also take place.

Before embarking on the development of a phytoprocess for the remediation of a contaminated site, it is important to identify the objectives or the desired level of contaminant removal as desired by the regulatory authorities. Next the site needs to be characterized to evaluate the extent of contamination both aerially and with depth. The toxicity and the potential for the movement of the contaminants to receptors should be examined. Biodegradation and bioavailability of these compounds should also be evaluated. The growth of the plants requires adequate soil fertility, which should be evaluated for the site. The addition of appropriate soil amendments can improve the fertility of soil if necessary. Soil water holding capacity is another property that needs to be evaluated. In most cases, amendments can be added to improve the water holding capacity of the soil if it is not adequate. Some guidance regarding local soil and agronomic conditions can be best obtained from local experts in horticulture and agriculture.

The plants selected for the treatment process depend very much on the type of contaminants present and the objectives of the remediation strategy. Climatic conditions must also be considered in the selection of plants. In arid areas, species must be able to withstand periods of drought and higher temperatures. In places where flooding occurs they must be able to survive periods of soil saturation. Colder weather may be a problem for some common species such as alfalfa or hybrid poplar trees. In some cases where data regarding plant germination under particular soil conditions are not available, greenhouse experiment can be conducted to provide this information. Irrigation may be necessary for crop maintenance, if the rainfall for the area does not satisfy the needs of the plants. Appropriate water management measures may need to be implemented to prevent excessive runoff and percolation through the soil.

Due to the site specific conditions which exist at each treatment location, it is difficult to specify plant types that may be successful in remediating a particular contaminant at a specific site. However some guidance is provided here.

For metal contaminants in soils, the selection of plant species for phytoremediation will depend on the location of the contaminants (depth and surficial area), the type of contaminants present and their chemical species and potential by-products. For surface contamination, a shallow-rooted species may be suitable, whereas, a deep-rooted species may be needed if the contamination runs deeper into the soil. Metal hyperaccumulator plants may not be practical for extraction of metals from soil as they have a slow growth and have a lower biomass production rate. Therefore, attention has been turned to other plants that have a higher growth rate and produce larger biomass. For example, several plant species such as hemp dogbane (*Apocynum cannabinum*), common ragweed (*Ambrosia artemisiifolia*) can accumulate large amounts of lead in their tissues (Berti and Cunningham, 1993). These non-hyperaccumulating plants can accumulate additional quantities of metals in presence of some amendments such as chelators. By adding a synthetic chelator (EDTA) at a rate of 10 mmol/kg soil, lead could be accumulated in the shoots of maize up to 1.6% (Blaylock et al., 1997). It is important to note that high metal concentrations in some plant tissues can be toxic to the plant causing death, so the chelator should be applied only after the plant has established large biomass base and should be harvested after a week of the treatment to avoid the loss of lead incorporated in shoots. Soil fertilization with nitrogen-rich fertilizers may be beneficial, but the addition of phosphorous compounds may reduce plant uptake of some metals (especially lead) by rendering the metals insoluble (Chaney et al., 2000).

For remediation of soils or shallow groundwater sites with organic contaminants, the selection of plants depends on the location of the contaminants. For contamination near the surface, plants are selected to treat up to a depth of 2 ft. The species selected should have fibrous and dense roots to maximize the bioactivity in the rhizospheric zone. Grasses such as fescue (*Festutca sp.*), ryegrass (*Lolium* sp.) and wheatgrass (*Agropyron sp.*), as well as clovers (*Trifolium sp.*) can be suitable for this type of remediation. However, it should be noted that specific site conditions will favor certain plant species to enhance the effectiveness of remediation. For example in a study involving the remediation of a site contaminated with polycyclic aromatic hydrocarbons in Texas, it was found that kleingrass (*Panicum coloratum*), native of Africa, was best suited for the site. For deeper contaminated areas, plants with deeper roots (Compass plant, Indian grass, Buffalo grass) may be used (Tsao, 2003).

For shallow aquifer or deeper soil (3-5 m) organic contaminant remediation, trees of the genus *Populus* have been successfully used (Hirsh et al., 2003). These trees are facultative phreatophytes that can extract contaminated water from the soil or groundwater. Cottonwood, aspen and hybrid poplar all belong to the genus *Populus*. Typical spacing in an open area could be 5-10 ft (1.5 – 2.6 m) with neighboring rows offset at a 60° angle. Initially, the water uptake rate should be conservatively designed and considerations should be given for an anticipated plant mortality rate of 10-20 % with additional retardation of growth due to adverse site conditions. Irrigation, fertilization and amendment addition for the site should be practiced to maintain the growth of the plants. Harvesting of the plant may be necessary to prevent the loss of contained materials during winter months. The harvested materials should be properly disposed if they pose any danger to the environment or human health.

Planting of trees or plants depends upon the type of the plant that is to be planted, the depth of the root system, and the type of soil. For planting phreatophytes in sandy or loose soils, a small hole can be poked with a rod or dowel and a plug or whip (for 1 to 2 ft (0.3 to 0.6 m) tall stock) placed into the hole, followed by hand packing of the soil around the hole. Experienced foresters can plant several thousand trees in a day. For clayey soils, a trench digger can be used to plant the whips, followed by manually filling the trench. The trees thus planted take quite some time to reach the shallow aquifer for transpiring the water and contaminants. For larger stocks (up to 10 ft (2.6 m) in length), individual boreholes can be drilled with an auger. The diameters of the holes are typically six to twelve inches and are drilled down to the groundwater table. The holes are backfilled with excavated soil and sand/peat mixture after planting. This mixture allows for a better root growth particularly in clayey soils. General fertilizer in powder form may also be added to the backfilled materials, which allows for a faster tapping of the groundwater by the plant roots. Sometimes breather tubes are inserted into the boreholes prior to the backfilling to allow an influx of air to the root zone (Tsao, 2003).

An extensive monitoring program must be included in phytoremedication projects, which should include surface runoff, groundwater, air sampling, soil sampling at the site both aerially and with depth and sampling of plant tissues (roots, shoots, leaves). This monitoring program can then be employed to validate the success of the process over time.

In summary, phytoprocesses may be applied successfully at selected sites where soil and/or shallow groundwater has been contaminated with inorganic or organic contaminants. The success of the process will depend upon factors such as climate, soil type, contaminant types and locations, remediation objectives, and selection of appropriate plant species for remediation.

References

Alexander, M., Hatzinger, P-B, Kelsy, J.W., Kottler, B.D., and Nam, K. (1997) Sequestration and realistic risk from toxic chemicals remaining after bioremediation. *Annals of New York Academy of Sciences*, 829: 1-5.

Bachmann, G., and Kinzel, H. (1992) Physiological and ecological aspects of the interactions between plant roots and rhizosphere soils. *Soil Biol. and Biochem.*, 24 (6): 543-552.

Baker, A.J.M., and Brooks, R.R. (1989)   Terrestrial higher plants which hyperaccumulate metallic elements- A review of their distribution, ecology and phytochemistry. *Biorecovery*, 1: 81-126.

Berti, W.R., and Cunningham, S.D. (2000) Phytostabilization of metals. *Phytoremediation of Toxic Metals: Using plants to clean up the environment*. Raskin, I., and Ensley, B. D., eds., John Wiley & Sons, New York, 53-70.

Blaylock, M. J., Salt, D.E., Dushenokov, S., Zakharova, O., Gussman, C., Kapulinik, Y., Ensley, B. D., and Raskin, I. (1997) Enhanced accumulation of Pb in Indian mustard by soil-applied chelating agents. *Envir. Sci. Technol.*, 31 (3): 860-865.

Blaylock, M. J., and Huang, J. W. (2000) Phytoextraction of metals. *Phytoremediation of Toxic Metals: Using plants to clean up the environment*. Raskin, I., and Ensley, B. D., eds., John Wiley & Sons, New York, 53-70.

Bowen, H. J. M., (1966) Trace Elements in Biochemistry, Academic Press, New York.

Carman, E. P. and Crossman, T. L. (2001) Phytoremediation. Chapter 2 *in In-Situ Treatment Technology*. 2nd Ed. E. K. Nyer, P. L. Palmer, E. P. Carman, G. Boettcher, J. M. Bedessem, F. Lenzo, T. L. Crossman, G. J. Rorech and D. F. Kidd (Eds.). Arcadis Environmental Science and Engineering Series. Lewis Publishers, Washington, D.C. pp. 392-435.

Carmichael, L. M., and Pfaender, F. K., (1997) Polynuclear aromatic hydrocarbon metabolism in soils: Relationship to soil characteristics and preexposure. *Envir. Toxic. Chem.*, 16 (4): 666-675.

Champagne, P. (2007) Chapter 10 - Phytoremediation *in Remediation Technologies for Soil and Groundwater Contamination*. A. Bhandari, R. Surampalli, P. Champagne, S. Ong, R. Demera and R. Tyagi Eds. ASCE.

Champagne, P. and Bhandari, A. (2007) Chapter 2 - Fundamental Processes in *Remediation Technologies for Soil and Groundwater Contamination*. A. Bhandari, R. Surampalli, P. Champagne, S. Ong, R. Demera and R. Tyagi Eds. ASCE.

Colmer, T. D. (2003) Long-distance transport of gases in plants: a perspective on internal aeration and radial oxygen loss from roots. *Plant, Cell and Environment*. 26: 17–36.

Cunningham, S.D., and Berti, W.R. (1993) Remediation of contaminated soils with green plants: An overview. *In vitro Cell. Dev. Biol.*, 29: 207-212.

Cunningham, S. D., Shann, J. R. Crowley, D. E. and Anderson, T. A. (1997) Phytoremediation of contaminated water and soil. *Phytoremediation of Soil and water Contaminants*, Kruger, E. L., Anderson, T. A. and Coats, J. R., eds., ACS Symposium Series 664, American Chemical Society, Washington, D.C., 2-17.

De Souza, M.P., Pilon-Smits, E. A. H., and Terry, N. (2000) The physiology and biochemistry of selenium volatilization by plants. *Phytoremediation of Toxic Metals: Using plants to clean up the environment*. Raskin, I. Ensley, B. D., eds., John Wiley & Sons, New York, 171-190.

Dushenkov, S., Nanda Kumar, P. B. A., Moto, H. and Raskin, I. (1995) Rhizofiltration: the use of plants to remove heavy metals from aqueous sources. *Envir. Sci. Technol.*, 29: 1239-1245.

Dushenkov, S., and Kapulnik, Y. (2000) Phytofiltration of Metals. *Phytoremediation of Toxic Metals: Using plants to clean up the environment*. Raskin, I. and Ensley, B. D., eds., John Wiley & Sons, New York, 89-106.

Gaudy, A., and Gaudy, E. (1980) Microbiology for Environmental Scientists and Engineers. McGraw-Hill, New York.

Gordon, M., Choc, N., Duffy, J., Ekuan, G., Heilman, P., Muiznieks, I., Newman, L., Ruszaj, M., Shurtleff, B. B., Strand, S, and Wilmoth, J. (1997) Phytoremediation of trichloroethylene with hybrid poplars. *Phytoremediation of Soil and water Contaminants*, Kruger, E. L. , Anderson, T., A. and Coats, J. R. (1997), 177-185.

Hirsch, S. R., Compton, H. R., Matey, D. H., Wrobel, J. G., and Schneider, W. H. (2003) Five-year pilot study: Aberdeen Proving Grounds, Maryland, *Phytoremediation: Transformation and Control of Contaminants*, McCutcheon, S. C. and Schnoor, J. L., eds., Wiley- Interscience, Hoboken, New Jersey, 635-659.

Hong, M. S., Farmayan , W. F., Dortch, I. J. and Chiang, C. Y. (2001) Phytoremediation of MTBE from a groundwater plume. *Envir. Sci. Technol.*, 35 (6): 1231-1239.

Huang, J. W., Chen, J., Berti, W. R., and Cunningham, S. D. (1997) Phytoremediation of lead contaminated soils: Role of synthetic chelates in lead phytoextraction. *Envir. Sci. technol.*, 31 (3): 800-805.

Jeffers, P. M., Ward, L. M., Woytowitch, L. M., and Wolfe, N.L. (1989) Homogeneous hydrolysis rate constants for selected chlorinated methanes, ethanes, ethenes and propanes. *Envir. Sci. Technol.*, 23 (6): 965-969.

Mathys, W. (1997) The role of malate, oxalate and mustard oil glucosides in the evolution of zinc resistance in herbage plants. *Physiol. Plant*, 40: 130-136.

McCutcheon, S. C. (1996) Phytoremediation of organic compounds: science validation and field testing, USEPA workshop on Phytoremediation of Organic wastes, Kovalick, W. W., and Oxlesey, R., eds., Fort Worth, TX.

McCutcheon, S. C., and Schnoor, J. L. (2003) Overview of Phytotransformation and control of Wastes. *Phytoremediation: Transformation and Control of Contaminants*, McCutcheon, S. C. and Schnoor, J. L. , eds., Wiley-Interscience, Inc., Hoboken, New Jersey. 3-58.

McIntyre, T. (2003) Phytoremediation of heavy metals from soils. *Advances in Biochemical engineering/Biotechnology, Phytoremediation*, ed., Tsao, D. T., Springer –Verlag, Berlin, 97-123.

Medina, V. F., Larson, S. L., McCutcheon, S. C. (2002) Evaluation of continuous flow-through phytoreactors for the treatment of TNT –contaminated water. *Envir. Progress*, 21 (1): 29-36.

Newman, L.A., Wang, X., Muiznicks, I.A., Ekuan, G., Ruszaj, M., Cortellucci, R., Domroes, D., Karscig, G., Newman, T., Crampton, R. S., Hashmonay, R. A., Yost, M. G., Heilman, P. E., Duffy, J., Gordon, M. P., and Strand, S. E. (1999) Remediation of trichloroethylene in an artificial aquifer with trees: a controlled field study. *Envir. Sci. Technol.*, 33: 2257-2265.

Nzangung, V. A., Wang, C., and Harvey, G. (1999) Plant-mediated transformation of perchlorate into chloride. *Envir. Sci. Technol.*, 33 (9): 1470-1478.

Oritz, D. F., Kreppel, L., Speiser, D. M., Scheel, G., McDonald, G., and Ow, D. W. (1992) Heavy metal tolerance in the fission yeast requires an ATP-binding cassette-type vacuolar membrane transport. *EMBO*, 11: 3491-3499.

Otten, A., Alphenaar, A., Pijls, C. Spuij, F. and deWi, H. (1997) In-situ soil remediation, Kluwer Academic Publishers, Boston, MA.

Qui, X., Leland, T.W., Shah, S. I., Sorensen, D. L., and Kendall, E.W. (1997) Field study: Grass remediation for clay soil contaminated with polycyclic aromatic hydrocarbons. Phytoremediation of Soil and Water Contaminants, Kruger, E. L., Anderson, T. A., and Coats, J. R., eds., ACS Symposium Series 664, American Chemical Society, Washington, D.C., 186-199.

Reynolds, C. M. and Wolf, D. C. (1999) Microbial based strategies for assessing rhizosphere –enhanced phytoremediation. *Proc. Phytoremediation Tech. Seminar*, Calgary, AB, Environment Canada, Ottawa, 125-135.

Rice, P. J., Andersoan, T. A., and Coats, J. R. (1997) Evaluation of the use of vegetation for rendering the environmental impact of deicing agents, *Phytoremediation of Soil and Water Contaminants*. Kruger, E. L., Anderson, T. A., and Coats, J. R., ACS Symposium Series 664, American Chemical Society, Washington, D. C., 162-176.

Salt, D. E., Thurman, D. A., Tomsett, A. B., and Sewell, A. K. (1989) Copper phytochelatins in *Mimulas gallatus*. *Proc, Royal Soc., London*, Series B, 236: 79-89.

Schnoor, J. L. (2000) Phytostabilization of metals using hybrid poplar trees, Phytoremediation of Toxic Metals: Using Plants to clean up the environment, Raskin, I. and Ensley, B. D., John Wiley & Sons, New York, 133-150.

Schwendinger, R. B. (1968) Reclamation of soil contaminated with oil, *J. Instit.of Petroleum*, 54: 182-197.

Siciliano, S. D., and Germida, J. J. (1997) Bacterial inoculants of forage grasses that enhance degradation of 2-chlorobenzoic acid in soil. *Envir. Toxic. And Chem.*, 16 (6): 1098-1104.

Simonich, S. L., and Hites, R. A. (1994) Importance of vegetation in removing polycyclic aromatic hydrocarbons from atmosphere. *Nature*, 370: 49-51.

Sogorka, D. B., Goswami, D., and Sogorka, B. (1998) Emerging technologies for soil contaminated with metals- Phytoremediation. *Proc. 30th Mid-Atlantic Ind. and Hazardous Waste Conference*, Suri, R. P. S., and Christensen, G. L., eds., Technomic Publs. Co., Inc., Lancaster, PA. 637-651.

Suthersan, S. S. (2002) *Natural and Enhanced Remediation Systems.* Lewis Publishers, Inc. Washington, D.C.

Thurman, D. A. (1981) Mechanism of metal tolerance in higher plants. *Effect of Heavy Metal Pollution on Plants*, Lepp, N. N., ed., Appl. Sci. Publs., London, 239-249.

Tsao, D. T. (2003) Overview of Phytotechnologies. *Advances in Biochemical Engineering/Biotechnology, Phytoremediation*, Tsao, D.T., ed., Springer-Verlag, Berlin, 1-50.

Uhlmann, D., (1979) Hydrobiology, John Wiley, New York.

USEPA, (1997) Technology alternatives for remediation of soils contaminated with As, Cd, Cr, Hg, and Pb. Document No. EPA/540/S-97/500, U.S. Environmental Protection Agency, Office of Emergency and Remedial Response, Washington, D. C.

USEPA, (1999) Phytoremediation Resources Guide. Document No. 542/B-99/003, U.S. Environmental Protection Agency, Office of Solid Waste and Emergency Response, Technology Innovation Office, Washington, D. C.

Van Steveninck, R. F. M., van Steveninck, M. E., Fernando, D. R., Horst, W. J., and Marschener, H. (1987) Deposition of Zn phytate in globular bodies in roots of *Deschampsia caesptosa* ecotype: A detoxification mechanism?. *J. Plant Physiol.*, 131: 237-257.

Vazquez, J. A. C., Poscharreider, C. Barcelo, J. Baker, A. J. M., Hatton, P., and Cope, G.H. (1994) Compartmentation of zinc in roots and leaves of the zinc hyperaccumulator *Thalspi caerulescens*, JC. Presl. Bot. Acta, 107: 243-250.

Verkleij, J. A. C., Lolkema, P. C., de Neeling, A. L., and Harmens, H. (1991) Heavy metal resistance in higher plants: Biochemical and genetic aspects. Ecological Response to Environmental Stress, Rozema, J. and Verkleij, J. A. C., eds., Kluwer Academic Publs., Dordecht, The Netherlands, 8-19.

Wang, X., Dossett, M. P., Gordon, M., and Strand, S. E. (2004) Fate of carbon tetrachloride during phytoremediation with poplar under controlled field conditions. *Envir. Sci. Technol.*, 38 (21): 5744-5749.

Wenzel, W. W., Adriano, D. C., Salt, D., and Smith, R. (1999) Phytoremediation: A plant-microbe- based remediation system. Bioremediation of Contaminated Soils, Adriano, D. C., Bollag, J. M., Frankenberg, Jr., W. T., and Sims, R. C., eds., Agronomy Monograph No.37, American Society of Agronomy, Madison, WI, 457-508.

Wright, A. L., Weaver, R. W., and Webb, J. W. (1997) Oil bioremediation in salt marsh mesocosms as influenced by N and P fertilization, flooding and seasons. *Water, Air and Soil Poll.*, 95: 179-191.

Xu, J. G., and Johnson, R. L. (1997) Nitrogen dynamics in soils with different hydrocarbon contents planted to barley and field pea. Canadian J. of Soil Sci., 77, 453-458

CHAPTER 7

Wetlands

PASCALE CHAMPAGNE

7.1 Introduction

The ability of natural wetlands to improve water quality through its physical, chemical and biological processes and their interactions has been recognized for many years. The processes and functions occurring within these natural systems have been modified and adapted for many different types of water quality improvement applications. These systems have proven effective in the mitigation of municipal, agricultural and industrial wastewaters, in reducing the nutrient concentrations of solid waste applied to land, in retaining heavy metals resulting from urban stormwater runoff, drainage and mining activities. In addition to improving water quality, wetlands are also able to store and slowly release surface water, rainfall runoff, groundwater and flood waters. Conversely, they can maintain stream flows during dry periods and replenish groundwater sources. Wetlands also provide a valuable aquatic habitat for a diverse species of flora and fauna.

7.2 Wetland System Definitions

In general terms, wetlands are lands where a depth of water covers the soil, or where water is present either at or near the surface of the soil or within the root zone, consistently or intermittently throughout the year, including during the growing season. The presence of water at or near the soil surface is at a frequency and duration sufficient to contribute to the formation of hydric soils which are characteristic of wetlands and the establishment of plants (hydrophytes) adapted for life in saturated soil conditions in their rhizosphere, and animal communities living in the soil and on its surface. In some cases, there are wetlands that lack hydric soils and hydrophytic vegetation, however, these wetlands support other organisms which are indicative of recurrent saturation and flooding (NAS, 1995).

7.2.1 Natural Wetlands

Natural wetlands are poorly drained, transitional areas between deeper open water and dry land. Often located in low-lying areas, wetlands receive runoff water

189

and overflow from rivers, streams and tides, and these areas of land are cyclically, intermittently or permanently inundated or saturated with fresh, brackish or saline water. Natural wetlands show an array of biota, including plants and animals that have adapted and are dependent on these permanent or intermittent inundations (NSW DLWC, 1998).

Natural wetlands are also known as swamps, marshes, bogs, fens, wet meadows, pocosins and sloughs, to name a few. These are classified on the basis of dominant plant types, hydrologic and geographic conditions which can vary considerably from wetland to wetland. The organic and inorganic materials entering the wetland form a complex substrate. This substrate along with the occurrence of gas/water exchanges promotes a varied community of microorganisms, which break down, transform or store a wide variety of natural and man-made constituents.

7.2.2 Constructed Wetlands

A constructed wetland is a man-made system designed using natural materials of soil, water, and biota to simulate and optimize the physical, chemical and biological processes and functions of a natural wetland to achieve a desired water quality or habitat objective. Kadlec and Knight (1996) state that wetlands are typically engineered and constructed to meet one of four objectives: (1) constructed habitat wetlands to offset the depletion of natural wetlands due to agricultural and urban development; (2) constructed treatment wetlands to improve water quality; (3) constructed flood control wetlands to provide flood control; and (4) constructed aquaculture wetlands used for the production of food and fiber.

Of particular interest in this chapter, are constructed treatment wetlands. If properly designed, built, maintained and operated, constructed treatment wetlands can effectively treat many pollutants associated with municipal effluents, industrial and commercial wastewater, agricultural runoff, stormwater runoff, animal wastes, acid mine drainage and landfill leachates. Such systems are especially efficient at removing contaminants such as organics, suspended solids, nitrogen, phosphorus, hydrocarbons, hazardous compounds, micropollutants and metals. Constructed treatment wetlands are typically divided into two basic types: free water surface wetlands (FWS) and subsurface flow wetlands (SSF).

7.2.2.1 Free Water Surface Wetlands (FWS)

Free water surface wetlands are shallow excavations or shallow earth banked lagoons, typically densely vegetated basins, underlain by a subsurface barrier, liner or compacted clay, to prevent seepage (Fig. 7.1). Soil or another suitable medium such as gravel or organic matter provides a growing medium to support roots of emergent vegetation, and water at a relatively shallow depth (3 to 24 inches) flowing through the unit (McCutcheon and Schnoor, 2003). A system of pipes or channels distributes the wastewater over the inlet end of the wetland, and a collection channel collects the treated effluent at the outlet. The wastewater flows along the surface, where it comes into contact with the bacterial populations on the surface of the media and plant stems. Treatment mechanisms are complex but include solids sedimentation, biogeochemical cycling of nutrients and other elements within the water column and contact with bacteria suspended in the water column and attached to surfaces. This type of constructed wetland is particularly efficient in removal of pathogens as a result of the high sunlight UV exposure of the generally shallow water column. Surface flow wetlands have a lower efficiency per unit surface area than sub-surface flow wetlands, and therefore require larger areas of land. Several of these systems may be provided in series for effective treatment. To avoid short-circuiting a good distribution of the wastewater must be provided at the inlet to the wetland, and similarly good collection facilities must be provided at the outlet, although very big wetlands or multiple wetlands in series may not need these at all.

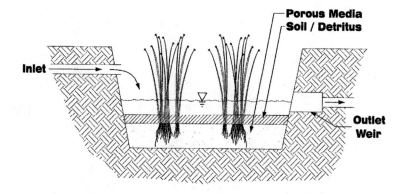

Figure 7.1 Free water surface wetland system

Aerobic wetland is a term employed for the treatment of wastewaters, drainages and discharges with elevated metal concentrations. They are generally used to collect water and provide residence time and aeration to allow metals in the water column such as iron, aluminum and manganese, to be oxidized or hydrolyzed and, subsequently, precipitated. Wetland plant species are generally incorporated to add organic matter and encourage a more uniform flow distribution, and hence, a more effective wetland treatment area. These types of wetlands are generally employed when water requiring treatment has a net alkalinity. The extent of metal removal depends on dissolved metal concentrations, dissolved oxygen content, pH and net alkalinity of the water, the presence of active microbial biomass, and detention time of the water in the wetland.

7.2.2.2 Subsurface Flow Wetlands

Sub-surface flow wetlands (SSF) are considered to be high rate wetlands (Fig. 7.2). The flow moves through a matrix (1 to 3 feet deep), usually soil, gravel, sand, organic substrate or a mixture that supports vegetation roots and biofilms (McCutcheon and Schnoor, 2003). Water is mitigated by physical and chemical processes, as well as microbiological degradation. Anaerobic and aerobic processes take place within the pores of the matrix media. The vegetation supports remediation processes by insulating the filter at colder temperatures and through the uptake of nutrients. The root zone also plays an important role by providing a surface for micro-organisms and by maintaining the hydraulic properties of the substrate through the growth of the roots. Sub-surface systems typically have an increased capacity for water treatment per unit area (NSW DLWC, 1998).

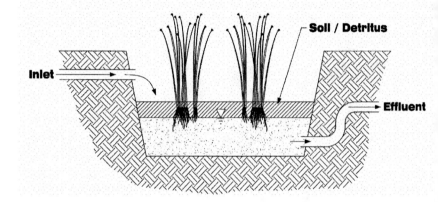

Figure 7.2 Subsurface horizontal flow wetland system

Horizontal Flow Systems
Horizontal flow wetlands are shallow excavations with a synthetic or clay liner. In a horizontal flow system the wastewater is fed through an inlet and passes the filter matrix under the surface of the bed in a relatively horizontal path until it reaches the outlet zone. The matrix is planted with marsh plants such as the common reed. Mitigation of the wastewater takes place through a complex mixture of aerobic and anaerobic processes. The primary removal mechanisms are filtration and biodegradation where dense populations of bacteria growing on the roots of the marsh plants contribute significantly to the treatment process. Due to the long retention time of the wastewater in the substrate matrix which acts as a filter, high pathogen and solids removal can be observed. The presence of a roots system slows down the rate at which the media clogs with the accumulated solid. The biodegradation of organics, as well as, significant nitrogen reduction can also be expected. However, a fully nitrified effluent is not possible due to the lack of oxygen to support the nitrification process.

Vertical Flow Systems

In vertical flow systems (VSF), the wastewater is applied over the entire surface area through a distribution system and the flow passes through the matrix in a relatively vertical path (Fig. 7.3). Marsh plants such as the common reed, *Phragmites australis*, are typically planted in the matrix. The wastewater is dosed on the bed in large batches, thereby flooding the surface to a depth of several centimetres. At the base of the excavation are drainage pipes which are usually turned up so they reach the surface at their ends, which allows air to move in and out of the wetland. The wastewater then slowly percolates downward through the substrate matrix, which acts as a filter, where high pathogen and solids removal can be observed. Between dosing times, oxygen can diffuse through the pores of the matrix. This enables nitrification processes to take place. Biodegradation of organics and significant nitrogen reduction can also be expected. In this case, due to the primarily aerobic matrix of the VSF, denitrification will not take place to a large extent. The treated water is then typically collected in a bottom drainage system and in an outlet well.

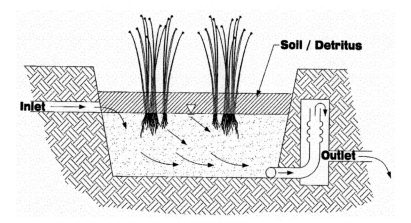

Figure 7.3 Subsurface vertical flow wetland system

Anaerobic Wetlands

Anaerobic wetlands encourage water passage through organic rich substrates, which contribute significantly to treatment of waters with elevated metal concentrations. The wetland substrate may contain a layer of limestone at the bottom of the wetland or mixed with the organic substrate. Wetland plants are transplanted into the organic substrate and serve several purposes including substrate consolidation, metal accumulation, adsorption of metal precipitates, and carbon source for microbial processes (Fernandes and Henriques, 1990). These systems are used when the water has net acidity. Hence, alkalinity must be generated in the wetland before dissolved metals will precipitate. Anaerobic wetlands promote metal oxidation and hydrolysis in the aerobic surface layer but also rely on subsurface chemical and microbial reduction reactions to precipitate metals and neutralize acid. The water infiltrates through thick permeable organic subsurface sediments and becomes anaerobic due to the high biological oxygen demand, where treatment mechanisms such as microbial generation of alkalinity, formation and precipitation of metal sulfides, as well as metal exchange and complexation reactions are enhanced.

7.2.3 Other Aquatic Systems

Other aquatic passive systems also exist which are continuously flooded and develop an anaerobic sediment and soil layer. The diffusion of oxygen occurs naturally and generally results in the development of aerobic and anaerobic mitigation processes, which take place simultaneously in a single, layered treatment system.

7.2.3.1 Facultative Ponds

Facultative ponds are low energy passive treatment systems that are designed to maintain a natural aerated surface layer over a deeper anaerobic layer. Aeration occurs as a result of diffusion through the pond surface and through photosynthesis by algae in the water column. Facultative ponds are effective in reducing the concentration of biodegradable organics through the activity of aerobic microorganisms. With the presence of aerobic and anaerobic layers in the pond, the potential for complete nitrification/denitrification exists. However, the achievement of low suspended solids effluent concentrations can be difficult due to algal growth.

7.2.3.2 Floating Aquatic Plant Systems

Floating aquatic plant systems are essentially ponds that have been inoculated with pleating aquatic plants such as water hyacinths or duckweed to provide wastewater treatment. In these systems, photosynthetic plants are located above the water surface as opposed to within the water column as is the case with algae. Oxygen is released above the water surface and the presence of the aquatic plants on the surface limits oxygen diffusion from the atmosphere to the pond. As a consequence, these systems are typically oxygen deficient (anaerobic) and aerobic processes are limited to the root zone of the plants. These systems are effective at reducing the level of biodegradable organics where facultative microorganisms present on plant roots in the water column and in the sediment are responsible for their removal as well as decreasing suspended solids through sedimentation. The largely anaerobic environment in the water column provides the opportunity for denitrification to occur, and consequently nitrate removal. However, total nitrogen and phosphorous removal can only be achieved if the plants are harvested regularly.

7.2.4 Natural vs. Constructed Wetland Systems

As the interactions of the many treatment mechanisms in wetland systems are often poorly understood, it can be difficult to predict the performance of a wetland system to a particular contaminant stream. Natural wetlands have typically had several decades to develop a complex level of functional interactions which enables them to adapt to a variety of contaminant streams. Constructed wetlands are less mature and generally lack this level of complex interactivity. Hence, the long-term performance of these constructed wetland systems is often unknown.

7.2.4.1 Natural Wetland System Characteristics

Natural wetlands typically exist when all of the necessary structural attributes are present for their formation to occur. These elements are highly variable and their characteristics depend upon climate, hydrology, underlying sediment and water quality (Kadlec and Knight, 1996).

(1) The underlying strata, located below the active rooting zone of the wetland vegetation, is generally composed of a saturated impervious material and does not contribute significantly to the mitigation processes of the wetland.

(2) The hydric soil, which can be a permanently or intermittently saturated mineral to organic soil layer depending on the type of wetland, provides a rooting environment for the wetland vegetation and a connection to the active surface environment.

(3) The detritus layer is formed through the accumulation of dead organic materials and provides a substrate for wetland invertebrates and microorganisms.

(4) The seasonally flooded zone which is a portion of the wetland that is seasonally or cyclically flooded and provides a habitat for aquatic organisms, submerged and floating vegetation depending on water buoyancy for support, as well as algae and microorganisms.

(5) The emergent vegetation are vascular, rooted plants that contain structural components which rise above the water surface and provide an attachment surface for microorganisms and a medium which reduces flow velocities enabling sedimentation.

7.2.4.2 Constructed Wetland System Characteristics

Constructed wetlands designs typically incorporate one or more of these structural attributes, however, these elements are generally less mature than in natural wetland systems which may impact the stability of the system or its ability to mitigate various wastewater streams over a long-term.

In terms of wastewater treatment, both natural and constructed wetlands have five functional components which contribute to the mitigation of pollutants (Hammer and Bastian, 1989).

(1) Wetland vegetation has adapted to growth in a hydric soil and the plants are able to efficiently transport atmospheric gases to the root zone enabling the roots to survive in an anaerobic environment. The transport of oxygen to the root zones creates an aerobic area termed rhizosphere which allows for a large microbial population to thrive and actively transform and/or biodegrade nutrients, metals, organic compounds and other contaminants.

(2) Microorganisms present in wetland system include bacteria, fungi and protozoa. Aerobic and anaerobic microorganisms uptake, transform and biodegrade contaminants in the wetland system to obtain nutrients and energy to conduct their life cycle activities. These microorganisms are responsible for a variety of complementary biological activities, have short generation times, reproduce rapidly, have considerable genetic plasticity, all of which enables them to adapt relatively quickly to new compounds or contaminants introduced to the wetland environment. Some microorganisms are also predatory and will prey upon pathogenic organisms.

(3) Vertebrates and invertebrates present in wetland system serve primarily to close the ecological cycle of wetland systems. Some will break down the organic material converting it to a more useable form for microorganisms. Others will burrow in the soil and substrate facilitating bulk transport to the root zone. While others still, may prey upon pathogenic organisms.

(4) The water column flowing in and above the substrate is primarily responsible for the surface and subsurface transport of nutrients, gases, organics and contaminants to the microorganisms as well as the removal of the by-products of metabolism. Many of the biogeochemical processes occur within the water column environment where exchanges can take place through bulk transport and diffusion.

(5) The soils and substrates of wetlands have a variety of hydraulic conductivities which can facilitate or restrict the transport of substances to the vegetative root zone and rhizosphere. They offer considerable reactive surface area for the complexing of ions and other compounds as well as provide an attachment surface for microorganisms.

7.2.4.3 Differences Between Natural and Constructed Wetland Systems

It is important to recognize that not all wetlands, whether natural or constructed, are similar in the treatment functions they can perform. The hydrologic regime of a wetland system (frequency, timing, depth, duration of flooding) is responsible for the regulation of its structural and functional characteristics. In general, natural wetlands tend to experience large fluctuations in water levels and the water quality will be based upon the watershed area flowing to the wetland system. On the other hand, constructed wetlands tend to maintain more uniform water levels and the water chemistry will be dependent upon the wastewater source (Gopal, 1999).

Another major difference between natural and constructed wetlands is biodiversity. Although the predominance of one or two macrophytes can be reported in a particular natural wetland area, these systems support a large biodiversity, which is not well established until after many years of operation in constructed wetlands. Furthermore, the periodic harvesting of macrophytes which may be necessary to enhance nutrient removal or metal uptake in some constructed wetlands may limit biodiversity (Gopal, 1999). At least three to four growing seasons are often required for constructed wetlands to develop the botanical and biogeochemical characteristics of a natural wetland and, hence, for the constructed wetland to achieve the desired treatment efficiencies for a particular waste stream (Sistani et al., 1999; Hammer and Bastian, 1989). The organic matter accumulation may take several more years to reach a level comparable to that of a natural wetland (Sistani et al., 1999). In assessing nutrient cycling, Hunt et al. (1997) reported spatial differences in water hydrogeochemistry between natural wetland sites as well as between constructed wetland sites located in close geographical proximity. In most cases, these spatial differences were more significant than temporal or seasonal differences which are typically observed and could mainly be attributed to dominant wetland vegetation (Hunt et al., 1997).

7.3   Wetland Structure and Function

The differences in wetland characteristics result from variations in soils, chemistry, climate, geology, hydrology and vegetation on a local and regional basis. Constituent removal and storage capacity in wetlands is a function of the interaction of a number of hydrologic, physical, chemical and biological processes in the soil and biota of the system. The net result of these processes determines the potential for a wetland to serve as a filter or sink for constituents of interest. Wetland microorganisms, vegetation and wildlife are an integral part of, and can therefore impact, the global wetland cycle (Fig. 7.4) for water, nutrients and other constituents.

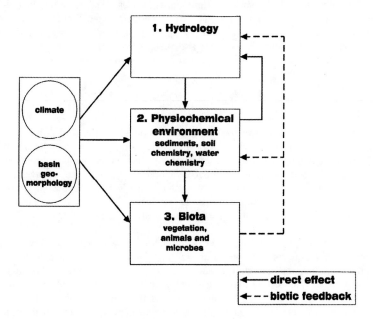

Figure 7.4 Global wetland cycle

7.3.1   Wetland Hydrology and Hydraulics

Wetlands are transitional areas between terrestrial and open water ecosystems and they play a critical role in regulating the movement of water within watersheds. They are typically characterized by water saturation in the root zone at or above the ground surface, for a particular hydroperiod creating the saturated conditions that lead to the development of hydric soils and the presence of hydrophytic plants. Wetlands can be transitional in the amount of water they store and process. The retention time in the wetland will ultimately impact the effectiveness of mitigation processes and the treatment performance of the system.

7.3.1.1 Wetland Hydrology

A hydroperiod is the pattern of the water level in a wetland. It is critical to the type of wetland that evolves and to the adaptive strategies used by its biota. The hydroperiod combines all inflows and outflows of water that are influenced by physical features of the land, proximity of other water bodies and subsurface conditions (Mitsch and Gosselink, 2000).

Water Budget

Groundwater flow, surface flow, ocean tides, periodic flooding and precipitation are the primary ways wetlands receive their water. Wetlands can store precipitation and surface water and then slowly release the water through seepage to ground water, transpiration, evaporation and surface flow from a surface outlet as given by the simplified water budget Equation 7.1 presented and shown in Figure 7.5.

$$\frac{\Delta S}{t} = \left(I_R + I_I + I_S + I_P\right) - \left(Q_o + Q_S + Q_P\right) + A\left(P - ET\right) \quad \text{(Eq. 7.1)}$$

where ΔS is the change in wetland water volume (m^3), t is the time (d), I_R is the overland runoff inflow to the wetland (m^3/d), I_I is the inflow through the inlet structure (m^3/d), I_S is the seepage from the groundwater (m^3/d), I_P is the pumped influent rate (m^3/d), Q_o is the outflow from the outlet structure (m^3/d), Q_s is the seepage to the groundwater (m^3/d), Q_P is the pumped extraction rate (m^3/d), A is the wetland surface area (m^2), P is the precipitation over the area of the wetland (m/d) and ET is the evapotranspiration from the wetland surface area (m/d).

This hydrologic model would be representative of a simple system for which the various components can be estimated or measured, which would include impoundments and wetlands with distinct inlets and outlets. However, obtaining an accurate representation of the hydrologic budget may be more difficult where some significant inflow and outflow elements are not easily quantified. Sheet flow across the surface of a wetland into or out of a stream or lake, tidal inputs, shallow horizontal groundwater flows through substrates of varying hydraulic conductivities are examples of hydrologic components which are often not readily quantifiable. The wetland hydrology budget is important in determining the capacity of the wetland for flood control, stream flow regulation, and its contribution to groundwater recharge (Owen, 1995).

Wetland types differ in their capacity to retain water based on a number of physical and biological characteristics, including landscape position, soil saturation, fiber content/degree of decomposition of the organic soils, vegetation density and type of vegetation (Taylor et al., 1990). Throughout the growing season, plants actively take up water and release it to the atmosphere through evapotranspiration, which reduces the amount of water in wetland soil and increases the capacity for absorption of additional precipitation or surface water flow. As a result, water levels and outflow from the wetland are less than when plants are dormant.

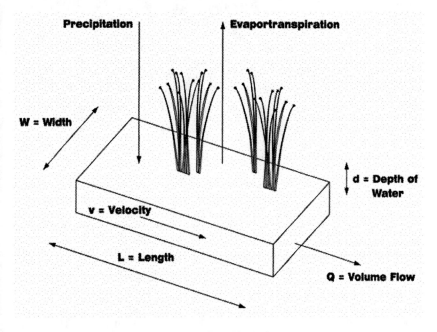

Figure 7.5 Wetland hydrology

Hydrologic Conditions

The performance of wetland systems in the mitigation of wastewater contaminants can also be impacted by wetland hydrology. Spieles and Mitsch (2000) compared the nitrate removal efficiencies of two constructed wetlands receiving river water and one constructed wetland receiving municipal wastewater over a two-year period. In all three wetland systems, nitrate retention efficiency was found to be considerably lower during flood events. In another study, Catallo and Junk (2003) investigated the impact of diurnal tidal compared to statically flooded or statically drained hydrologic wetland conditions on the transformation of N- and S-heterocycles (NSH). In this case, the diurnal tidal hydrologic condition transformed the NSH contaminants at a faster rate and to a lower concentration than the two static hydrologic regimes. This was attributed to the diurnal changes in sediment Eh which enhanced NSH transformation under the diurnal tidal hydrologic regime. These studies illustrate the importance of understanding wetland hydrology and its impact on wetland ecology which can, in turn, affect the treatment efficiency of the wetland system for various contaminants.

Hydrologic conditions are extremely important for the development and maintenance of wetland structure and function. Hydrology transports sediments, nutrients, organics and contaminants into wetlands, which influences the chemical and physical properties of the environment. Hence, wetland hydrology is the main driving force in establishing and maintaining the functionality of wetland ecology. Small changes in hydrology can have a significant affect on abiotic factors, including soil anaerobiosis and redox state, water chemistry and nutrient availability, and in costal wetlands, salinity. This can then, in turn, result in changes in the development of wetland biota leading to distinctive vegetation composition that can enhance or limit species richness. When soils become waterlogged, there is less available oxygen for the vegetation to use, which can limit the number and types of rooted plants that can survive. Finally, to close the cycle, the biotic component can then actively impact and alter wetland hydrology and other physicochemical processes.

7.3.1.2 Wetland Hydraulic Retention Time

Removal processes in a wetland are generally dependent on the length of time the wastewater is held in the system. Hence, one of the key design parameters is the hydraulic retention time (HRT) or, conversely, the hydraulic loading rate of the system. If land area is readily available, wetlands offer passive mitigation strategies which can generally achieve a desired treatment efficiency at relatively low maintenance and operational costs when compared active treatment systems due to the simplicity of these passive treatment systems. In existing treatment wetland systems, HRTs typically range between 1 to 10 days for a particular wetland cell, which corresponds to a hydraulic loading rate of 1 to 10 cm/day (Suthersan, 2002).

Other important design parameters which require consideration is establishing the design HRT required for treatment or which may impact the design HRT as the wetland system matures:

- Water budget or wetland system hydrology, flood protection
- Climatic and operational variables, temperature, pH and DO profiles
- Wetland configuration and associated operational parameters
- Surface water elevation and anticipated seasonal and precipitation-based fluctuations
- Flow patterns, flow velocities, mixing and expected dead zones and short-circuiting
- Reactor models: plug flow vs. completely mixed design
- Biological and chemical reaction rates, mass transfer rates and mechanisms

- Suspended solids
- Nutrient demand and cycling
- Vegetation selection, maturity of system
- Influent constituent concentrations, desired and/or regulatory effluent criteria.

7.3.2 Wetland Soils

Soils are a complex mixture of organic and inorganic materials that are formed as a result of physical, chemical and biological modifications of rocks and sediments. Soils are a fundamental part of wetlands because they provide the necessary elements that define the basic structure of a wetland. They also provide the support for plant growth and sustain microbial communities which are crucial in the mitigation of wastewaters constituents.

Hydric soils are formed where land has been flooded and soil saturated, and the consequent lack of oxygen and development of anaerobic conditions, inhibit the decomposition of plant and animal remains, which in turn favors the growth and regeneration of hydrophytic vegetation. There are two categories of hydric soils: mineral soils and organic soils or histosols. The physical characteristics of the soil will determine the flow of water through the wetland system and, the chemical and biological properties will govern the treatment processes that can occur to mitigate various wastewater streams. The physicochemical properties of wetland mineral soils and histosols are summarized in Table 7.1.

7.3.2.1 Physical Properties of Wetland Soils

In general, wetland histosols are composed of 20 to 35% (dry weight basis) of organic material, while wetland mineral soils contain less than 20 to 35% organic material. Histosols typically have 10 to 20% of the density of mineral soils, and their bulk densities increase with the degree of decomposition. Because of their fibrous composition, histosols tend to shrink, oxidize and subside upon drainage (Suthersan, 1997). Mineral soils will absorb 20 to 40% of their dry weight in water, while histosols can hold from 200 to 400% of their dry weight in water (Mitsch and Gosselink, 2000).

Table 7.1 Summary of general physicochemical properties of wetland soils

Properties	Mineral Soils	Histosols
Organic content	< 20% to 35%	> 20% to 35%
Organic carbon content	< 12% to 20%	> 12% to 20%
pH	Circum-neutral	Acidic
Bulk density	High	Low
Porosity	Low (45-55%)	High (80%)
Hydraulic conductivity	High (except clays)	Low to very high
Water holding capacity	Low	High
Nutrient availability	Generally high	Often low
Cation exchange capacity	Low, governed by major cations	Low, dominated by hydrogen ions

The hydraulic conductivity of the soil will be governed by the pore size and degree of compaction of the soil material. Mineral soils are typically light to dark grey in wetland areas that are permanently flooded or light tan to brown in areas that are intermittently saturated. On the other hand, the range in color for wetland organic soils is from black mucks to brown peats (Kadlec and Knight, 1996).

7.3.2.2 Chemical Properties of Wetland Soils

After flooding, the pH in wetland soils typically tends toward neutrality (pH 6.7 to 7.2). This is a result of ferric iron reduction which occurs after some time under saturated soil conditions. In certain highly organic soils such as peatlands, the pH can remain fairly acidic which is likely due to the oxidation of organic sulfur compounds to sulfuric acid in the presence of humic acids. Wetland soils generally have a high cation exchange capacity (CEC) for a variety of chemical constituents. Mineral soils can undergo isomorphic substitutions of Mg^{2+} for Al^{3+} or Al^{3+} for Si^{4+} at any solution pH. Wetland organic soils contain humic substances, which are a product of the undecomposed vegetation in the wetland soils, contain a large number of phenolic (C_6H_4OH), carboxyl (-COOH) and hydroxyl (-OH) functional groups, which are hydrophilic and serve as pH dependent ion exchange sites. Wetland soils may remove wastewater contaminants by adsorption, ion exchange, precipitation or complexation.

When wetland soils are not saturated, microbial and chemical processes can generally occur using oxygen as the electron acceptor. However, when wetland soils are saturated, microbial respiration and biological and chemical reactions consume available oxygen. This then shifts the soil from an aerobic to an anaerobic state. As conditions become increasingly reduced, other electron acceptors other than oxygen must be used for reactions. These acceptors are, in order of microbial preference, nitrate, ferric iron, manganese, sulfate, and organic compounds. This property makes wetlands ideal environments for chemical transformations because of the range of oxidation states that occur in their soils (Faulkner and Richardson, 1989). Under saturated soil conditions, the rapid use of free dissolved oxygen as microorganisms consume organic materials, leads to oxygen depletion and an increasingly negative redox (Eh) potential with increasing soil depth, as illustrated in Figure 7.6.

When the Eh is greater than +400 mV the conditions are considered to be aerobic. At an Eh of less than -100 mV they are considered to be anaerobic, while the intermediate conditions are termed anoxic (Kadlec and Knight, 1996). In the Eh range between above +320 mV to +340 mV, oxygen depletion is generally considered to be complete and nitrate (NO_3^-) reduction can begin to take place until and Eh of approximately +225 mV is reached. Manganese reduction from manganic (Mn^{4+}) to manganous (Mn^{2+}) is prominent from an Eh of +225 mV until an Eh of +120 mV is obtained. Then ferric (Fe^{3+}) to ferrous (Fe^{2+}) reduction will take place from an Eh of +120 mV to between -75 to -150 mV, followed by sulfate (SO_4^{2-}) reduction to hydrogen sulfide (H_2S) from -150 mV to -250 mV, and, finally, carbon dioxide (CO_2) is reduced to methane (CH_4) from -250 mV to -350 mV.

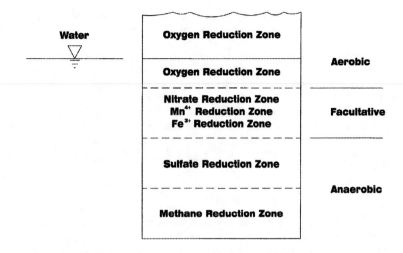

Figure 7.6 Oxidation-reduction zones with increasing soil depth

The chemical characteristics of the wetland soil play an important role in the remediation potential of the wetland. Hence, the variability in mitigation performance observed in natural and constructed wetland systems is likely due to the differences in wetland soils. These will ultimately impact the establishment and diversity of wetland vegetation, as well as wetland microbiology and, subsequently, wildlife.

7.3.2.3 Biological Properties of Wetland Soils

Once a natural or constructed wetland is established, the microorganisms of the wetland make up a significant portion of the organic carbon and nutrients found in wetland soils. Wetland vegetation also contributes to the physical, as well as chemical soil characterization and diversification of the soil through the growth and decay of roots, algae, generation and accumulation of plant litter. It is important to note that biological, chemical and physical properties of the wetland soils are interdependent and are instrumental in establishing the structural and functional foundation of the wetland ecosystem.

7.3.3 Role of Wetland Vegetation and Microbial Communities

Wetland ecosystems offer a diversified range of biological activities. The primary function of algae and hydrophytic plants is the uptake dissolved nutrients and contaminants from the water and, which they employ to produce new plant biomass. The interaction of wetland soil, sediment vegetation and aquatic constituents also provides a desirable environment for the growth and reproduction of microorganisms, which can play an active role in contaminant mitigation.

Besides playing an instrumental role in the reduction and cycling of nutrients and mitigation of metals and other contaminants, wetland vegetation influences wetland hydrology and hydraulics through the system. Macrophytes reduce flow velocities which enables sedimentation, physical filtering of sediment particles and decreases erosion. They also provide shade which limits light availability for algal photosynthesis, and reduces the impact of the diurnal effect of this process (NSW DLWC, 1998). The vegetation also has a positive aesthetic component in terms of waste treatment systems and provides shelter for wetland microorganisms and wildlife.

7.3.3.1 Algae

Algae are unicellular and multicellular photosynthetic bacteria and plants that are highly diverse and can adapt to a wide range of aquatic habitats, including wetlands. They are dependent on light, carbon, typically CO_2, and other nutrients for their energy, growth and reproduction. When these elements are not limiting, algae can generate large populations and contribute significantly to the overall food chain, the transformation, storage and cycling of wetland nutrients, the fixation of contaminants through sorption and settling as well as contaminant mitigation. However, when they are shaded by macrophytes, algae tend to play a lesser role in the processes of the wetland system.

In wetland areas with few emergent macrophytes, filamentous algal mats can develop which tend to control DO and CO_2 concentrations in the wetland water column. DO concentrations can be high saturating the water during the day when photosynthetic rates are high, and decrease to zero at night due to algal respiration. Changes in dissolved CO_2 concentrations will vary inversely proportional to DO concentration, because it is generally utilized when photosynthetic oxygen production is high. Correspondingly, the pH may rise by 2 to 3 units during the day, as dissolved CO_2 concentrations decrease and rise again at night as CO_2 is produced as a product of algal respiration (Kadlec and Knight, 1996).

When algae growth is substantial, they can play a significant role in the short-term fixation and immobilization of nutrients and hydrocarbons in wetland systems, followed by a gradual release through algal death and decomposition. From a wetland function perspective, this form of nutrient cycling can lead to the immobilization and transformation of some contaminants which can be used for growth by the algae, where the gradual release would result in a reduction in discharge concentration when compared to inflow concentrations. Long-term algal fixation of contaminants into wetland sediments via sorption, settling and burial may also be significant, particularly in low-nutrient wetlands where the microbial decomposition rates of algal biomass is low.

7.3.3.2 Macrophytes

In macrophytes, nutrients and contaminants generally move to underground storage in the roots when the plants senesce, and are deposited as litter in the bottom sediments when the plants die. This creates additional substrates and ion exchange sites, as well as provides a source of carbon, nitrogen and phosphorous to fuel microbial processes. Macrophytes also create microenvironments and attachments sites for communities of microorganisms to thrive, along stems and in the root zone or rhizosphere.

Macrophytes are the dominant vegetative structural component of most wetland systems. They include vascular plants with tissues which result from specialized cells. As oxygen is often limiting in flooded environments, macrophytes have adapted to saturated conditions of wetland systems where they thrive, despite the challenges of an anaerobic environment, and contribute to its treatment capacity. These wetland plants have adopted other strategies such as long, oxygen transporting tubes (e.g., emergent reeds), the ability to float on shallow water (e.g., lilies), or buttresses trunks (e.g., Cypress trees) to obtain and transport oxygen to areas of the plant where it is required for various plant functions. Wetland plants are adapted to changing redox conditions. They often contain arenchymous tissue (spongy tissue with large pores) in their stems and roots that allows air to move quickly between the leaf surface and the roots. Oxygen released from wetland plant roots oxidizes the rhizosphere and allows processes requiring oxygen, such as organic compound breakdown, decomposition, and denitrification, to occur (Steinberg and Coonrod, 1994).

The types of plants involved in wetland systems can be classified as: emergent, floating or submerged plants based on their predominant growth form. In emergent plant species, most of the above ground part of the plant emerges above the waterline and into the air. They are rooted in the sediment soil where they have an extensive root and rhizome structure. Emergent plants generally have structural components which allow them to be self-supporting and to transport oxygen to the roots and rhizomes. This type of oxidizing environment surrounded by anoxic sediments enhances the presence of heterotrophic and nitrifying microorganisms, and their ability to biodegrade organic and nitrogenous compounds. The presence of macrophytes in wetlands is also important because they provide surface area for microbial attachment and growth both in the water column and in the root zone or rhizosphere.

Submerged species are also rooted in the wetland soil, but have buoyant stems and leaves with that cover large surface areas that fill the niche in the water column above the wetland sediment layer. Floating species have been adapted to float on the surface of the wetland due to buoyant leaves and stems. They are not generally rooted in the wetland soil, and roots typically dangle below the surface within the water column (Guntenspergen et al., 1989).

In a study by Collins et al. (2004), aerobic wetlands with plants had lower pH levels and NH_4^+-N concentrations, as well as higher Fe and Mn concentrations, especially during the growing season, than their non-vegetated counterparts. The vegetated wetlands were also found to affect bacterial assemblages in the rhizosphere and the water column which utilize these elements in their metabolisms. Vegetated vertical flow wetland microcosm systems were found to provide a more rapid rate of *E. coli* removal than their non-vegetated counterpart (Decamp and Warren, 2000). Two macrophytes-based constructed wetlands, *Cyperus papyrus* and *Miscanthidium violaceum,* were compared for wastewater treatment efficiency (Kyambadde et al., 2004). It was found that the *Cyperus papyrus*-based constructed wetlands exhibited a higher treatment efficiency than the *Miscanthidium violaceum*-based wetlands because *Cyperus papyrus* plants possess root structures for a greater number of microbial attachment sites, increased residence time for suspended solids trapping and settlement, and greater surface area for adsorption, uptake and organic and inorganic constituent oxidation in the rhizosphere. Engelhardt and Ritchie (2001) reported on the importance of macrophyte species diversity in wetland systems on the establishment of algal populations and the consequent increase in the phosphorous removal efficiency of the system.

7.3.3.3 Rhizosphere

In natural plant ecosystems, indigenous soil microorganisms present in the rhizosphere are found in mutual relationships with plants. Rhizosphere is a zone of increased microbial activity at the root-soil interface that is under the influence of the plant root. In emergent and some submerged macrophyte species, oxygen is transported by diffusion or convective air flow to the roots and rhizomes (Guntenspergen et al., 1989). A portion of the oxygen leaks from the root system into the rhizosphere which creates an oxidizing environment surrounded by anoxic sediments. The translocation of oxygen in the plant root zone is shown in Figure 7.7. Studies have shown that the redox state of the rhizosphere can have a significant effect on the intensity of oxygen release through the roots (Stottmeister et al., 2003). Plants also release exudates into the surrounding soil which include sugars, alcohols, amino acids, and enzymes. The exudates and enzymes enhance microbial growth and the growth of mycorrhizal fungi. These fungi grow in symbiotic association with the plant and have unique enzymatic pathways that help degrade organics that could not be degraded by bacteria alone (Suthersan, 1997). The overall effect is an increase in the microbial biomass surrounding the roots by an order of magnitude of 100 to 1000-fold greater, compared with microbial populations in the bulk soil (Carman and Crossman, 2001; Crowley et al., 1997; Anderson and Coats, 1995).

The microbes and mycorrhizal fungi promote degradation and co-metabolism of organics and, hence, are integral to phytoremediation. The plant is then affected by the microbial population it has stimulated, since the root zone is the area from which mineral nutrients are obtained. The rhizosphere is divided into the inner rhizosphere, at the root surface, and the outer rhizosphere, embracing the adjacent soil. The microbial population is typically largest and most active in the inner zone where the biochemical interactions are most pronounced and root exudates are most concentrated. The rhizosphere microbial community composition is dependent on root type, plant age, plant species, environmental conditions, substrates, root exudates and soil type, as well as factors such as plant root exposure history to various contaminants. Generally, the primary root microbial population is determined by the habitat created by the plant. The secondary microbial population, however, depends upon the activities of the initial population (Anderson and Coats, 1995).

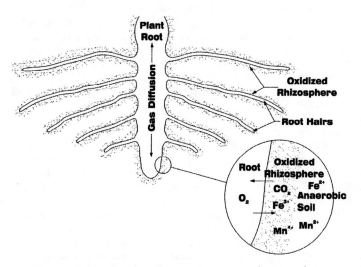

Figure 7.7 Translocation of oxygen to root zone in macrophytes

The fate of organic and inorganic constituents in the rhizosphere and the tendency for these constituents to be taken up by plants can be predicted using the log of the octanol-water partitioning coefficient (K_{ow}) of a constituent. Moderately hydrophobic organic constituents (log K_{ow} ranging from 0.5 to 3) such as VOCs are efficiently removed by direct plant uptake. Constituents with log K_{ow} of less than 0.5 are generally too water soluble to be taken up by plant roots while constituents with log K_{ow} values of greater than 3 are too large or relatively immobile (Carman and Crossman, 2001).

Wetland plants, in combination with microbial activities, alter redox conditions, pH and organic matter content of sediments and can, therefore, impact the chemical speciation and mobility of some constituents, particularly metals. The biogeochemistry of the rhizosphere has a tendency to change on a seasonal basis (Jacob and Otte, 2003). Continual changes at the root-soil interface, both physical and chemical, produces constant alterations in the soil matrix and microbial environment (Crowley et al., 1997; Anderson and Coats, 1995). Constituents may be mobilized or immobilized depending on the actual combination of factors and it is extremely difficult to predict which effects plants will have on metal mobility under a particular set of conditions.

7.3.3.8 Microorganisms

Bacteria and fungi are generally the first organisms to colonize and begin the sequential decomposition of organic solids and are the first access readily available dissolved constituents. Wetland microorganisms remove soluble organic matter, coagulate colloidal material, stabilized organic matter and convert soluble and colloidal organic matter. These microbial populations can have a significant influence on the soil and water chemistry of the wetland ecosystem. Different microorganisms have a variety of tolerances and requirements for dissolved oxygen, temperature ranges, pH and nutrients, which will results in important transformations of nitrogen, iron, manganese, sulfur and carbon depending on the redox potential and pH of the local environment.

In wetland systems, bacteria are often found on solid plant surfaces, decaying organic matter and in the soil and sediment environment. Most bacteria are heterotrophic which implies that they obtain their food and energy requirements for growth from organic compounds and a few are autotrophic and they synthesize organic molecules from inorganic carbon (CO_2). Fungi are heterotrophic organisms. They are also abundant in wetland environments and they play an important role in wastewater treatment. Most aquatic fungi are molds and are associated with the detritus and sediment layers where they can degrade the dead organic matter. Fungi are ecologically important in wetlands because they mediate a significant proportion of the recycling of carbon and other nutrients. If fungi were inhibited through the action of toxic metals and other chemicals in the wetland environment, nutrient cycling of scarcer nutrients would be reduced, greatly limiting the primary productivity of algae and higher plants (Kadlec and Knight, 1996).

7.4   Constituent Cycling and Removal in Wetlands

Wetlands are characterized by a range of properties which make them attractive for the management of contaminants in water. The treatment mechanisms of wetland systems involve physical, chemical and biological processes occurring in the soil-water matrix, in the plant and in the rhizosphere. Wetlands are valuable because they can greatly influence the quality of water and its flow. They improve water quality by intercepting surface runoff and removing or retaining inorganic constituents, processing organic matter and reducing suspended sediments.

A wetland is considered to be a sink if it has a net retention of a specific constituent, which occurs when constituent inputs are greater than the outputs. The opposite is true for a source wetland, where if a wetland exports more constituent to an adjacent system than would occur without the wetland, it is considered a source. When a wetland transforms a constituent, (e.g., dissolved form to particulate form), but the total amount of this constituent present in the wetland does not change, it is a transformer. A wetland can be a sink for an inorganic form of a constituent and still be a source for the organic form of the same constituent (Mitsch and Gosselink, 2000). Although they often present an inexpensive, low-maintenance alternative to wastewater treatment, wetland may not be a final sink for contaminants, and may only retain them for a period of days, months or years before they are re-released to the environment. Thus, proper design, operation and maintenance of wetlands should aim remove accumulated contaminants from the system on a periodic basis.

7.4.1 Abiotic Factors Influencing Wetland Processes

Chemical and biological wetland processes occur at rates dependent upon environmental factors including temperature, oxygen, pH and salinity. At lower temperatures, process rates may be slower, cease altogether or even reverse, releasing nutrients, under certain environmental conditions. This is especially true for biological activity in wetland vegetation, in the rhizosphere and in biofilms. Wetland processes occur under the influence of interrelated and constantly changing environmental conditions, which will ultimately impact the performance of a wetland treatment system. Two of the most important cyclic changes in ambient conditions which impact wetland treatment and function are diurnal and seasonal changes.

Diurnal changes usually result in variations in wetland temperature profiles and dissolved oxygen concentrations. Lower night temperatures slow various chemical reactions and microbial activity, in comparison to warmer daytime temperatures. Photosynthesis adds oxygen, which determines the direction of many wetland processes, to the water column by day, but this oxygen is reduced and possibly depleted overnight due to microbial respiration.

Seasonal changes cause mainly variations in daylight hours from season to season (photoperiod), and temperature. The growth and reproductive activities (flowering) of wetland plants are stimulated by changes in day length. During the growing season, emergent and submerged vegetation from water and sediments uptake nutrients at high rates. By the time the vascular plants die, they have translocated a portion of nutrient material to the roots and rhizomes. Cold weather reduces plant and microbial activity in the water and sediments, affects community selection, and slows down biological and chemical processes as well as their resultant biogeochemical nutrient cycling kinetics (Werker et al., 2002; Newman et al., 2000). In the spring, excessive runoff combined with cool temperatures leads to high flow rates and reduced nutrient and metal retention, and hence diminished nutrient retention from the fall to early spring (Mitsch and Gosselink, 2000). Dry seasons tend to accentuate organic matter decomposition while wet seasons lead to contaminant dilution. Wind, rain, frost and snow are other factors that add variations to wetland activity and, hence, overall treatment performance.

Fluctuating water levels that are characteristic in wetlands effectively control oxygen concentrations in the wetland water column and sediments, and the oxidation-reduction (redox) chemistry conditions of the wetland system. Oxygen diffuses slowly in water, and is often used by microbial activity faster than it can be replenished. This affects root respiration, and impacts nutrient availability. The redox conditions are, therefore, governed by hydroperiods and play a significant role in nutrient cycling and availability, pH, solubility or insolubility of nutrients and metals within the wetland system, vegetation composition, sediment and organic matter accumulation, degradation, and inorganic constituent availability. As an example, a decreased redox potential, will affect the pore and column water chemistry and may impact the composition of metal complexes and release the metal ions, bound to sediments or organic matter, into the overlying water column (Connell et al., 1984).

The pH of water and soils in wetlands also exerts a strong influence on the direction of many biological and chemical reactions and processes. Organic wetland soils tend to be acidic, while mineral wetland soils are more neutral or sometimes alkaline. The consequence of flooding previously drained wetland soils is usually to push the pH toward neutrality, whether it was formerly acidic or alkaline. Many processes are pH dependent, and are less effective if pH is out of their respective optimal range including biological transformations, partitioning of ionized and unionized acids and bases, cation exchange, solid and gas solubilities (NSW DLWC, 1998). For example, a lower pH increases the competition between metal and hydrogen ions for binding sites on soils and organic substrates. A decrease in pH may also dissolve metal-carbonate complexes, releasing free metal ions into the water column. Another factor that must be taken into consideration is that while abiotic chemical oxidation reaction rates may decrease with a decrease in pH, some microbial oxidation kinetics may increase with the same decrease in pH, thereby enhancing the overall oxidation process.

The salinity of water within the wetland can increase as water levels decrease and, the contaminants may become concentrated. This will depend on wetland size and its design. Change in salinity may impact some of the removal and retention processes in wetlands. For instance, elevated salt concentrations create increased competition between cations and metals for binding sites on soils and organic substrates. Often, inorganic contaminants such as metals will be desorbed into the overlying water column (Connell et al., 1984).

7.4.2 Wetland Processes

Wetlands provide a diversity of niches and micro-environments, all of which play important roles in the mitigation of contaminants. Various processes occur within the vegetation, the water column, on the wetland substrate, in the soil and in concentrated areas of microbial activity on plant stems and roots, in the water column and within the soil and substrate matrixes. These processes may be biological, chemical or physical, although considerable overlap and interaction occurs between them. General wetland processes involved in the mitigation of organic and inorganic contaminants is illustrated in Figure 7.8.

In natural or constructed wetland treatment systems both biotic and abiotic mechanisms are involved. Biological mechanisms can be vegetative and microbiological in nature, with the remainder of the mitigation occurring via chemical and physical processes, principally at the interfaces of the water and soil, the soil and the root, or the vegetation and the water. The effectiveness of all of these processes varies with residence time.

Longer residence times having beneficial effects by permitting more contaminants to be removed. However, if too long it can also have detrimental effects as the redox potential of the sediments may change, allowing the release of nutrients and pollutants back into the water column.

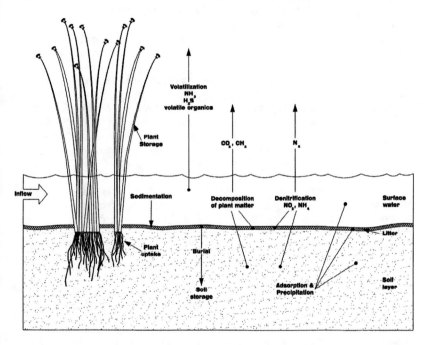

Figure 7.8 Wetland processes in the mitigation of contaminants

7.4.2.1 Physical Processes

In wetland systems, the physical removal of contaminants associated with particulate matter in wastewater are able to occur efficiently primarily via sedimentation and filtration processes. Water typically moves very slowly through wetlands due to the resistance provided by rooted and floating plants in free water surface systems and due to the soil, sediment and substrate matrix in subsurface flow systems, and is consequently settled, filtered and retained in the wetland system.

7.4.2.2 Chemical Processes

A wide range of chemical processes are involved in the removal and mitigation of contaminants in wetland treatment systems. Exposure to light and atmospheric gases can lead to the breakdown of organic compounds and kill pathogens. Chemical reactions between constituents leading to their transformation and subsequent precipitation from the water column as insoluble compounds are important chemical processes within wetland treatment systems. For instance, phosphate can precipitate with iron and aluminum oxides to form new mineral compounds (iron- and aluminum-phosphates), which can be very stable in the soil, allowing for the long-term storage of phosphorus. Another important precipitation reaction that occurs in wetland soils is the formation of metal sulfides, which are highly insoluble compounds and represent an effective means for immobilizing many metals in wetlands treatment systems.

Volatilization involving the diffusion of a dissolved compound from the water into the atmosphere is another potential means of constituent removal in wetlands. Ammonia (NH_3) volatilization can result in significant removal of nitrogen, if the pH of the water is greater than 8.5 (Reddy and Patrick, 1984). However, at a pH lower than 8.5, ammonia nitrogen exists almost exclusively in the ionized ammonium (NH_4^+) form, which is not volatile. Many other organic compounds are volatile, and are readily lost to the atmosphere from wetlands.

One of the most important chemical removal processes in wetland soils, sediments, substrates and plant surfaces is sorption, which results in short-term retention or long-term immobilization of several types of contaminants. Sorption describes a group of processes, which include adsorption and precipitation reactions for the transfer of ions from the water phase to the solid phase. Adsorption is a surface process referring to the attachment of ions to soil particles by either ion exchange or chemisorption. Ion exchange involves the physical attachment of ions, generally cations, to the surfaces of clay particles and organic matter in the soil. It is a much weaker attachment than chemical bonding, therefore the ions are not permanently immobilized in the soil. Many constituents of agricultural, industrial and municipal wastewaters and runoff exist as cations, including ammonium and metals. The capacity of soils for the retention of cations, expressed as the cation exchange capacity (CEC), generally increases with increasing clay and organic matter content. Chemisorption represents a stronger and more permanent form of bonding than ion exchange. A number of metals and organic compounds, such as phosphate, can be immobilized in the soil via chemisorption with clays, and organic matter, as well as iron and aluminum oxides.

7.4.2.3 Biological Processes

In addition to physical and chemical mitigation processes, biological processes involving wetland microorganisms and vegetation also contribute significantly to the mitigation effectiveness of treatment wetlands. The most widely recognized and utilized biological process for contaminant removal in wetlands is biological uptake. Contaminants that are present in the form of essential plant nutrients such as nitrate, ammonium and phosphate are readily taken up by wetland plants. However, many wetland plant species take up and transform a variety of organic and inorganic constituents. The rate of contaminant removal by plants varies depending on plant growth rate and concentration of the constituent in plant tissues. Algae may also provide a significant amount of nutrient uptake, but are the storage of nutrients and other organic and inorganic constituents may be short-term, due to the short life-cycle of algae. Microorganisms in the soil also provide uptake and short-term storage of nutrients and other contaminants.

As a result of vegetation and microbial death, detritus and litter accumulates at the soil surface, of wetland systems. Some of the nutrients, metals or other elements previously removed from the water by wetland biota are then recycled back to the water and soil by leaching and decomposition. Leaching of water-soluble constituents may occur rapidly upon the death of the plant or plant tissue, while a more gradual loss of constituents occurs during the decomposition of detritus by bacteria and other organisms. Recycled contaminants may be flushed from the wetland in the surface water, or may be removed again from the water by biological uptake or other phytoprocesses.

Although microorganisms may provide a measurable amount of contaminant uptake and storage, it is their metabolic processes that play the most significant role in removal of organic compounds. Many of the microbiological processes can be facilitated by or occur in conjunction with plants and algae present in the system, and other phytoprocesses and the biological processes are often part of larger element cycles occurring within the wetland system. Microbial activity will depend on the concentration and nature of the substrate undergoing transformation, as well as the presence and availability of the suitable enzymes.

Microbial decomposers, primarily soil bacteria, utilize the carbon in organic matter as a source of energy, converting it to carbon dioxide (CO_2) or methane (CH_4) gases. This provides an important biological mechanism for removal of a wide variety of organic compounds. The efficiency and rate of organic compound degradation by microorganisms is highly variable for different types of organic compounds. Aerobic respiration using O_2 as an electron acceptor dominates under aerobic conditions while under the anaerobic conditions often found in wetlands, organic decomposition can occur in the presence of other terminal electron acceptors, including NO_3^-, Mn^{2+}, Fe^{3+}, SO_4^{2-} and CO_2. The redox potential decreases through the sequence of electron acceptors, as O_2 is the acceptor at +400 to +600 mV. Nitrate becomes an acceptor at approximately +250 mV, manganese at +225 mV, iron at +120 mV, and sulfate between -75 to -150 mV. Carbon dioxide will become the terminal electron acceptor below -250 to -350 mV (Mitsch and Gosselink, 2000). The biological degradation of organic compounds usually occurs rapidly in the presence of oxygen, and more slowly for other electron acceptors. It should be noted that other abiotic factors such as temperature and pH are also crucial in establishing the rate at which these biological processes will proceed in the wetland.

7.4.2.4 Phytoremediation Processes

Phytoremediation is a diverse and emerging technology that uses plants and their rhizosphere for the remediation of inorganic and organic contaminants (Fig. 7.9). The vegetation-based treatments make use of natural cycles within the plant and its environment. In general, it is possible to characterize the primary processes within macrophyte systems as the uptake and transformation of contaminants by microorganisms and vegetation, and their subsequent biodegradation and biotansformation. In the both emergent and submerged vegetation, the surface of the plants can also provide a structural support for the attached growth of bacteria and other microbial biomass. The processes of phytoremediation that directly address the removal of organic and inorganic constituents from wetland soils and wastewaters, and their subsequent retention are summarized in Table 7.2 and addressed in the following sections.

Phytoaccumulation
Phytoaccumulation, also called phytoextraction, phytoconcentration, phytotransfer, phytomining or hyperaccumulation, is the use of plants to uptake, translocations and accumulate constituents from the plant root zone into above-ground plant tissue (Suthersan, 1997). Constituents include: metals, metalloids, radionuclides, perchlorate, BTEX, PCP and organic compounds not tightly bound to soils. The constituents may be extracted from the soil, as well as in the dissolved form in pore water or shallow groundwater, by cation pumps, absorption and other mechanisms (McCutcheon and Schnoor, 2003). Certain plants exhibit the ability to accumulate and tolerate very high concentrations (2 to 5%) of metals in their biomass, and are called hyperaccumulators. (Carman and Crossman, 2001). Hyperaccumulation is a special case of phytoaccumulation and defined to be 100 times the normal plant accumulation of elements and is 0.01% by dry weight for cadmium and other rare elements, 0.1% for most other heavy metals and 1% for iron, manganese and other more common elements (McCucheon and Schnoor, 2003).

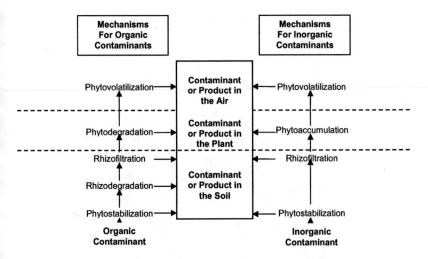

Figure 7.9 Potential fate of organic and inorganic contaminants during phytoremediation.

Phytodegradation

Phytodegradation, also referred to as phytotransformation, phytoreduction, phytooxidation and phytolignification, is the biochemical degradation or transformation of organic constituents taken up and stored by plants via metabolic processes within the plant or external to the plant through the effect of enzymes produced by the plants. Contaminants are degraded to products used as nutrients and incorporated into the plant tissues or by-products that can be broken down further by microorganisms to less harmful compounds (McCutcheon and Schnoor, 2003). Plant transformation pathways will differ depending on the plant species, tissue type and nature of the contaminant. The pathways can be generally categorized as reduction, oxidation, conjugation, and sequestration. Plants synthesize a large number of enzymes as a result of primary and secondary metabolisms forming enzymes that are useful for phytodegradation including nitroreductase (munitions and pesticides), dehalogenases (chlorinated solvents and pesticides), phosphatases (pesticides), peroxidases (phenols), laccases (aromatic amines), cytochrome P-450 (pesticides and chlorinated solvents), nitrilase (herbicides). Contaminant degradation by these plant-formed enzymes can occur in an environment free of microorganisms, even in environments that are not ideal for biodegradation (Suthersan, 2002)

Table 7.2 Summary of phytoremediation processes.

Mechanism	Process Goal	Media	Contaminants	Plant Type
Phytostabilization	Containment erosion control	Soils, sediments Sludges	Metals: As, Cd, Cr, Cu, Pb, Zn	Herbaceous species, grasses, trees, wetland species
Rhizodegradation	Transformation mineralization	Soils, sediments sludges, groundwater	Organic compounds (TPH, PAHs, BTEX, chlorinated solvents, pesticides, PCBs)	Herbaceous species, grasses, trees, wetland species
Phytoaccumulation	Extraction accumulation	Soils, sediments Sludges	Metals: Ag, Au, Cd, Co, Cr, Cu, Hg, Mn, Mo, Ni, Pb, Zn, Radionuclides: ^{90}Sr, ^{137}Cs, ^{239}Pu, $^{234, 238}U$	Herbaceous species, grasses, trees, wetland species
Phytodegradation	Transformation mineralization	Soils, sediments, sludges, groundwater surface water	Organic compounds, chlorinated solvents, phenols, pesticides, munitions	Algae, herbaceous species, trees, wetland species
Phytovolatilization	Extraction and release to air	Soils, sediments, sludges, groundwater	Chlorinated solvents, MTBE, metals: (Se, Hg, As)	Herbaceous species, trees, wetland species
Evapotranspiration	Containment erosion control	Groundwater, surface water, stormwater	Water soluble organics and inorganics	Herbaceous species, grasses, trees, wetland species

Phytostabilization

Phytostabilization is the use of plants to immobilize organic and inorganic constituents in the soil and groundwater through adsorption and accumulation by roots, adsorption onto roots, or precipitation within the rhizosphere. It is also known as biomineralization, phytosequestration and lignification and can be applied to metals, phenols, and chlorinated solvents. Plant exudates released to the root zone can increase the local soil pH up to 1.5 pH units, as well as increase the soil oxygen content, having a significant effect on the redox conditions of the soil and promoting oxidation, causing speciation, and adsorption to form stable mineral deposits, thereby reducing the mobility and bioavailability of metals (Carman and Crossman, 2001).

Humification, lignification and irreversible binding of some organic compounds can also occur as a result of plant functions.

Phytovolatilization

In this process volatile inorganic and organic compounds are taken up, often respeciated and transpired by the plant. It is also referred to as biovolatilization and phytoevaporation and applies to selenium, tritium, arsenic, mercury and chlorinated solvents. In some cases, organic compounds that are extracted, and subsequently either degraded or transformed within the plant and form catabolic intermediates or end-products that can be volatilized from the plant tissue. Inorganic contaminants that are extracted and accumulate can be methylated and subsequently volatilized from the plant tissue (Evans and Furlong, 2003).

Rhizodegradation

Rhizodegradation is also called phytostimulation, rhizosphere biodegradation, enhanced rhizosphere biodegradation or plant-assisted bioremediation and can be applied to BTEX, petroleum hydrocarbons, PAHs, PCP, perchlorate, pesticides, PCBs and other organic compounds (McCrutcheon and Schnoor, 2003). The process involves the breakdown of organic constituents in the soil through microbial activity that is enhanced by plant-mediated processes within the rhizosphere. As much as 60% of the oxygen transported to the root zone of the plants can be transferred to the rhizosphere generating aerobic conditions for the thriving microbial community associated with the root zone (Evans and Furlong, 2003). Plant exudates (amino acids, carbohydrates, nucleic acid derivatives, growth factors, carboxylic acids and enzymes), root necrosis and other processes provide the environment, organic carbon and nutrients necessary to spur soil bacteria growth to as much as 20 times that normally found in non-rhizosphere soil. The conditions also stimulate enzyme induction and cometabolic enzyme degradation by mycorrhizal fungi and other microorganisms in the rhizosphere. Plants can also enhance biodegradation by physical rooting mechanisms which loosen the soil in the root zone and facilitate the transport of oxygen and water.

Rhizofiltration

Rhizofiltration involves the uptake, absorption, adsorption, or precipitation of constituents that are in solution surrounding the root system onto or within the plant root, young shoots, fungi, algae and bacteria (Carman and Crossman, 2001). Due to the various adsorption media in a variety of configurations and application, it is also referred to as phytofiltration, blastofiltration, phytosorption, biosorption, biocurtain, biofilter, and contaminant uptake for the removal of metals, radionuclides, organic compounds, nitrate, ammonium, phosphate and pathogens (McCrutcheon and Schnoor, 2003). Macrophytes and algae are known to accumulate metals and other toxic elements from solution. Plants that are naturally immobilized, such as attached algae and rooted plants, and those that can be easily separated from suspension, such as filamentous algae, macroalgae and floating plants have been found to have high adsorption capacities (Suthersan, 1997). The process of rhizofiltration first involves the containment of the constituent via immobilization or accumulation on or within the plant or microorganism. Plant or microbial uptake, accumulation and translocation may follow depending on the contaminant. Plant root exudates might lead to precipitation of some metals (Suthersan, 2002).

Phytocontainment

Phytocontainment or hydraulic containment is the use of plants, particularly phreatophytes which can transpire large quantities of water, to control the migration and flow of pore water, shallow groundwater and wetlands and their associate contaminants dissolved in the water (Carman and Crossman, 2001). This process is often applied in conjunction with rhizodegradation and phytodegradation processes leading to the mitigation of water-soluble contaminants such as chlorinated solvents, MTBE, explosives, other organic contaminants, salts and some metals (McCrutcheon and Schnoor, 2003).

7.4.3 Solids Cycling and Removal

Suspended solids enter wetlands in runoff with inflow from associated water bodies, waste streams requiring treatment, as well as detritus and dead organic matter formed in the wetland system. Other contaminants that can impact water quality such as nutrients, organic compounds, metals, metalloids and radionuclides are often adsorbed onto suspended solids. Thus, the removal of suspended solids, to which contaminants are adsorbed, also removes these constituents from the water and can provide multiple benefits to water quality (Hupp and Bazemore, 1993).

Sedimentation of suspended solids depends upon water velocity, flooding regimes, vegetated area of the wetland, and water retention time. Mats of floating plants and filamentous algae may also serve, as sediment traps. Wetland vegetation typically traps 80-90% of sediment from runoff (Gilliam 1994; Johnston 1991). Filtration of suspended solids also occurs in wetlands as water comes into contact with wetland vegetation. Stems and leaves provide friction for the flow of the water, thus allowing settling of suspended solids and removal of related constituents from the water column. Flocculation, precipitation and the uptake of fine sediments onto biofilms also operate to assist the removal of suspended solids by filtration. Sediment deposition and retention is variable depending on the wetland and wetland type, because deposition depends on flow rate, particulate size, and vegetated area of the wetland (Johnston 1991; Hemond and Benoit 1988).

With time, contaminants will accumulate in successively buried layers on the wetland soil, sediment and substrate as a result of physical removal processes. Sedimentation is generally considered to be an irreversible process, resulting in accumulation of solids and associated contaminants on the wetland soil surface. However, some resuspension may occur, on occasion, during periods of higher flows and correspondingly higher velocities in the wetland, from wind-driven turbulence, bioturbation resulting from animal activity, and gas lift. Gas lift results from production of gases such as oxygen, methane, hydrogen sulfide and carbon dioxide, produced by algae and plants as a result of photosynthesis and microorganisms in the sediment during decomposition of organic matter. If the amount of suspended or colloidal material in the water column becomes too great, sediment overloading of the wetland may occur, leading to the smothering of wetland plants and adversely affecting effluent water quality. Periodic sediment removal form the wetland will reduce the impacts of bioaccumulation and resuspension.

7.4.4 *Pathogen Removal*

Bacteria, viruses, fungi, protozoa and helminthes, diseases-causing organisms in humans, enter wetlands mainly through municipal sewage, urban stormwater, leaking septic tanks, agricultural runoff or animal feces. As some pathogens are charged they can attach to suspended solids as well as organic matter, and be entrained in the wetland system (Hemond and Benoit 1988). Once they have entered the wetland system, pathogen numbers can be effectively reduced by up to 5 orders of magnitude from wetland inflows (Reed et al., 1995).

The reduction of pathogens in treatment wetlands is due to a combination of physical entrapment, filtration, sedimentation, natural die-off, ultra-violet light ionization, excretion of antibiotics and toxins excreted plant roots and other microorganisms, biofilm adsorption flowed by natural die-off, predation by other organisms (protozoan and viral), and unfavorable water chemistry (Stott and Tanner, 2004; Werker et al., 2002; Newman, et al., 2000; Hemond and Benoit, 1988). Outlet fecal coliform concentrations were reported to be significantly greater in the winter than during the summer, which followed influent trends during winter and spring season (Newman et al., 2000). Increases in flow rates through the wetland system were also found to negatively affect fecal coliform and fecal streptococci removal (Perkins and Hunter, 2000)

Vegetated wetlands are generally more effective at removing pathogens than unvegetated ponds. This is may be due to the fact that wetlands provide a habitat for a diverse range of microorganisms which may be predators for pathogens. Vegetated wetland systems also provide greater surface areas for potential biofilm development along the stems and leaves of plants which can lead to pathogen removal via biofilm adsorption and filtration (Stott and Tanner, 2004).

7.4.5 *Nutrient Cycling and Removal Processes*

The complex cycles of carbon, nitrogen, phosphorous and sulfur involve a number of conceptual compartments which include the water column, soil, sediments, substrate/litter, plant roots, biofilms, plant stems and leaves. Often, a large fraction of these nutrients may be attached to other particles or are particulates. Hence, sedimentation and filtration processes can be very effective for the removal of nutrients from the flow and their retention within the wetland system. However, this is not always the case, as there are other physical, chemical and biological processes that can transform these nutrients, making them available again in a different form.

Generally, both biotic and abiotic processes are involved in the cycling of these nutrients within the wetland system.

7.4.5.1 Carbon Cycle

Wetlands are generally considered to act as a sink for carbon, since organic mater decomposition is stable and relatively slow, leading to the storage of carbon within wetland soil, sediment and litter. Storing carbon is an important function in the carbon cycle, considering that carbon dioxide is considered to be a green house gas, and its release increases concerns about global warming. Biological oxygen demand (BOD) is a measure of the oxygen required for the decomposition of organic matter and oxidation of some inorganic constituents such as sulfide. BOD is introduced into surface water through inputs of organic matter such as sewage effluent, surface runoff, and natural biotic processes (Hemond and Benoit 1988).

Algae are important in wetland systems and can contribute significantly to the long-term storage of carbon. The algae utilize dissolved carbon in the water during photosynthesis, incorporating it into their biomass and increasing the inflow gradient form the atmosphere, when the algae die, they sink, locking up transient carbon and removing it from the aerobic rapid biodegradation cycle to the slower anaerobic cycle which is bounded by less efficient anaerobic biological activities within the sediments.

The use of algal effluent wetland treatment systems as carbon sinks is a simple process which works on the basis of functional eutrophication and rely on an ecological equilibrium between the algae and aerobic bacteria of the wetland system. In essence, organic constituents are biologically decomposed by the aerobic bacteria which make use of the oxygen provided by algal synthesis, while the algae use the nutrients produced by the bacterial biodegradation for growth and photosynthesize producing more oxygen. The process is self-sustaining, as well as self-limiting and will result in characteristics of eutrophication leading to the eventual death of all component organisms, if the dynamics of the system are not well established or maintained. Hence, in this case, the removal of excess algal and bacterial biomass is crucial in maintaining wetland treatment system efficiency. Elements of the carbon cycling and removal are presented in Figure 7.10.

Photosynthesis and respiration are the two basic life processes responsible for maintaining the recycling of carbon and oxygen in nature. Carbon is transformed both under aerobic and anaerobic conditions. Green plants and algae are unique in that they have the ability to produce their own food using energy from the sun in a process called photosynthesis. The key component in the photosynthetic process is chlorophyll found in the chloroplast of algae, and of leaves and stems of higher plants. It does not become involved in the chemical reactions, but rather functions as a catalyst by

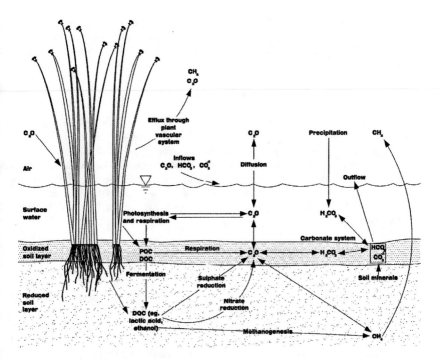

Figure 7.10 Carbon cycling and removal in a wetland system

capturing light energy from the sun. Equation 7.2 outlines the photosynthetic process.

$$6\ CO_2 + 6\ H_2O \xrightarrow{\ LightEnergy\ } C_6H_{12}O_6 + 6\ O_2 \qquad \text{(Eq. 7.2)}$$

Photosynthesis occurs in two stages. Light is required for the first stage, where the chlorophyll traps the light energy from the sun and this energy splits the water molecule. The hydrogen atom remains in the chloroplasts, and the oxygen atom is released. During the second stage, the hydrogen atoms combine with carbon dioxide, producing a new glucose (sugar) molecule. The glucose molecule stores the chemical potential energy which can be released during respiration.

All living organisms require a continuous energy supply to maintain their life functions. In aerobic environments, organisms are able to maintain this supply of energy through a process called respiration. Respiration occurs in the cytoplasm and mitochondria of cells in all living organisms. It is a series of chemical reactions where energy is released from the breakdown of glucose molecules. Plants store this energy in roots, stem, and leaves and eventually utilize it for its own metabolic processes or it is passed on to the animals that forage on the plants within the wetland system.

In respiration, chemical reactions occur that break down glucose molecules and transfer energy. Respiration requires oxygen, which combines with the atoms in the glucose releasing carbon dioxide and water. The following equation outlines respiration:

$$C_6H_{12}O_6 + 6\ O_2 \rightarrow 6\ CO_2 + 6\ H_2O \qquad \text{(Eq. 7.3)}$$

Organic carbon is biodegraded by facultative or obligate anaerobic microorganisms by fermentation under anaerobic conditions. This process does not release as much energy as aerobic respiration, and generally results in a less efficient assimilation of organic matter than under aerobic respiration. In fermentation, larger carbohydrates serve as the terminal electron acceptor, producing lower molecular weight, energy rich compounds, including alcohols such as ethanol (Eq. 7.4) and volatile fatty acids such as lactic acid (Eq. 7.5) (Lehninger, 1982). These can then be utilized by other microorganisms.

$$C_6H_{12}O_6 \rightarrow 2\ CH_3CH_2OH + 2\ CO_2 \qquad \text{(Eq. 7.4)}$$
$$C_6H_{12}O_6 \rightarrow 2\ CH_3CHOHCOOH \qquad \text{(Eq. 7.5)}$$

Other terminal electron acceptors such as nitrate (Eq. 7.6), iron (Eq. 7.7) and sulfate (Eq. 7.8) can be employed by facultative and obligate anaerobes instead of oxygen. The selection of the electron acceptor will depend upon the redox potential conditions of the local environment as illustrated in Figure 7.6. In nitrate reduction or denitrification, shown in Equation 7.6, carbohydrates are oxidized to produce carbon dioxide, water and nitrogen gas (vanLoon and Duffy, 2005).

$$5\ \{CH_2O\} + 4\ NO_3^- + 4H_3O^+ \rightarrow 5\ CO_2 + 11\ H_2O + 2\ N_2 \qquad \text{(Eq. 7.6)}$$

Equation 7.7 shows the reduction of ferric iron as the electron acceptor to ferrous iron which oxidizes carbohydrates.

$$\{CH_2O\} + 4\ Fe^{3+} + 5\ H_2O \rightarrow 4\ Fe^{2+} + CO_2 + 4\ H_3O^+ \qquad \text{(Eq. 7.7)}$$

Similarly, Equation 7.8 demonstrates the use of sulfate as the electron acceptor for the oxidation of acetate to carbon dioxide, hydrogen sulfide and gas (vanLoon and Duffy, 2005).

$$2\ \{CH_2O\} + SO_4^{2-} + H_3O^+ \rightarrow 2\ CO_2 + 3\ H_2O + HS^- \qquad \text{(Eq. 7.8)}$$

Methane can also be formed in wetlands due to the action of methanogens (Eq. 7.9), which are obligate anaerobes and can use carbon dioxide as an electron acceptor under very low redox potential conditions.

$$4\,H_2 + CO_2 \rightarrow CH_4 + 2\,H_2O \hspace{3cm} \text{(Eq. 7.9)}$$

The wetland gases produced via methanogenesis and sulfate reduction are generally released to the atmosphere when sediments are disturbed. The term "swamp gas" refers to the production of methane and hydrogen sulfide from carbon transformations in wetland systems.

7.4.5.2 Nitrogen Cycle

Nitrogen is the most common atmospheric gas, and a major constituent of living tissue. Sources of nitrogen in wetlands include organic matter, sediments, wastewater, fertilizers, and stormwater and agricultural runoff. Nitrogen is present in a variety of compounds with varying stability and oxidation states. The most important nitrogen species in wetlands are dissolved ammonia, nitrite, and nitrate. Other forms include nitrous oxide gas, nitrogen gas, urea, amino acids and amines. Nitrogen has a very complex cycle within wetland systems, involving multiple biotic and abiotic transformations as illustrated in Figure 7.11. In order to be permanently removed from the wetland system, organic nitrogen must undergo a series of reactions, which occur in spatially separate compartments of wetlands due to their respective transformation requirements.

Organic nitrogen in the form of amino acids or urea is biologically transformed to ammonium through the process of mineralization which results as a consequence of organic matter decomposition that results from both aerobic and anaerobic microorganisms. In aerobic substrates, organic nitrogen may mineralize to ammonium (ammonification) as shown in Equations 7.10 and 7.11 which plants and microbes can then utilize, adsorb to negatively charged particles, or diffuse to the surface.

$$NH_2CONH_2\ (organic\ nitrogen, urea) + H_2O \xrightarrow{\ microorganisms\ } CO_2 + 2\,NH_3 + H_2O$$
$$\text{(Eq. 7.10)}$$
$$NH_3 + H_2O \rightarrow NH_4^+ + OH^- \hspace{2cm} \text{(Eq. 7.11)}$$

Emergent macrophytes are efficient in the removal and storage of nitrogen in their root systems. Ammonium can become bound to organic matrices since the soils have negative charges and the ammonium ion is positively charged. Ammonium is a reduced compound and in the absence of oxygen it could not be transformed. It should be noted that high concentrations of ammonia are toxic to plants and would have a detrimental impact on wetland vegetation.

Nitrification is a two-step aerobic process catalyzed by *Nitrosomonas* and *Nitrobacter* bacteria which are thought to be the main nitrifying bacteria. This process generally takes place in the rhizosphere and in biofilms. In the first step, ammonium is oxidized to nitrite in an aerobic reaction catalyzed by *Nitrosomonas* bacteria as shown in Equation 7.12.

$$NH_4^+ + O_2 \xrightarrow{\text{Nitrosomonas}} H^+ + NO_2^- + H_2O \tag{Eq. 7.12}$$

In order to perform nitrification, the *Nitrosomonas* bacteria must compete with heterotrophic bacteria for oxygen. The BOD of the water must be less than 20 mg/L before significant nitrification can occur (Reed et al., 1995). The nitrite produced is then oxidized aerobically by *Nitrobacter* bacteria forming nitrate as per Equation 7.13.

$$NO_2^- + O_2 \xrightarrow{\text{Nitrobacter}} NO_3^- \tag{Eq. 7.13}$$

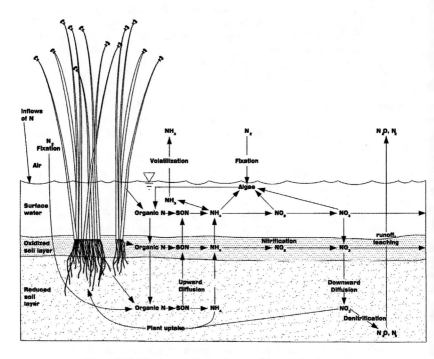

Figure 7.11 Nitrogen cycling and removal in a wetland system

Plants or microorganisms can assimilate nitrate. Alternatively, as the nitrated diffuses into an anoxic zone in soils and at the soil/water interface and under the appropriate redox potential conditions (Fig. 7.6), anaerobic bacteria may reduce nitrate to nitrogen gas in a process called denitrification.

Denitrification is a process performed by facultative anaerobes where nitrate is reduced to gaseous nitrogen. Equation 7.14 demonstrates the general reaction for denitrification. These organisms flourish in an aerobic environment but are also capable of breaking down oxygen-containing compounds such as nitrate to obtain oxygen in an anoxic environment. Examples include fungi and the bacteria *Pseudomonas* (Brodrick et al., 1988)

$$NO_3^- + Organic-C \xrightarrow{\text{Denitrifying Bacteria}} N_2 (NO \& N_2O)_{(G)} + CO_{2(G)} + H_2O \quad \text{(Eq. 7.14)}$$

Denitrification requires nitrate, anoxic conditions and a readily biodegradable carbon source. The nitrate must diffuse from the aerobic to the anaerobic sediments before denitrification can occur which can represent a limiting factor for the denitrification process. The gaseous nitrogen volatilizes and the nitrogen is eliminated as a water pollutant. Thus, the alternating reduced and oxidized conditions of wetlands complete the needs of the nitrogen cycle and maximize denitrification rates.

To close the cycle, nitrogen fixation is the conversion of nitrogen gas to ammonia or nitrate. Nitrate is the product of high-energy fixation by lightning, cosmic radiation, and meteorite trails, where atmospheric nitrogen and oxygen combine to form nitrate, which are carried to the earth's surface in rainfall as nitric acid. High-energy fixation accounts for little (10%) of the nitrate entering the nitrogen cycle. In contrast, biological fixation accounts for 90% of the fixed nitrogen in the cycle. In biological fixation, nitrogen gas is split into two free nitrogen molecules, which combine with hydrogen molecules to yield ammonium. The nitrogen fixation process is accomplished by a series of different microorganisms. The symbiotic bacteria *Rhizobium* is associated with the roots in the rhizosphere. To a lesser extent, some root-noduled non-leguminous plants also exhibit symbiotic relationships with bacteria. Some free-living aerobic bacteria, such as *Azobacter* and *Clostridium*, freely fix nitrogen in the soil. Finally, blue-green algae (cyanobacteria) such as *Nostoc* and *Calothrix* can fix nitrogen both in the soil and in water, yielding ammonia as the stable end product.

Wetland systems generally have a natural background level of organic nitrogen (1-2 mg/L) as a result of biomass decomposition (Kadlec and Knight, 1996). This background level is usually higher during the winter months and the nitrogen is generally present as an amine group in organic compounds. Wetlands are vertically stratified into zones or micro-environments which promote different nitrogen reactions. For example, plant roots, detritus/litter, soil, water column, biofilms on plant stems and leaves all foster particular nitrogen reactions. Nitrogen is preferentially taken up by wetland plants and microorganisms in the form of ammonium or nitrate, a portion is leached into the subsoil, while another fraction is volatilized as nitrogen gas to the atmosphere. Nitrogen may also be redistributed within the wetland by the decomposition of detritus and litter.

Nitrogen removal in wetlands is accomplished primarily by physical settlement, denitrification and plant and microbial uptake. The biological processes of nitrification/denitrification in the nitrogen cycle transform the majority of nitrogen in wetland systems, resulting in a removal rate of between 70% and 90% (Knight et al., 1993; Gilliam 1994). Plant uptake does not represent permanent removal unless plants are routinely harvested. The conservation of nitrogen is generally controlled by biological processes that may change with ecological succession (Craft, 1997). Operational and climatic conditions such as vegetation, constituent and hydraulic loading rate, media depth, DO concentrations, pH and temperature can play a significant role in these transformations which will ultimately affect the HRT design requirements of the system (Werker et al., 2002; Spieles and Mitsch, 2000; Raisin and Mitchell, 1995). In a study involving the synthetic wastewater addition to 28 microcosms over a period of 132 days, Hunter et al. (2001) noted that the average percent removal of NH_4^+-N was significantly greater in microcosms containing plants (67%) than in those without plants (29%). Average removal was significantly lower in microcosms subjected to variable hydraulic and constituent loadings where NH_4^+-N removal ranged from 51% to 83% in systems under varying and constant loading rates, respectively. Percent NH_4^+-N removal was significantly greater in microcosms with a 6-day retention time (80%) than in those with 2-day retention (53%). Elevated metal concentrations were also found to impact nitrogen uptake, where nitrogen cattail uptake rates were inhibited in the presence of metals (Lim at al., 2003). Research performed by Bachand and Horne (2000) demonstrated that water temperature and organic carbon availability affect denitrification rates and that longer retention times resulted in enhanced settlement of particulate organic N within the wetland.

7.4.5.3 Phosphorous Cycle

Phosphorus is one of the most important limiting constituents, where excess phosphorous can trigger algal blooms in a wetland system, leading to eutrophication. The phosphorus cycle is governed by geochemical processes which tend to act separately form biological processes (Craft, 1997). At any given time, a significant fraction of the phosphorus in wetlands is tied up in organic detritus and litter, as well as in inorganic sediments (Fig. 7.12). Phosphorus occurs as soluble and insoluble complexes in both organic and inorganic forms. Phosphorus retention is considered on to the most crucial attributes of wetland systems (Mitsch and Gosselink, 2000).

Phosphorus typically enters wetland systems with suspended solids or as dissolved phosphorus. Phosphorus removal in wetlands occurs as a result of plant, as well as soil and biofilm microbial uptake; adsorption by aluminum and iron oxides and hydroxides; precipitation of aluminum, iron, and calcium phosphates; and burial of phosphorus adsorbed to sediments or organic matter (Walbridge and Struthers 1993). In order to allow sufficient time for chemical and physical processes to take place, one of the most important design factors in phosphorous removal in wetland systems is the HRT (Raisin and Mitchell, 1995). Manipulating the hydrologic regime to increase phosphorous removal may be beneficial in wetland systems (Wang and Mitsch, 2000). Dissolved phosphorus is processed by wetland biota such as soil and water column bacteria, algae, and macrophytes, as well as geochemical mechanisms (Kadlec, 1997a; Walbridge and Struthers, 1993). The vegetative and microbial removal of phosphorus from wetland soils or water column is rapid and efficient, however, it only represents a short-term storage, because 35 to 75% of the phosphorous stored in this fashion will eventually be released back into the water upon dieback of algae, plants and microorganisms (Richard and Craft, 1993; Kadlec, 1997a; White et al., 2000). The potential for long-term storage of phosphorus through adsorption to wetland soil is greater than the maximum rates of phosphorus accumulation possible in biomass of the wetland biota (Walbridge and Struthers, 1993).

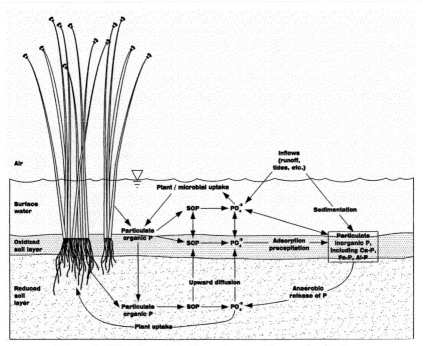

Figure 7.12 Phosphorous cycling and removal in a wetland system

Lake and reservoir sediments have been shown to act as phosphorous sinks (Sundby et al., 1992; Richardson and Craft, 1993; White et al., 2000) and similar cycling and removal processes are involved in wetland systems. A significant fraction of phosphorus associated with the suspended solids is deposited in wetlands (Walbridge and Struthers, 1993). Phosphorus-containing particles settle to the soil and are covered continuously by sediment, detritus and litter, which will, in time, place a fraction of the phosphorus too deep within the substrate to be reintroduced to the water column (Shatwell and Cordery, 1999). In this case, a fraction of the phosphorous will be permanently removed from biocirculation (Holtan et al., 1988). Wetland soils can, however, reach a state of phosphorus saturation, after which phosphorus may be released from the system (Richardson, 1985). The efficiency of long-term peat storage is a function of the loading rate and also depends on the amount of native iron, calcium, aluminum, and organic matter in the substrate (Shatwell and Cordery, 1999).

The effects of vegetation, wastewater level variations, hydraulic retention time (HRT) and media depth on phosphorous removal was investigated by Hunter et al. (2001) using microcosms for a period of 132 days. The percent removal of PO_4^{3m}-P was found to be significantly greater in microcosms with plants (42%) than in microcosms without plants (20%). Average PO_4^{3m}-P removals were significantly lower in microcosms with wastewater level fluctuations than in those without wastewater fluctuations (14% versus 71%), while media depth did not have a significant effect on nutrient removal. The percent PO_4^{3m}-P removal was significantly greater in microcosms with a 6-day retention time (55%) than in those with 2-day retention (29%). From their study, Hunter et al. (2001) concluded that the required design parameters were different depending on the nutrient being removed in systems simulated by these microcosms. In a study examining the operational HRT requirements of horizontal-flow and vertical-flow constructed wetland cells (hybrid systems), Cui et al. (2006) showed that removal efficiencies of total phosphorous in horizontal-flow and vertical-flow cells were improved significantly with the extension of HRT under a given temperature. The percent phosphorous removal was higher using a 3-day HRT compared to that of a 1 day HRT. Similarly, removal efficiencies were higher using 5-day HRT, reaching 95% to 98% average phosphorous removal efficiencies. However, this increase in removal efficiency was not significant in the 3-day to 5-day HRT increment. Hence, a 3-day HRT was recommended in the operation of the horizontal-flow/vertical-flow hybrid systems based on operation and maintenance requirements and economics.

7.4.5.4 Sulfur

Wetlands have been identified as net sinks for atmospheric sulfur and sulfur occurs in various states of oxidation within the wetland system. Like nitrogen, it is transformed through several pathways which are often mediated by microorganisms, as illustrated in Figure 7.13. Most of the sulfur in wetland soils is present as organic sulfur. Sulfur is rarely a limiting element for plants and microbial growth in wetlands because it is usually found in higher concentrations than required for biological processes. The dissolution of sulfide minerals under acidic conditions, the precipitation of metal sulfides under anaerobic conditions, the adsorption of metals by bacteria or algae, and the formation and destruction of organometallic complexes are all examples of microbiologically mediated sulfur processes that can take place in wetland systems.

Sulfate can be reduced to sulfide by anaerobic bacteria of the genera *Desulfovibrio* and *Desulfotomaculum* in wetland systems. Sulfate reducing bacteria utilize low molecular weight organic compounds such as lactate as their carbon source during anaerobic respiration which produces acetate as per Equation 7.15 (Cork and Cusanovich, 1979). The sulfate serves as an electron acceptor in the absence of free oxygen at low redox potentials (Fig. 7.6) forming sulfides, which are subsequently released to the atmosphere as hydrogen, methyl and dimethyl sulfides, or are bound in insoluble complexes with phosphate and metal ions in wetland sediments.

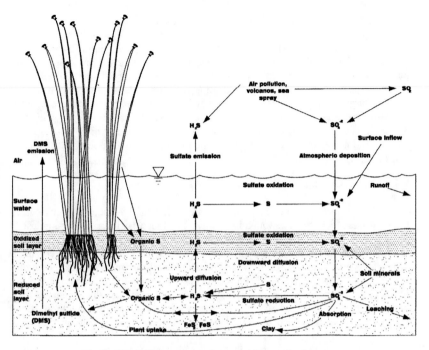

Figure 7.13 Sulfur cycling and removal in a wetland system

$$2\ CH_3CHOHCOO^- + SO_4^{2-} \rightarrow 2\ CH_3COO^- + 2\ HCO_3^- + H_2S \qquad (7.15)$$

The process of sulfate reduction seems to be important in wetland systems, where it benefits plants by regenerating nutrients and buffering soil pH. The metal sulfide complexes formed enable the precipitation of dissolved heavy metals in treatment wetlands. Although iron sulfides are the most abundant metal sulfides in most sediments, sulfides form highly insoluble minerals with a variety of other metals, including lead, copper, cadmium, zinc, silver, and mercury (Framson and Leckie, 1978). The precipitation maintains concentrations of soluble sulfides relatively low, which can be important because soluble sulfides are toxic to a number of organisms, including wetland plants, and can inhibit nutrient uptake (Kappelmeyer et al., 2004; Vamos and Koves, 1972; Goodman and Williams, 1961). However, zinc, copper and ferrous iron may become limiting as they precipitate with sulfides and are no longer available for plant uptake. Sulfide can be released when wetland sediments are disturbed, which causes the 'rotten egg' odor which is sometimes associated with wetlands.

Under aerobic wetland conditions, the oxidation of sulfur or sulfides by bacteria of the genera *Thiobacillus, Thiomicrospira, Sulfolobus* results in the formation of sulfuric acid, sulfur and sulfate as metabolic products. Oxidation reactions reduce sulfide concentrations, which will increase the ability of wetland plants to take up nutrients. However, it will promote nutrient immobilization by aerobic bacteria and phosphorus precipitation with iron oxides resulting in decreased overall nutrient concentrations. In the presence of pyretic and pyrrhotitic rocks, sulfur oxidation bacteria have been reported to accelerate the generation of acid mine drainage, where sulfide oxidation catalyzed by these bacteria can have reactions rates 1,000,000 times greater than the same reaction in an abiotic environment (Evangelou and Zhang, 1995).

7.4.5.5 Metal Cycling and Removal

Metals in wetland systems can originate from natural or anthropogenic sources. Many of these metals can be found in municipal and industrial wastewaters, industrial processing waters, as well as stormwater runoff and enter wetland systems in surface or groundwater flow. Hence, currently, anthropogenic inputs of metals exceed natural inputs. Trace amounts of some metals are required for animals (cobalt, chromium, copper, iron, manganese, molybdenum, nickel, selenium, tin, vanadium, zinc) and plants (boron, copper, iron, manganese, molybdenum, selenium, zinc). Some of these micronutrients, however, can be toxic to the organisms at higher concentrations. Other metals that have no known biological benefits included arsenic, barium, cadmium, lead and mercury. Many metals can be toxic to living organisms at relatively low concentrations (antimony, arsenic, beryllium, cadmium, cobalt, copper, lead, mercury, nickel, selenium, silver, tin and zinc). Furthermore, certain metals have a tendency to biomagnify and become concentrated at higher levels of the food chain.

In wetland systems, metals exist in colloidal, particulate, and dissolved phases. Colloidal and particulate metals may be found in 1) hydroxides, oxides, silicates, or sulfides; or 2) adsorbed onto clay, silica, or organic matter. Soluble metals are generally present in the form of ions or unionized organometallic chelates or complexes (Connell and Miller, 1984). The removal of metals in wetlands can occur through a number of processes including: precipitation; soil, sediment, substrate and biomass adsorption; as well as plant and microbial uptake.

The behavior of metals in natural waters is dependent on substrate sediment composition, suspended sediment composition, and water chemistry. The solubility of metals is primarily a function of water pH, the oxidation state of the mineral components and the redox environment of the system. Metals also have a high affinity for humic acids, organo-clays, and oxides coated with organic matter, thus the type of substrate and the number of sites on which the metal can adsorb is also significant in their removal from wetland systems (Connell and Miller, 1984).

Data on wetland performance for metal removal are highly scattered among various sites and highly variable in reported treatment efficiencies. From an operational point of view, metal removal efficiencies are highly dependent upon mass loading, influent constituency and concentrations, as well as the HRT (Goulet and Pick, 2001b; Stark and Williams, 1995). Removal efficiencies will also be affected by the water biogeochemistry of the wetland, wetland maturity, the surface area available for adsorption processes to take place, and competition between metal constituents for available sorption sites.

Aerobic Wetland Conditions

Under aerobic wetland conditions, iron and manganese can be chemically oxidized. Bacteria can transform reduced iron and manganese to oxidized forms, where oxygen is utilized as an electron acceptor. The pH and net acidity/alkalinity of the wetland are particularly important because pH influences both the solubility of the subsequent metal hydroxide precipitates and the kinetics of metal oxidation and hydrolysis. Abiotic oxidation reaction rates decrease a hundred-fold with each unit drop in pH, but microbial oxidation is usually accelerated by acidic conditions. Following iron oxidation, abiotic hydrolysis reactions precipitate iron hydroxides. Metal hydrolysis produces H^+ ions, but the alkalinity in the water can buffer the pH and allow metal precipitation to continue. Abiotic manganese oxidation occurs at pH levels higher than 8, while microorganisms are thought to catalyze this reaction at pH levels higher than 6 (Wildeman et al., 1993). Manganese oxidation occurs more slowly than iron oxidation and is sensitive to the presence of ferrous iron (Fe^{2+}), which generally prevents or reverses manganese oxidation. As a consequence, iron and manganese tend to precipitate sequentially, not simultaneously.

Anaerobic Wetland Conditions

Wetland soils are potentially effective sinks for metals, due to the relative immobility of most metals in the reducing wetland soil environment. Under the appropriate redox conditions (-150 mV to -250 mV), sulfate is reduced to sulfides by anaerobic bacteria (Eq. 7.15), which are subsequently bound in insoluble metal complexes. The metal sulfide complexes formed enable the precipitation of dissolved metals including cadmium, copper, iron, nickel, lead, manganese, mercury, silver and zinc (Framson and Leckie, 1978). It should be noted that slightly higher redox potentials than sulfate reduction, the microbial reduction of manganese (+225 mV to +120 mV) and iron (+120 mV to -150 mV) also takes place. Both elements are more soluble and more readily available to organisms in reduced forms which may lead to the release of dissolved iron and manganese and their associated ions from anaerobic sediments (Kadlec and Knight, 1996). Reduced iron (ferrous) causes a grey-green coloration of mineral soils, instead of the normal red of brown color of oxidized (ferric) conditions (Mitsch and Schnoor, 2000).

Adsorption

The adsorption of metals onto wetland soils, sediments, substrates and biomass is highly variable and dependent on the nature of the metal. The water chemistry of the wetland system is often the primary factor controlling the rate of adsorption and desorption of the metals. Adsorption removes the metal from the water column and stores the metal onto the solid surface, while desorption returns the metal to the water column. Metals can be desorbed from the solid surface as a result of increases in salinity, decreases in redox potential, or decreases in pH within the wetland system (Connell and Miller, 1984). Metals, such as chromium, copper, lead and zinc, form stronger chemical complexes (chemisorption) with organic matter present in the soil or water. Some metals such as chromium and copper may also be chemically-bound to clays and oxides of manganese, aluminum and iron. Nickel also binds with organic matter, iron and manganese, but may become re-mobilized under certain abiotic wetland conditions. Microorganisms, algae and other vegetative surfaces in constructed wetland systems also compete with cellulose and chitin as chelating surfaces for metal adsorption in a process often referred to as biosorption. The biosorption of metal ions and radionuclides by biomass is due to the non-specific binding or physical-chemical complexation reaction between dissolved metal species and charged cell surface-associated or extracellular polysaccharides and proteins (Kosolapov et al., 2004; Mullen et al., 1989; Volesky, 1990) as illustrated in Figure 7.14. The metal ions are, thus, bound to extracellular material or the cell wall. Bacterial cell walls and envelopes, and the walls of fungi, yeasts, and algae are efficient metal biosorbents that bind charged groups. Metallic ions can be removed by alive or dead biomass and the biosorption of some metals can be described by the linearized Freundlich adsorption equation (Eq. 7.16):

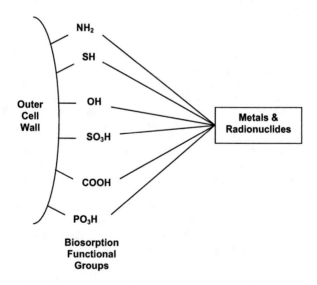

Figure 7.14 Potential extracellular or cell wall functional sites for metal and radionuclide biosorption

$$\log S = \log K + n \log C \qquad \text{(Eq. 7.16)}$$

where S is the amount of metal absorbed in $\mu mol/g$, C is the equilibrium solution concentration in $\mu mol/L$, K and n are the Freundlich constants (Mullen et al. 1989). Microorganisms are also able to remove heavy metals from solution by via metabolism dependent uptake into the cell. Some potentially toxic divalent ions are micronutrients at low concentrations and microorganisms possess the active uptake system to bind these ions.

Phytoremediation Processes
Phytoprocesses have the ability to use certain plant species for the removal or stabilization metals via phytoaccumulation, rhizofiltration or phytostabilization (Fig. 7.9, Table 7.2). Phytoaccumulation involves the uptake of metal contaminants from within the soil by the roots and their translocation into the above-ground plant tissues. Certain species, hyperaccumulators, can absorb large amounts of metals compared to most ordinary plants 50-100 times as much. Currently the best candidates for metal removal by phytoaccumulation are copper, nickel and zinc, since these are the metals most readily taken up by the majority of hyperaccumulator plants (Evans and Furlong, 2003). Rhizofiltration of metals involves the absorption into or the adsorption or precipitation onto plant roots of the metals present in the soil pore water. For phytoaccumulation and rhizofiltration, harvesting should follow once the plant tissues or roots have become saturated with metals and the collected biomass will require further treatment. The phytostabilization of metals is similar to both phytoaccumulation and rhizofiltration in that it also makes use of the uptake and accumulation by adsorption onto or precipitation around the roots of plants. What distinguishes this particular phytoremediation strategy is that unlike phytoaccumulation and rhizofiltration, harvesting the grown plants is not required to complete the process. Hence, it does not remove the metals, but immobilizes them by concentrating and containing them within a living system, thereby reducing their bio-availability and migration offsite.

The biomethylation of mercury, lead, tin, arsenic, selenium and tellurium resulting in the formation of volatile derivatives is also a possible phenomenon occurring in wetland systems. The methylation of metals and metalloids involves the enzymatic transfer of methyl groups to the metals, which can be mediated by a range of aerobic and anaerobic bacteria as well as fungi, algae and plants. The methylated compounds formed may then be eliminated from the system by evaporation. Wetland system conditions are favorable for the biomethylation of mercury. However, methylmercury is an organic form of mercury that is lipophilic, extremely toxic and readily bioaccumulated by aquatic organisms (Kosolapov et al., 2004).

Plant bioaccumulation rates and tolerance of metals varies considerably among plant species. Deng et al. (2004) investigated the accumulation of lead, zinc, copper and cadmium in 12 emergent rooted wetland plant species. They found that metal accumulation by wetland plants differed among species, populations and tissues. Populations grown in substrates with elevated metal concentrations generally accumulated significantly higher metals within the plants. Some species could accumulate relatively high metal concentrations in their shoots which suggested that internal detoxification metal tolerance mechanisms in the plants were involved. Matthews et al. (2004) also reported that the accumulation of zinc differed between populations of floating sweetgrass grown on different soil concentrations, but that these differences in accumulation did not appear to affect plant tolerance. Sparling and Lowe (1998) reported that that emergent plant species generally exhibited lower metal accumulations in their tissues than submerged or floating species. Metal accumulation in the emergent plants was also more frequently influenced by soil type. A comparison of heavy metal uptake by cattail and duckweed (Debusk et al., 1998) demonstrated that duckweed accumulated lead and cadmium more effectively than cattail, on a whole plant basis. Duckweed also appeared to be a more suitable candidate for use in a biomass harvesting process for metal removal than cattail because of its high productivity and its ease of harvesting. Similarly, the accumulation of zinc and lead in the pondweed and duckweed was found to be greater than in common reed, when compared on a whole plant basis by Peltier et al. (2003).

Many plant species that are able to accumulated metals exhibit similar metal uptake patterns in concentrating metals primarily in root tissues (Weis and Weis, 2004; Karathanasis and Johnson, 2003; Peltier et al., 2003; Polprasert et al., 1996), suggesting that an exclusion strategy for metal tolerance widely exists in plants (Deng et al., 2004). Weis and Weis (2004) found that some plant species retained more of their metal accumulation in below ground plant structures, while other species redistribute a greater proportion of metals into above ground tissues. They proposed that the storage of metals in root tissues would be most beneficial for the phytostabilization of the metals which tend to be least biologically available when concentrated below ground.

Sparling and Lowe (1998) noted that different metals were influence by different removal mechanisms in wetland systems. All species of wetland plants accumulated manganese and boron, which are essential elements and can be actively assimilated. They were also found to accumulate strontium, which is not a required mineral for plants but is chemically similar to calcium and often assimilated when calcium availability is low. Submerged plants were also reported to accumulate barium, which is also in the same chemical group as strontium and calcium and may be assimilated in a similar manner. Karathanasis and Johnson (2003) noted that aluminum and iron were primarily in the roots of the plants, while manganese was more mobile throughout the entire plant. In another study by Walker and Hurl (2002), zinc lead and copper exhibited significant reductions in dissolved concentrations due to metal sulfide formation. Chromium removal by sedimentation was found to be strongly dependent on its association with organic matter. Arsenic concentrations initially decreased, but showed an increase in concentration as the flow progressed through the wetland, which was attributed to the fact that arsenic is strongly associated with sulfur and carbon in organic compounds and, thus, forms relatively stable complexes with organic matter.

Many studies have been conducted to assess the metal removal processes of various metals within different types of wetland systems. Several researchers have reported that although plants bioaccumulated many of the metals introduced in the system, biological accumulation was often only a small fraction of the total metal removal compared to metal retention in wetland substrates and sediments. (Karathanasis and Johnson, 2003; Peltier et al., 2003; Walker and Hurl, 2002; Goulet and Pick, 2001a; Mays and Edwards, 2001; Debusk et al., 1998; Mungur et al., 1995). In wetlands containing organic sediments, most metals were typically present in sediments as immobile residual forms of metal sulfides, limiting the bioavailability and toxicity of these elements due to the neutral to alkaline pH and reducing biogeochemistry of the treatment system in the sediment layer (O'Sullivan et al., 2004a; O'Sullivan et al., 2004b; Debusk et al., 1998). O'Sullivan et al. (2004b) found that the high iron content of both sediment and suspended solids suggest that removal of zinc an lead from the water column could likely be due to contact of these metals with iron precipitates either through co-precipitation or adsorption, which was support by the presence of dissolved zinc and lead in sediment pore waters during iron reduction.

Macrophyte foliage can attenuate the flow of surface discharges and facilitate the sedimentation of metal contaminated particles and the uptake of dissolved metal species. Rhizomes and roots can reduce the resuspension of particles during intermittent discharges (Mungur et al., 1995). These findings would suggest that the main plant contribution in wetland systems may be through substrate stabilization, microbial attachment and rhizosphere oxidation rather than phytoaccumulation. Therefore, plant selection criteria for high metal load wetlands should mainly be based on metal tolerance and rhizosphere surface area rather than metal phytoaccumulation (Karathanasis and Johnson, 2003).

7.4.5.6 Organic Compound Removal

In addition to the easily-degradable organic compounds collectively referred to as BOD, there are a number of xenobiotic and/or toxic natural and anthropogenic organic compounds that may be present in wastewater or runoff entering wetland systems, including many types of pesticides, solvents and lubricants. Quantitative data on the removal of organics, such as pesticides and petrochemicals, are limited. An increasing number of wetland sites are being designed for the treatment of various target organic chemicals. These may present a difficult set of problems due to the potential toxicity to plants and the receiving environment, as well as limitations in aerobic and anaerobic degradation due to compound complexity. Many wetland soils are organic in nature and possess an affinity for introduced organics via sorption and other binding mechanisms. The removal mechanisms for organic compounds from wetland systems are phytovolatilization, volatilization, sedimentation, sorption, biodegradation, phytodegradation and rhizodegradation (Table 7.2).

Both mineral and organic soils may be able to contribute to the adsorption of organic compounds via chemisorption or physical adsorption. Microorganisms are capable of degrading most classes of organic pollutants, but the rate of degradation varies significantly, depending on the chemical and structural properties of the organic compound, and the chemical and physical environment of the soil. After xenobiotics enter the wetland system, the partitioning of compounds from the aqueous phase into biota and/or the partitioning of xenobiotics from the aqueous phase into the sediment phase must take place to achieve some degree of treatment. Not all organic compounds are equally accessible to plant roots. The ability of the roots to take up organic compounds can be described by the hydrophobicity or lipophilicity of the organic compounds of interest. This parameter is often expressed as the log of the octanol-water partitioning coefficient, K_{ow}. Direct uptake of organics by plants is a surprisingly efficient removal mechanisms for moderately hydrophobic organic compounds, where, generally, the higher the K_{ow} of a particular compound, the greater the potential for plant root uptake. Organic compounds that are highly water soluble (log K_{ow} less than 0.5) are generally not highly absorbed in the roots or actively transported through plant membranes (Suthersan, 1997).

Once an organic compound has been taken up, the plant can store (sequestration) the chemical and its fragments in new plant materials via a process known as lignification, or it can volatilize, metabolize or mineralize the chemical to carbon dioxide, water and by-products. Phytovolatilization involves the uptake of the organic contaminants by plants and their release into the atmosphere, typically in a modified form, via transpiration. Phytodegradation occurs either internally, having been taken up by the plant, or externally, using enzymes secreted by the plants, and involves the biological breakdown of the organic compound. The complex organic molecules are biodegraded into simpler substances that can, subsequently, be incorporated into the plant tissues. Rhizodegradation is the biodegradation of contaminants in the soil by microorganisms enhanced by the character of the rhizosphere. This region supports high microbial biomass and a high level of microbial activity which tends to increase the rate and efficiency of the biodegradation of organic compounds within the rhizosphere compared with other wetland soil and water column communities. In addition, mycorrhizae fungi associated with the roots may also play a significant role in metabolizing organic compounds. They have unique enzymatic pathways that enable the biodegradation of organic substances that could not be otherwise transformed by simple bacterial pathways.

Each year, there is an increasing number of organic compounds that are introduced to the environment, that have the potential to become environmental issues. Many of these have not been studied with respect to their treatability in wetland systems. The addition of these anthropogenic organic compounds to wetland systems for remediation is contributing to an increasingly complex array of hydrocarbons and organic chemical reactions. Hydrocarbon molecules are susceptible to fragmentation and chemical conversion in wetland environments, predominantly via microbiologically mediated pathways. Partial conversion may occur via hydrolysis, dealkylation and ring cleavage or the removal of amino, nitro, chlorine, hydroxyl, acid or thio groups from the parent molecule. Oxidative processes ultimately produce carbon dioxide and water while anaerobic processes will enhance reductive dechlorination and may, ultimately, result in methane generation (Suthersan, 2002).

Polycyclic Aromatic Hydrocarbons (PAHs)

Wetland systems have all the basic elements necessary the remediation of chlorinated alkenes and alkanes. These include high organic carbon content in the sediments to bind the contaminants, high microbial density and diversity in the sediments to biodegrade contaminants, and both anaerobic and aerobic conditions to ensure that contaminants can be fully degraded without the accumulation of potentially toxic intermediates. Aliphatics can also be degraded, but more slowly when molecular weights are high. Aromatics follow a similar pattern, with PAHs degrading more slowly than benzene, and those with more than 3 rings cannot support microbial growth (Zander, 1980). The existence of the extracellular plant enzymes has allowed for the wetland treatment to be extended to chemicals such as chlorinated solvents, explosives and herbicides. This process depends on the direct uptake of contaminants from soil water and the accumulation of resultant metabolites within the plant tissues. It is important that the metabolites which accumulate be nontoxic, or significantly less toxic than the original contaminant. In the rhizosphere, pesticides, trichloroethylene, PAHs and petroleum hydrocarbons can be degraded at accelerated rates because of the enhanced rhizosphere environment.

Polprasert et al. (1996) achieved an overall phenolic compound removal of 99% from free water surface (FWS) constructed wetland operating at a HRT of 5-7 days, where biodegradation, plant uptake and adsorption accounted 76-78% of the removal and volatilization approximately 20%. The phenolic compounds were found to accumulate primarily in the root zone of cattail plants. Three pilot-scale subsurface flow (SSF) constructed wetlands (2 vegetated, 1 unvegetated) were utilized to determine their potential use in the removal of complex halogenated and nonhalogenated organic compounds represented by benzoic acid (Zachritz et al., 1996). At higher influent benzoic acid concentrations the double stage vegetated reactor did not exhibit increases in effluent benzoic acid, while the single stage vegetated and double stage unvegetated systems showed marked increases in effluent concentrations at this higher influent concentrations. Comparison of microbial dominance/diversity of the systems indicated a rich microbial population containing facultative anaerobes and fermenters, as well as strict anaerobes and aerobes that could potentially adapt to a variety of organic substrates, but little difference in microorganisms was noted between the vegetated and unvegetated systems. The performance of these systems suggested that the staging of treatment components might be advantageous in certain treatment schemes involving chlorinated organics and other complex compounds, to allow for the reaeration of substrates and to mediate the sequence of anoxic, aerobic and anaerobic environments which would enhance the treatment of these complex organic compounds.

An experimental oil spill was applied to an impacted wetland to evaluate bioremediation as a remedial action in natural wetland systems (Mills, 1997). The removal of target petroleum compounds such as PAHs (naphthalene, phenanthrene), n-C18, n-C17, phytane and pristine was monitored, and the results indicated a significant biodegradation of petroleum compounds. The addition of an inorganic nutrient (N) and inorganic nutrient (N) + alternative electron acceptor (nitrate) were found to enhance the degradation. The nutrient + alternative electron acceptor demonstrated a greater rate of degradation, but all treatments generated similar overall removals. In another study, Giraud et al. (2001) investigated the possible role of fungi in the remediation of PAHs (fluoranthene and anthracene) in contaminated wetland systems. Using pilot-scale constructed wetlands, they noted an increase in fungal activity in site contaminated with PAHs compared to a control wetland. Forty fungal species were isolated and identified from sediments of the contaminated and control wetlands. Fluoranthene was degraded by 33 species but only 2 were able to degrade anthracene. The study showed that fungal species were able to adapt to the contaminated environment, and that the biodegradation ability of fungi was not related to their extracellular phenoloxidase. Rather, they were thought to act synergistically with wetland bacteria and plants to remove PAHs by adsorption and biodegradation.

Chlorinated Compounds

Experiments to determine whether natural attenuation of TCE in the wetland was possible in microcosm study were undertaken by Bankston et al. (2002). A C^{14} tag was employed to identify the fraction of TCE bound to soil and water as a result of adsorption and sedimentation, volatized TCE, CO_2 produced by mineralization of TCE, and as carbon incorporated into the plant material. Volatized TCE accounted for more than 50% of the recovered label. In microcosms without plants, CO_2 represented 3.2 to 15.6% recovery, in organic and sandy soils respectively, indicating that TCE was mineralized by indigenous microorganisms. Microcosms vegetated with cattail exhibited an increased production of CO_2 to 5.3 in organic soil, which may be due to the presence of the rhizosphere in vegetated systems leading to enhanced mineralization. In laboratory microcosm experiments, Lorah and Voytek (2004) assessed the biodegradation pathways for 1,1,2,2-tetrachloroethane and 1,1,2-trichloroethane and the associated microbial communities in anaerobic wetland sediments. They observed greater 1,1,2-trichloroethane dichloroelimination in microcosms constructed with sediment that was initially iron-reducing and subsequently iron-reducing and methanogenic. These generated 2 times as much vinyl chloride as microcosms constructed with sediment that was only methanogenic. The microcosms with higher vinyl chloride production also showed a substantially higher rate of vinyl chloride degradation. Measurements of redox-sensitive constituents 1,1,2,2-tetrachloroethane and its anaerobic degradation products in field wetland pore water also showed greater production and degradation of vinyl chloride with concurrent methanogenesis and iron reduction. Microcosm and field results also demonstrated that complete reductive dechlorination could take place under methanogenic conditions without ethane accumulation in the wetland sediments. This was attributed to the presence of organic matter in wetland sediments providing a sustainable source of electron donors for reductive dechlorination. Kowles and Stein (2004) applied three polar organic solvents (acetone, tetrahydrofuran and 1-butanol) to experimental microcosm SSF constructed wetlands. Solvent removal was typically higher in vegetated systems. There was a noted seasonal variability in remediation where removal rates of all solvents were slower at colder temperatures, but the seasonal effects were more pronounced in the unvegetated systems. The vegetation and seasonal effects were believed to be due to plant and season induced variation in available rhizosphere oxygen.

Pesticides and Herbicides
    The transformation of alachlor and atrazine has been show to be catalyzed by inorganic sulfur present in anoxic pore water of salt marsh sediments. The rates of sulfhydrolysis in these environments may greatly exceed the rates of other removal processes (biotic or abiotic) for these compounds and may represent and important sink for herbicide compounds in salt marsh environments (Sutherson, 2002). Anderson et al. (2002) evaluated the mineralization of atrazine in two wetlands; a natural wetland which does not directly receive agricultural runoff and constructed wetland which was exposed to agricultural runoff. Atrazine is typically associated with suspended solids which include soil particles from runoff and microorganisms. The constructed wetland showed 70-80% mineralization of atrazine, while the natural wetland showed a low potential for atrazine mineralization. The highest levels of mineralization were localized in the top 5 cm zone of the wetland sediment and sediment activity near the outflow was similar to that near the inflow, which would indicate a distribution of atrazine-mineralizing organisms by depth with the highest activity located at the sediment-water interface. However, overall mineralization activity was found to be more pronounced near the inlet, which would imply that the mineralization of atrazine in the wetland could be attributed to the actions of microorganisms introduced from the river inflow as well as by microbial populations already established in the sediment zone. Other results also suggested that the level of activity may be depended upon seasonality of atrazine runoff entering the wetland, temperature, flow rate, suspended solids, plant and microbial community diversity and sediment organic carbon content.

    In a surface flow wetland study at a container nursery, Runes et al. (2003) examined the treatability of atrazine. Sediment-water column experiments indicated that sorption was an important mechanism for atrazine retention in the constructed wetland. They subsequently determined organic carbon adsorption coefficients for atrazine degradation products from equilibrium adsorption isotherms and reported that the coefficients decreased in the order of hydroxyatrazine > deisoproplyatrazine > atrazine > deethylatrazine. MPN assays demonstrated the existence of a low-density population of microorganisms with the potential to mineralize the atrazine ethyl side chain. The stimulation of simazine degradation via nitrate or sulfate reduction in a wetland microcosm was investigated by Slem et al. (2004). Vegetated microcosms significantly enhanced the removal of simazine because of higher levels of organic matter and organic acids which enhance adsorption and microbial degradation. Denitrification and sulfate reduction microbial processes were not found to have a significant effect on simazine. Measured metabolites indicated that aerobic microbial degradation of simazine was likely taking place in the rhizosphere, where microorganisms in the vegetated microcosms initially cleaved an ethyl group from simazine to form deisoproplyatrazine which would be an aerobic process.

Nitrogenous Compounds

Contamination of soil and water with explosives, particularly TNT, is a widespread problem on military sites, explosives producing plants and ammunition factories. A study conducted by Best et al. (1997) indicated that the presence of plants enhanced the removal of explosives, TNT and TNB, in water. Reed canary grass, coontail and pondweed were found to be most effective. The plants degraded TNT to reduced by-products and other metabolites. Submerged plants were identified as having the highest explosives removal activity. However, the biomass production of submerged plants was reported to generally be 5 to 10 times less than that of emergent plants per unit area. Thus, in plant selection for wetland construction, consideration of explosives removal potential and biomass production were considered essential. Sikora et al. (1997) reported that the anaerobic degradation of TNT and TNB in anaerobic SSF constructed wetland systems was rapid and similar to removal rates in parrot feather wetlands. Vegetated and unvegetated anaerobic SSF systems were found to be the only systems that provided significant reduction of RDX and HMX.

Vegetated systems with parrot feather had no significant impact on removal rates of explosives in the anaerobic SSF systems. The effectiveness of anaerobic SSF systems indicated that anaerobic SSF constructed wetlands could be established as the primary treatment for remediation with the initial addition of a carbon source. With time, it was surmised that the need for carbon supplementation would be reduced due to the carbon exudates and redox lowering potential of certain plants like canary grass. Results of a study employing 1 m^3 constructed wetland microcosms for the remediation of TNT-contaminated water were presented by Gerth et al. (2003). The constructed wetland systems required an acclimatization period of more than 8 weeks before they reached a steady state. The biological degradation efficiencies of microorganism for TNT differed depending on the carbon source added to the system. Molasses was found to be a superior carbon source for TNT cometabolism compared to wood chips. Reduction of TNT concentrations of 45% were observed without the addition of a carbon source, and 80% for 2,6 DNT. The addition of molasses increased the efficiency to 95% with outflow concentration meeting target effluent criteria.

7.5   Case Studies

7.5.1 Wetland Treatment of Explosives at the Iowa Ammunition Plant

The Iowa Army Ammunition Plant (IAAP) is located in Middletown, Iowa. IAAP is a 20,000 acre (8,094 ha) site, of which 8,000 acres (3,237 ha) are used for both munitions production and testing (Best et al. 1998; Best et al. 1997a; Best et al. 1997b). In 1948, a concrete impoundment was constructed near Line 1 to contain explosives production wastewater in a lagoon having an area of about 3.6 acres (1.5 ha). The five-acre Line 800 Pink Water Lagoon Area was surrounded by an earthen berm and at one time was also used for sludge disposal. The lagoon basins and underlying groundwater had been contaminated with 2,4,6-trinitrotoluene (TNT) and hexahydro-1,3,5-trinitro-1,3,5-triazine (RDX) as well as many other compounds and heavy metals. The action levels were determined by the explosives contamination. Soil action levels were set at 47 mg/kg for TNT and 1.3 mg/kg for RDX. Action levels for aqueous concentrations were set at existing EPA health advisory lifetime levels of 2 mg/L for both TNT and RDX.

Preliminary studies investigating the potential of phytoremediation and active remediation at the site were conducted as a coordinated effort by the U.S. Army Corps of Engineers, Waterways Experiments Station, and the University of Iowa. Excavation was used initially to remediate the most highly contaminated areas. Approximately 100,000 yd^3 (76,455 m^3) of soil were excavated from the Line 1 (24,000 yd^3 (18,350 m^3)) and Line 800 (76,000 yd^3 (58,106 m^3)) areas in early 1997. Following excavation, the potential impact on the local groundwater supplies was expected to have been greatly diminished and the U.S. Army Corps of Engineers anticipated a level of residual contamination equal to or less than the action levels. As a precaution, residual contamination would require additional treatment. Studies were then conducted to assess the potential use of phytoremediation biotechnologies with poplars as a final polishing step for soil and the use of wetland systems to treat residual contamination in surface and ground waters. The process of selecting the wetland treatment approach and obtaining construction and operational approvals included many steps and investigations. A full-scale wetlands treatment system is currently in place at the IAAP site.

Research was conducted on aquatic systems at Waterways Experiments Station in Vicksburg, Mississippi (Best et al., 1997; Larson, 1997; Best et al., 1998). These screening studies also investigated the potential of phytoremediation, from the perspective of aquatic-based systems. The goals of the research were as follows:
- Quantify the ability of aquatic and wetland plants to treat TNT and RDX containing surface water or groundwater,
- Investigate the transport and fate of the explosives in plants, and
- Determine the degradation by-products and distribution in sediments and potential phytoremediation species.

From the initial wetland research involving the screening of plants, a number of species were determined to have the capability of removing TNT and RDX from aqueous systems. Plants that exhibited the greatest potential for use in wetland systems were the submerged plants coontail and American pondweed, as well as the emergent plants common arrowhead, reed canary grass, and fox sedge. TNT removal occurred at a higher rate than RDX removal. In the laboratory studies, TNT was not detected in plant tissues and TNT degradation products were also below detection limits. RDX was found to bioaccumulate in a number of plants, reaching concentrations up to 1000 mg/kg dry weight (Best et al., 1997). RDX had also been detected in plant tissues from the IAAP site prior to the initial excavation in arrowhead roots (7 mg/kg dry weight) and in reed canary grass shoots (10 mg/kg dry weight) (Schneider et al., 1995). The detection of RDX in plant tissues was a concern going into implementation at the IAAP.

Following excavation of the Line 1 and Line 800 sites at the IAAP, the excavation pits were converted to constructed wetlands. The wetlands were considered to be a means to remediate the collected groundwater, provide ecological enhancement to the area, and avoid the costs associated with backfilling of the area with clean material. Sediment/seed material from nearby Stump Lake was used in each wetland to introduce indigenous species. Following the addition of the sediment, the wetlands were allowed to fill with the infiltrating groundwater.

Desorption of RDX from the contaminated soils tainted the groundwater that filled the wetland resulting in a rebound effect in the RDX concentration. In January 1998, the Line 1 wetland RDX concentration reached 760-830 mg/kg. The RDX concentration declined in the wetland over the following months, reaching 1.6 mg/kg in August 1998. The following December, RDX reached 22 mg/kg in the Line 1 wetland. However, the concentrations were non-detect (< 0.25 mg/kg) the following summer. The water from the wetland was only released when RDX concentrations were below the 2 mg/kg action level. The seasonal fluctuations in RDX concentrations were also observed in the Line 800 impoundment wetland. The Line 800 wetland reached 190-760 mg/kg in January 1998. RDX concentrations were below detection limits during the following summer months, and rebounded to 1.9 mg/kg in November 1998.

Possible reasons for the observed RDX increases in concentration during the winter month include decreased plant activity and biomass, decreased rates of enzymatic reactions, decreased physical-chemical reaction rates, decreased photolytic reactions as incident sunlight decreases, and/or increased concentration in the infiltrating groundwater. Residual soil and sediment concentrations of RDX are greater than 1.3 mg/kg soil, which would suggest that this residual concentration could contribute to ground and surface water concentrations greater than 2 mg/L.

Bioaccumulation of RDX in plant tissues was a concern at the site. Thus, a sampling program of the plant material was undertaken in June 1999. Duplicate testing was conducted by the environmental engineering laboratories at the University of Iowa in Iowa City (Just, 1999) and the testing laboratories at the U.S. Army Corps of Engineers Waterways Experiments Station, Vicksburg Mississippi (Larson, 1999). The testing revealed that RDX and TNT were both below detection limits (< 1 mg/kg) in the tissues of the plant species tested, which included reed canary grass, arrowhead water plantain, and coon tail. 1,3,5-trinitrobenzene, a photodegradation product, was detected in some plant tissues at < 30 mg/kg (dry weight), which suggested that reactions other than those involved in plant-mediated processes are contributing to the degradation of the explosives.

7.5.2 Wetland Treatment of BTEX at the Former BP Refinery, Casper, Wyoming

An engineered wetland system implemented by British Petroleum (BP) in Casper, Wyoming is currently one of the largest and most recent remediation wetland implemented in the United States (Wallace and Kadlec, 2004; Wallace et al., 2001). The site was one of the oldest and largest Amoco refineries in the West, and began operating in 1912. It was the largest refinery in North America during the 1920s and continued its operation until 1991. As a result of common operating practices during the first 50 years of operation, much of the site is underlain with residual hydrocarbons. Remediation efforts were complicated because the North Platte River serves as the northern border of the 346-acre (140-hectare) site and historically, water table elevations have fluctuated in response to the water level of the North Platte River. These fluctuations have generated a "smear zone" containing residual hydrocarbons that extends from 5 ft (1.5 m) above the water table to 10 ft (3 m) below the water table. Since 1981, over 1,306,650 ft^3 (37,000 m^3) of LNAPL have been removed from the groundwater. Faced with the rising cost of environmental cleanup, The Amoco Oil Company decided to close the refinery in 1993. In 1998, the Wyoming Department of Environmental Quality finalized a Consent Decree establishing a framework for site remediation, where British Petroleum (BP) and the City of Casper agreed to convert the former refinery site into a golf course and office park with a trail system along the North Platte River. Because of the time required for remediation (50 to 100 years), BP became very interested in biological treatment processes due to the potential cost savings. A constructed wetland was identified as a low-maintenance system compatible with the intended golf course use of the property. The WDEQ permit requires that effluent benzene levels be less than 0.05 mg/L.

A pilot system was designed in 2002. It was estimated that a 3-tanks-in-series (3TIS) areal rate constant (k_A) of 377 ft/yr (115 m/yr) would be applicable for the treatment of BTEX and benzene. For the initial design criteria of 282,520 ft^3/d (8,000 m^3/d), this would have required 27 acres (11 ha) of subsurface-flow wetlands, which could not be accommodated at the site. This initial feasibility study also recognized the need for cascade oxygenation for iron control, together with a settling basin for iron precipitate collection. Further, all examples available in the literature at that time were for non-aerated wetlands, and since aeration is believed to improve treatment of petroleum hydrocarbons, a pilot system was constructed to assess aerated wetland performance.

Four subsurface-flow treatment cells (operating in parallel) were established. Each cell was 5.6 ft (1.7 m) wide by 23 ft (7 m) long by 3.6 (1.1 m) deep. Cells were loaded at a nominal flow rate of 190 ft^3/day (5.4 m^3/day), resulting in a nominal hydraulic retention time of one day. In each pilot cell, influent was introduced across the bottom area of the wetland, flowed upward through the (lower) gravel and (upper) sand bed, and then across the upper portion of the sand bed to the outlet. Various species of wetland plants were transported from the University of Wyoming-Laramie greenhouse for the pilot systems, including species of willows (*Salix*), reed (*Phragmites*), bulrush (*Schoenoplectus*), rush (*Juncus*), and dogwood (*Cornus*). Two of the four cells were vegetated using a 6-inch (15-cm) thick sod layer consisting of plant detritus interlocked with roots and rhizomes harvested from a nearby wetland (Soda Lake) which had a mature assemblage of wetland vegetation adapted to alkaline conditions.

The pilot system was operated between August and December 2002. During the course of operation, all four pilot cells were operated with aeration for at least part of the study period. The presence of wetland sod and aeration both improved treatment performance. The water temperature decreased throughout the period of pilot operation, however no impact on removal rates was observed. Mean rate constants based on assumed 3TIS flow were established for benzene, BTEX, TPH, and methyl *tert*-butyl ether (MTBE) as summarized below.

The full-scale system was designed to treat 211,890 ft³/day (6,000 m³/day) of contaminated groundwater. Because potential fouling of the SSF wetland media was identified during the pilot operation, a cascade aeration system (for iron oxidation) and FWS wetland (for iron precipitation) was added to the system. Federal regulations require that the benzene concentration of the water released to the wetland treatment system be less than 0.5 mg/L. An enclosed, ventilated cascade aerator was designed to reduce benzene levels in the oil/water separator effluent. Volatile organics stripped from the water column are routed to a soil-matrix biofilter for degradation. Average benzene reduction in the cascade aerator in 2004 was on the order of 54%.

The primary function of the FWS wetlands was to precipitate and remove iron to prevent fouling of the SSF wetland media. Reduced iron present in the groundwater is oxidized in the cascade aerator, which due to the pH of the water (8.3), forms ferric oxyhydroxide precipitates. The FWS wetland is divided into two parallel treatment cells with a combined surface area of 1.5 acre (0.6 ha). The water depth in each cell can be independently adjusted from 0 to 2 ft (0 - 60 cm), with a typical operating depth of 1 ft (30 cm). The FWS wetland cells were planted with Hardstem Bulrush (*Schoenoplectus acutus*), Cattail (*Typha angustifolia*) and Bur-reed (*Sparganium eurycarpum*).

The thousand-fold scale-up from the pilot until to the full-scale treatment system presented a number of design challenges, first and foremost of which was flow distribution. In order to apply the rate constants developed from the pilot, the full-scale system had to be designed such that the degree of short-circuiting and dispersion was less than that observed in the pilot study. The vertical flow path was abandoned in favor of a center-feed, horizontal, radial-flow configuration. In order to operate ice-free in the winter, a 6-inch (15 cm) insulating mulch layer was installed on top of the SSF wetland cells. The SSF wetland cells were also constructed with an aeration system capable of uniformly distributing air throughout the wetland basins. Air distribution to the wetland bottom was via aeration tubing on 2 ft (60 cm) spacing.

The full-scale system was commissioned in May 2003 and has been hydraulically loaded at approximately 95,350 ft³/day (2,700 m³/day) (45% of design). In 2004, the system was producing non-detect levels of petroleum hydrocarbons, with a BTEX mass load of about 15% of the design mass load. This project demonstrated, through the application of both pilot-scale and full-scale wetland treatment systems, that removal rates for petroleum hydrocarbons in aerated subsurface flow wetlands is considerably higher than in non-aerated wetlands. Areal rate constants (3TIS k_A values) BTEX degradation was measured in the pilot system to be 800 ft/yr (244 m/yr) for cells operating without aeration and mulch, and increased to 1,170 ft/yr (356 m/yr) for cells with aeration and mulch. The full scale system, which uses aeration and mulch, has a 3TIS k_A value of ~1,150 ft/yr (350 m/yr). Based on data from the pilot and the full-scale system, there appears to be little, if any, temperature effect on petroleum hydrocarbon degradation rate constants.

7.5.3 Wetland Treatment of Runway Runoff at the Heathrow Airport in Middlesex, UK

To ensure wintertime flight safety, de-icers are used to prevent ice formation on aircraft and runways. During the winter frost, ice and snow can accumulate on aircraft as well as on runways and taxiways and some method of removal is necessary before aircraft take-off. A widely used method of removal is to apply some form of de-icer and anti-icer. Glycol-based de-icer contains mainly propylene glycol or ethylene glycol, while anti-icers often consist of urea and a variety of acetate and formate-based products. Hence, the runoff from airfields containing de-icers and anti-icers has the potential to impose a significant oxygen demand on receiving waters, leading to degradation of the resource. A treatment system was commissioned by BAA (formerly the British Airports Authority) to attenuate airfield runoff contaminated with de-icers and anti-icers, as well as other contaminants such as heavy metals, nutrients, suspended solids, faecal coliform and hydrocarbon based oils and lubricant for the London Heathrow Airport (Richter et al., 2004, Richter et al., 2003; Revitt et al., 2001; Chon et al., 1999).

A wetland system for the treatment of airport runoff was developed with the construction of a gravel subsurface flow reed bed system at a small experimental scale at the Heathrow Airport in 1994. The primary requirement of the treatment system was to control the concentrations of glycols following their use as de-icers and anti-icers within the airport. The ability of reed beds to contribute to this treatment role was tested in the pilot scale system via on-site experiments over a 2 year period. The average reductions in runoff BOD concentrations achieved by pilot scale surface flow and sub-surface flow reed beds were 30.9% and 32.9%, respectively. One of the concerns with the application of wetland systems is their performance at colder temperature. During the winter months, the application of large quantities of glycol based anti- and de-icers to aircraft and runways poses a serious threat to receiving waters because of their toxicity and BOD effect. Results from the pilot-scale study indicated that most of the constructed wetland plants and substrate microbial populations throughout the beds were not adversely affected by airport runoff or exposure to shock-loads of glycols at initial total concentrations of 1,180 mg/L and 632 mg/L in the subsurface and surface flow reed beds respectively. Apart from Typha spp, the aquatic macrophytes have adapted well to the exposure to airport runoff. Glycol removal efficiencies have improved as the beds have matured, and average removal efficiencies of 78% for the sub-surface system and 54% for the surface system have been recorded. Complimentary monitoring of substrate micro-organism populations prior to and after glycol dosings have shown that aerobic microbial groups of bacteria, fungi and actinomycetes, are present in higher numbers (10^5-10^7 CFU/g substrate dry weight) than their anaerobic counterparts (10^3-10^5 CFU/g substrate dry weight). In the laboratory, studies have shown fungi and bacteria to be most tolerant of glycol, with several strains able to utilize these compounds. These treatment performances were then used to predict the required full scale constructed wetland surface areas necessary to attain the desired effluent water quality. The treatment system also incorporated aeration, storage and, combined with reed bed technology, was designed to reduce a mixed inlet BOD concentration of 240 mg/L to less than 40 mg/L for water temperatures varying between 6°C and 20°C.

The promising pilot-scale results of the experimental system lead to the construction of a full-scale treatment system, the Heathrow Constructed Wetlands, designed as an integrated wetland system to receive and treat the runoff form the Eastern and Southern Catchments of Heathrow Airport. The integrated system incorporates of a number of hydraulic balancing and aerated water bodies, feeding a polishing reed bed system prior to discharge into a receiving water body. The site consists of a 2.5 acre (1 ha) rafted reed bed canal system and a 3 acre (2 ha) sub-surface flow gravel reed bed. The subsurface treatment wetland on the Heathrow constructed wetland consist of 12 beds of varying size ranging from 27,026 ft^3 (765.3 m^3) to 80,906 ft^3 (2291.0 m^3), where each bed is hydraulically isolated from the adjacent bed in the terrace by concrete dividing walls. A 65 ft (20 m) x 6.5 ft (2 m) open water channel is situated at the front and at the end of the bed and between the cells of a bed. While the front and end channels distribute and collect the water flowing through the beds, the intermediate channels reduce the incidence of channelization and short-circuiting along the entire length of the bed. Each cell is filled to a depth of 2 ft (60 cm) with gravel with varying degree of limestone. An impermeable bentonite liner underlies each bed to prevent loss from or ingress to the system. The design hydraulic loading rates for the entire system is 20 L/s/ha, based on a design flow rate of 40 L/s. Due to the different surface areas of each bed, the design inflow rate fore each bed is different such that a constant hydraulic loading rate across the system is maintained. The beds were vegetated with *Phragmites* (common reed) in the summer of 2001.

In November 2002, the subsurface flow reed beds of the Heathrow Constructed Wetlands were conditioned prior to the de-icing season. A volume of 920 ft^3 (26 m^3) of industrial glycol product was used for the purpose, which contained approximately 28 vol-% mono-propylene glycol. It was transferred into the main reservoir, where it was mixed and diluted with runoff water before being transferred to the subsurface flow system. The system was operated in a closed loop configuration, where water exiting the subsurface reed beds was pumped back to the main reservoir. Aeration systems in the main reservoir and balancing pond were used to oxygenate the water. The temporal observations of glycol removal measured as COD demonstrated treatment efficiencies in the range of 30% to 70%, with absolute reductions of down to 45 mg/L. These observed efficiencies and total reductions of COD were lower than reported efficiencies or reduction of COD in various other applications. However, since the measurements were undertaken at the beginning of the de-icing season, in a pre-conditioned system, these figures could be expected. The removal of COD per unit area and flow rate were typically found to be highest in the smaller beds with the greatest flow rate. Thus, it was concluded that the absolute removal of COD was higher with higher influent concentrations of COD into the wetland system. This would also support observations that the main removal of COD takes place within the first half of the beds. Overall, the results of the full-scale system to date have shown that subsurface reed beds are generally able to attenuate glycol-based pollutants and have shown a relatively good performance without a long period of adaptation.

7.6   Conclusions

Wetland systems have shown a great deal of potential for the remediation of hazardous compounds in wastewaters. The unique advantage of these passive treatment systems is that, when adequately designed, they can provide the relatively low-cost and low-maintenance treatment of large volumes of contaminated water, when compared to conventional physical-chemical approaches. Wetlands provide multiple use benefits ranging from aesthetic enjoyment to enhanced biodiversity. Natural wetlands are generally less efficient than constructed wetlands, which have been designed for specific constituent removals and can generally deliver reliable treatment and meet strict discharge limits.

The combination of higher plants, algae and bacteria make wetland systems an exciting prospect for phytoremediation and bioremediation. Wetland phytoremediation generally differs from terrestrial phytoremediation in that bacterial transformations rather than plant uptake dominate remediation processes. Wetland plants provide the litter later that provides microbial habitants with a source of labile organic carbon for bacterial processes. The anoxic soils that are often characteristic of wetland systems can immobilize organic and metal contaminants, while oxidized soils in the rhizosphere or found in shallow free surface wetland systems can mobilize certain constituents into plant tissues in wetlands where is present. Combinations of oxic and anoxic sites and wet and dry cycles aid in the remediation of recalcitrant compounds. With heavy metals such as copper or lead, or metalloids such as selenium, emphasis is on creating conditions for immobilization in the highly reduced selenide or metallic form. To date, less is known regarding the treatment of toxic organics or pesticide removal in wetland system. However, recent studies have demonstrated that wetlands can efficiently remove some chlorinated compounds present at low concentrations that are generally difficult to remove by other treatment approaches.

One of the greatest issues associated with the treatment efficiency and design of wetland systems, particularly in temperate regions, is contaminant release due to seasonal biotic cycles or when the wetlands is fully loaded. This is often addressed by increasing the surface area of the wetland system or designing for colder temperatures when vegetation growth and bacterial activity would be expected to be minimized, which results in larger hydraulic retention times than in comparatively warmer climates.

The remediation of contaminants requires large energy inputs and wetland systems have become particularly interesting alternatives that are competitive with other remediation approaches because they are largely driven by free solar energy. Care must be taken to design these systems specifically for the target compounds of interest, as the treatment mechanisms involved in wetland systems differ with each class of contaminant. The regular harvesting and disposal of pollutant accumulating plants in vegetated wetlands can considerably increase the operational and maintenance costs of treatment. However, these generally remain below the operational and maintenance requirements of more traditional physical-chemical treatment systems with associated sludge dewatering, conditioning and disposal. Hence, constructed wetland systems are likely to continue to evolve as attractive, efficient and cost-effective remediation alternatives for the mitigation of hazardous compounds in wastewaters.

References

Anderson, K.L., Wheeler, K.A., Robinson, J.B. and Tuovinen, O.H. (2002) Atrazine Mineralization Potential in Two Wetlands. *Wat. Res.* 36:4785-4794.

Anderson, T.A. and Coats, J.R. (1995) An Overview of Microbial Degradation in the Rhizosphere and its Implications for Bioremediation. Chapter 8 *in Bioremediation: Science and Applications*. H.D. Skipper and R.F. Turco Eds. SSSA Special Publication No. 43. Soil Science Society of America, American Society of Agronomy, Crop Science Society of America. Madison, Wisconsin.

Bachand, P.A.M. and Horne, A.J. (2000) Denitrification in constructed wetland free-water surface wetlands: II. Effects of vegetation and temperature. *Ecol. Eng.* 14:17-32.

Bankstone, J.L., Sola, D.L., Komor A.T. and Dwyer, D.F. (2002) Degradation of Trichloroethylene in Wetland Microcosms Containing Broad-Leaved Cattail and Eastern Cottonwood. *Wat. Res.* 36:1539-1546.

Best, E.H.P., Miller, J.L., Fredrickson, H.L., Larson, S.L., Zappi, M.E. and Streckfuss, T.H. (1998) *Explosives Removal from Groundwater of the Iowa Army Ammunition Plant in Continuous-Flow Laboratory Systems Planted with Aquatic and Wetland Plants*. Omaha, NE, U.S. Army Engineer Waterways Experiment Station: 26.

Best, E.H.P., Zappi, M.E., Fredrickson, H.L., Sprecher, S.L., Larson, S.L. and Miller, J.E. (1997a) *Screening of Aquatic and Wetland Plant Species for Phytoremediation of Explosives-Contaminated Groundwater from the Iowa Army Ammunition Plant*. Omaha, NE, U.S. Army Engineer Waterways Experiment Station: 47.

Best, E.H.P., Zappi, M.E., Fredrickson, H.L., Sprecher, S.L., Larson, S.L. and Ochman, M. (1997b) Screening of Aquatic and Wetland Plant Species for Phytoremediation of Explosives-Contaminated Groundwater form the Iowa Army Ammunition Plant. *Bioremediation of Surface and Subsurface*

Contamination. R.K. Bajpai and M.E. Zappi (Eds.). Annals of the New York Academy of Sciences Volume 829. pp. 202-218.

Brodrick, S.J., Cullen, P. and Maher, W. (1988) Denitrification in a Natural Wetland Receiving Secondary Treated Effluent. *Wat. Res.* 22(4):431-439.

Burgoon, P.S., Kadlec, R.H. and Henderson, M. (1999) Treatment of Potato Processing Wastewater with Engineered Natural Systems. *Wat. Sci. Technol.* 40(3):211-215.

Carman, E.P. and Crossman, T.L. (2001) Phytoremediation. Chapter 2 *in In-Situ Treatment Technology*. 2nd Ed. E.K. Nyer, P.L. Palmer, E.P. Carman, G. Boettcher, J.M. Bedessem, F. Lenzo, T.L. Crossman, G.J. Rorech and D.F. Kidd (Eds.). Arcadis Environmental Science and Engineering Series. Lewis Publishers, Washington, D.C. pp. 392-435.

Catallo, W.J. and Junk, T. (2003) Effects of Static vs. Tidal Hydrology on Pollutant Transformation in Wetland Sediments. *J. Environ. Qual.* 32:2421-2427.

Chong, S., Garelick, H., Revitt, D.M., Shutes, R.B.E., Worrall, P. and Brewer, D. (1999) The Microbiology Associated with Glycol Removal in Constructed Wetlands. *Wat. Sci. Technol.* 40(3):99-107.

Collins, B, McArthur, J.V. and Sharitz, R.R. (2004) Plant Effects on Microbial Assemblages and Remediation of Acidic Coal Pile Runoff in Mesocosm Treatment Wetlands. *Ecol. Eng.* 23:107-115.

Connell, D.W. and Miller, G.J. (1984) *Chemistry and Ecotoxicology of Pollution*. John Wiley & Sons, NY.

Cork, D.J. and Cusanovich, M.A. (1979) Continuous Disposal of Sulphate by Bacterial Mutualism. *Devel. Ind. Microbiol.* 20:591-602.

Cui, L.H., Liu, W., Zhu, X.Z., Ma, M. and Huang, X.H. (2006) Performance of Hybrid Constructed Wetland Systems for Treating Septic Tank Effluent. *J. Environ. Sci. (China)* 18(4):669.

Craft, C.B. (1997) Dynamics of Nitrogen and Phosphorus Retention during Wetland Ecosystem Succession. *Wetland Ecol. Manag.* 4:177-187.

Crowley, D.E., Alvey, S. and Gilbert, E.S. (1997) Rhizosphere Ecology of Xenobiotic Degrading Microorganisms. Chapter 2 *in Phytoremediation of Soil and Water Contaminants*. E.L. Kruger, T.A. Anderson and J.R. Coats (Eds.). American Chemical Society. Washington. D.C.

Debusk, T.A., Laughlin, R.B. and Schwartz, L.N. (1996) Retention and Compartmentalization of Lead and Cadmium in Wetland Microcosms. *Wat. Res.* 30(11):2707-2716.

Decamp, O. and Warren, A. (2000) Investigation of *Eschrichia coli* Removal in Various Designs of Subsurface Flow Wetlands Used for Wastewater Treatment. *Ecol. Eng.* 14:293-299.

Deng, H., Ye, Z.H. and Wong, M.H. (2004) Accumulation of Lead, Zinc, Copper and Cadmium by 12 Wetland Plant Species Thriving in Metal-Contaminated Sites in China. *Environ. Poll.* 132:29-40.

Engelhardt, K.A.M and Ritchie, M.E. (2001) Effects of Macrophyte Species richness on Wetland Ecosystem Functioning and Services. *Nature* 411:687-689.

Evangelou, V.P. and Zhang, Y.L. (1995) A Review: Pyrite Oxidation Mechanisms and Acid Mine Drainage Prevention. *Crit. Rev. Environ. Sci. Technol.* 25(2):141-199.

Evans, G.M. and Furlong, J.C. (2003) *Environmental Biotechnology: Theory and Application.* John Wiley & Sons, Inc.

Faulkner, S.P. and Richardson, C.J. (1989) Physical and Chemical Characteristics of Freshwater Wetland Soils. Chapter 4 *in Constructed Wetlands for Wastewater Treatment: Municipal, Industrial and Agricultural.* Lewis Publishers. Michigan. pp. 41-72.

Fernandes, J.C., and Henriques, F.F. (1990) Metal levels in soil and cattails (*Typha latifolia L.*) plants in a pyrite mine, Lousal Portugal. *Int. J. Environ. Stud.* 36:205-210.

Foth, H. (1990) *Fundamentals of Soil Science.* 8th Edition. John Wiley & Sons, Inc. New York.

Framson, P.E. and Leckie, J.O. (1978) Limits of Coprecipitation of Cadmium and Ferrous Sulfides. *J. Am. Chem. Soc.* 12:465-469.

Gazsó, L.G. (2001) The Key Microbial Processes in the Removal of Toxic Metals and Radionuclides from the Environment. *Central European J. Occup. Environ. Med.* 7(3-4):178-185

Gerth, A., Hebner, A. and Thomas, H. (2003) Case Study: Natural Remediation of TNT-Contaminated Water and Soil. *Acta Biotechnol.* 23(2-3):143-150.

Gilliam, J.W. (1994) Riparian Wetlands and Water Quality. *J. Environ. Qual.* 23: 896-900.

Goodman, P.J. and Williams, W.T. (1961) Investigations into 'dieback' in *Spartina townsendii. J. Ecol.* 49:391-398.

Giraud, F., Guiraud, P., Kadri, M., Blake, G. and Steinman, R. (2001) Biodegradation of Anthracene and Fluoranthene by Fungi Isolated from an Experimental constructed Wetland for Wastewater Treatment. *Wat. Res.* 35(17):4126-4136.

Gopal, B. (1999) Natural and Constructed Wetlands for Wastewater Treatment: Potentials and Problems. *Wat. Sci. Technol.* 40(3):27-35.

Goulet, R.R. and Pick, F.R. (2001a) The Effects of Cattails (*Typha latifolia L.*) on Concentrations and Partitioning of Metals in Surficial Sediments of Surface-Flow Constructed Wetlands. *Wat. Air Soil Poll.* 132:275-291.

Goulet, R.R. and Pick, F.R. (2001b) Changes in Dissolved and Total Fe and Mn in a Young Constructed Wetland: Implications for Retention Performance. *Ecol. Eng.* 17:373-384.

Gross, B., Montgomery-Brown, J., Naumann, A. and Reinhard, M. (2004) Occurrence and Fate of Pharmaceuticals and Alkylphenol Ethoxylate Metabolites in an Effluent-Dominated River and Wetland. *Environ. Toxicol. Chem.* 23(9):2074-2083.

Guntenspergen, G.R., Stearns, F. and Kadlec, J.A. (1989) Wetland Vegetation. Chapter 5 *in* Constructed Wetlands for Wastewater Treatment: Municipal, Industrial and Agricultural. Lewis Publishers. Michigan.

Hammer, D.A. (1989) *Constructed Wetlands for Wastewater Treatment: Municipal, Industrial and Agricultural.* Lewis Publishers. Michigan.

Hammer, D.A. and Bastian, R.K. (1989) Wetland Ecosystmes: Natural Water Purifiers? Chapter 2 *in* Constructed Wetlands for Wastewater Treatment: Municipal, Industrial and Agricultural. Lewis Publishers. Michigan. pp. 5-19.

Hemond, H.F. and Benoit, J. (1988) Cumulative Impacts on Water Quality Functions of Wetlands. *Environ. Manag.* 12(5):639-653.

Holtan, H., Kamp-Nielson, L., and Stuanes, A.O. (1988) Phosphorus in Sediment, Water, and Soil: An Overview. *Acta Hydrobiol.* 170:19-34.

Hunt, R.J., Krabbenhoft, D.P. and Anderson, M.P. (1997) Assessing Hydrogeochemical Heterogeneity in Natural and Constructed Wetlands. *Biogeochem.* 39:271-293.

Hunter, P.G., Combs, D.L., George, D.B. (2001) Nitrogen, Phosphorous and Organic Carbon Remval in Simulated Wetland Treatment Systems. *Arch. Environ. Contam. Toxicol.* 41(3):274

Hupp, C.R. and Bazemore, D.E. (1993) Spatial and Temporal Aspects of Sediment Deposition in West Tennessee Forested Wetlands. *J. Hydrol.* 141:179-196.

Jacob, D.L and Otte, M.L. (2003) Conflicting Processes in the Wetland Plant Rhizosphere: Metal Retention or Mobilization? *Wat. Air Soil Poll.* 3:91-104.

Johnston, C.A. (1991) Sediment and Nutrient Retention by Freshwater Wetlands: Effects on Surface Water Quality. *Crit. Rev. Environ. Control* 21:491-565.

Kadlec, R.H. (1997a) An Autobiotic Wetland Phosphorus Model. *Ecol. Eng.* 8:145-172

Kadlec, R.H. (1997b) Integrated Natural Systems for Treating Potato Processing Wastewater. *Wat. Sci. Technol.* 35(5):263-270.

Kadlec, R.H. and Knight, R.L. (1996) *Treatment Wetlands.* Lewis Publishers, Washington, D.C.

Kappelmeyer, U., Wiebner, A., Kuschk, P. and Kastner, M. (2004) Interactions of C-N-S Transformation Processes and Microbial Community Structure in a Model Reactor for Constructed Wetlands. *9th Int. Conf. Wetland Systems for Water Pollution Control.* Avignon, France. 521-525.

Karathanasis, A.D. and Johnson, C.M. (2003) Metal Removal Potential by Three Aquatic Plants in an Acid Mine Drainage Wetland. *Mine Wat. Environ.* 22:22-30.

Knight, R.L., Ruble, R.W., Kadlec, R.H. and Reed, S. (1993) Wetlands for Wastewater Treatment Performance Database. Chapter 4 *in* Constructed Wetlands for Water Quality Improvement. G. Moshiri (Ed.). Lewis Publisher. Boca Raton, FL.

Kowles, J.L. and Stein, O.R. (2004) Polar Organic Solvent Removal in a Microcosm Constructed Wetland System. *9th Int. Conf. Wetland Systems for Water Pollution Control*. Avignon, France. September 26-30. pp. 343-351.

Kosolapov, D.B., Kuschk, P., Vainshtein, M.B., Vatsourina, A.V., Wiebner, A., Kastner, M. and Muller, R.A. (2004) Microbial Processes of Heavy Metal Removal from Carbon-deficient Effluents in Constructed Wetlands. *Eng. Life Sci.* 4(5):403-411.

Kruger, E.L., Anderson, T.A. and Coats, J.R. (1997) *Phytoremediation of Soil and Water Contaminants*. American Chemical Society. Washington. D.C.

Kyambadde, J., Kansiime, F., Gumaelius, L. and Dalhammar, G. (2004) A Comparative Study of *Cyperus papyrus* and *Miscanthidium violaceum*-Based Constructed Wetlands for Wastwater Treatment in Tropical Climate. *Wat. Res.* 38:475-485.

Lehninger, A.L. (1982) *Principles of Biochemistry*. Worth Publishers.

Lim, P.E., Tay, M.G., Mak, K.Y. and Mohamed, N. (2003) The Effect of Heavy Metals on Nitrogen and Oxygen Demand Removal in Constructed Wetlands. *Sci. Total Environ.* 301(1-3):13-21.

Lorah, M.M. and Voyek, M.A. (2004) Degradation of 1,1,2,2,-tetrachloroethane and Accumulation of Vinyl Chloride in Wetland Sediment Microcosms and In-Situ Porewater: Biogeochemical Controls and Associations with Microbial Communities. *J. Contam. Hydrol.* 70:117-145.

McCutcheon, S.C. and Schnoor, J.L. (2003) *Phytoremediation: Transformation and Control of Contaminants*. Wiley and Sons, Inc.

Maehlum, T. (1995) Treatment of Landfill Leachate in On-Site Lagoons and Constructed Wetlands. *Wat. Sci. Technol.* 32(3):129-135.

Martin, C.D., Johnson, K.D. and Moshiri, G.A. (1999) Performance of a Constructed Wetland Leachate Treatment System at the Chunchula Landfill, Mobile County, Alabama. *Wat. Sci. Technol.* 40(3):67-74.

Masi, F., Conte, G., Lepri, L., Martellini, T. and Del Bubba, M. (2004) Endocrine Disrupting Chemicals (EDCs) and Pathogens Removal in a Hybrid CW System for a Tourist Facility Wastewater Treatment and Reuse. *9th Int. Conf. Wetland Systems for Water Pollution Control*. Avignon, France. September 26-30. pp. 461-469.

Matamoros, V., Garcia, J., and Bayona, J.M. (2004) Behaviour of Pharmaceuticals in a Subsurface Flow Constructed Wetland Pilot Plant. Preliminary Results. *9th Int. Conf. Wetland Systems for Water Pollution Control*. Avignon, France. September 26-30.

Matthews, D.J., Moran, B.M., McCabe, P.F. and Otte, M.L. (2004) Zinc Tolerance, Uptake, Accumulation and Distribution in Plants and Protoplasts of Five European Populations of the Wetland grass *Glyceria fluitans*. *Aquat. Bot.* 80:39-52.

Mays, P.A. and Edwards, G.S. (2001) Comparison of Heavy Metal Accumulation in a Natural Wetland and Constructed Wetlands Receiving Acid Mine Drainage. *Ecol. Eng.* 16:487-500.

Mills, M.A. (1997) Bioremediation of an Experimental Oil Spill in a Wetland. *In Situ and On-Site Bioremediation: Volume 3.* 4th Int. In situ and On-Site Bioremediation Symposium. New Orleans, April 28-May 1. pp. 355-359.

Mitsch, W.J. and Gosselink, J.G. (2000) *Wetlands.* 3rd Edition. John Wiley & Sons Inc.

Moshiri, G.A. (1993). *Constructed Wetlands for Water Quality Improvement.* Lewis Publishers, Boca Raton, FL.

Mullen, M.D., Wolf, D.C., Ferris, F.C., Beveridge, T.J., Flemming, C.A. and Bailey, F.W. (1989) Bacterial sorption of heavy metals. *Appl. Environ. Microbiol.* 55:3143-3149.

Mungur, A.S., Shutes, R.B.E., Revitt, D.M. and House, M.A. (1995) An Assessment of Metal Removal from Highway Runoff by a Natural Wetland. *Wat. Sci. Technol.* 32(3):169-175.

National Academy of Sciences (NAS) (1995) *Wetlands: Characteristics and Boundaries.* National Academy of Sciences Press: Washington, D.C.

Newman, J.N., Clasen, J.C. and Neafsey, J.A. (2000) Seasonal Performance of a Wetland Constructed to Process Dairy Milkhouse Wastewater in Connecticut. *Ecol. Eng.* 14:181-198.

NSW Department of Land and Water Conservation (DLWC). (1998) *The Constructed Wetlands Manual.* Volume 1 & 2. Land and Water Conservation. Australia.

Nyer, E.K., Palmer, P.L., Carman, E.P., Boettcher, G., Bedessem, J.M., Lenzo, F., Crossman, T.L., Rorech, G.J. and Kidd, D.F. (2001) *In-Situ Treatment Technology.* 2nd Ed. Arcadis Environmental Science and Engineering Series. Lewis Publishers, Washington, D.C.

O'Sullivan, A.D., Murray, D.A. and Otte, M.L. (2004a) Removal of Sulfate, Zinc and Lead from Alkaline Mine Wastewater Using Pilot-Scale Surface-Flow Wetlands at Tara Mines, Ireland. *Mine Wat. Environ.* 23:58-65.

O'Sullivan, A.D., Moran, B.D. and Otte, M.L. (2004b) Accumulation and Fate of Contaminants (Zn, Pb, Fe and S) in Substrates of Wetlands Constructed for Treating Mine Wastewater. *Wat. Air Soil Poll.* 157:345-364.

Owen, C.R. (1995) Water Budget and Flow Patterns in an Urban Wetland. *J. Hydrol.* 169:171-187.

Peltier, E.F., Webb, S.M. and Gaillard, J.-F. (2003). Zinc and Lead Sequestration in an Impacted Wetland System. *Adv. Environ. Res.* 8:103-112.

Perkins, J. and Hunter, C. (2000) Removal of Enteric Bacteria in a Surface Flow Constructed Wetland in Yorkshire, England. *Wat. Res..* 24(6):1941-1947.

Polprasert, C., Dan, N.P. and Thayalakumaran, N. (1996) Application of Constructed Wetlands to Treat Some Toxic Wastewaters Under Tropical Conditions. *Wat. Sci. Technol.* 34(11):165-171.

Portier, R.J. and Palmer, S.J. (1989) Wetlands Microbiology: Form, Function, Processes. Chapter 6 <u>in</u> *Constructed Wetlands for Wastewater Treatment: Municipal, Industrial and Agricultural.* Lewis Publishers. Michigan.

Raisin, G.W. and Mitchell, D.S. (1995) Use of Wetlands for the Control of Non-Point Source Pollution. *Wat. Sci. Technol.* 32(3):177-186.

Reddy, K.R., and Patrick, W.H. (1984) Nitrogen transformations and loss in flooded soils and sediments. *Crit. Rev. Environ. Control* 13:273–309

Reed, S.C., Crites, R.W. and Middlebrookes, E.J. (1995) *Natural Systems for Waste Management and Treatment.* 2nd Edition. McGraw-Hill Inc., New York.

Richardson, C.J. (1985) Mechanisms Controlling Phosphorus Retention Capacity in Freshwater Wetlands. *Science* 228:1424-1427.

Richardson, C.J. and Craft, C.B. (1993) Efficient Phosphorus Retention in Wetlands: Fact or Fiction. *In Constructed Wetlands for Water Quality Improvement*, G. Moshiri (Ed.). Lewis Publishers. Boca Raton, FL.

Richter, K.M, Guymer, I., Worrall, P. and Jones, C. (2004) Treatment Performance of Heathrow Contrcuted Wetlands. *6th International Conference on Waste Stabilization Ponds & 9th International Conference on Wetland Systems for Water Pollution Control*. Avignon, France. September 26-30. pp. 125-132.

Richter, K.M, Margett, J.R., Saul, A.J., Guymer, I. and Worrall, P. (2003) Baseline Hydraulic Performance of the Heathrow Constructed Subsurface Flow System. *Wat. Sci. Technol.* 47(7-8):177-181.

Rivett, D.M., Worrall, P. and Brewer, D. (2001) The Integration of Constructed Wetlands Into a Treatment System for Airport Runoff. Wat. Sci. Technol. 44(11-12):469–476.

Runes, H.B., Jenkins, J.J., Moore, J.A., Bottomley, P.J. and Wilson, B.D. (2003) Treatment of Atrazine in Nursery Irrigation Runoff by a Constructed Wetland. *Wat. Res.* 37:539-550.

Scholes, L., Shutes, R.B.E., Revitt, D.M., Forshaw, M. and Purchase, D. (1999) The Removal of Urban Pollutants by Constructed Wetlands During Wet Weather. *Wat. Sci. Technol.* 40(3):333-340.

Scholes, L., Shutes, R.B.E., Revitt, D.M., Forshaw, M. and Purchase, D. (1998) The Treatment of Metals in Urban Runoff by Constructed Wetlands. *Sci. Total Environ.* 214:211-219.

Shatwell, T. and Cordery, I. (1999) Nutrient storage in urban wetlands. *In Impacts of urban growth on surface water and groundwater quality*. B. Ellis (Ed.). Publ. 259. IAHS, Wallingford, UK. pp. 339–347.

Sikora, F.J., Behrends, L.L., Phillips, W.D., Coonrod, H.S., Bailey, E. and Bader, D.F. (1997) A Microcosm Study on Remediation of Explosives-Containing Groundwater Using Constructed Wetlands. *Bioremedaiton of Surface and Subsurface Contamination* R.K. Bajpai and M.E. Zappi (Eds.). Annals of the New York Academy of Sciences Volume 829. pp. 202-218.

Sistani, K.R., Mays, D.A. and Taylor, R.W. (1999) Development of Natural Conditions in Constructed Wetlands: Biological and Chemical Changes. *Ecol. Eng.* 12:125-131.

Skipper, H.D. and Turco, R.F. (1995) *Bioremediation: Science and Applications.* SSSA Special Publication No. 43. Soil Science Society of America, Inc.,

American Society of Agronomy, Inc., Crop Science Society of America, Inc. Madison, Wisconsin.

Slemp, J.D., Norman, J.L., George, D.B., Stearman, G.K. and Wells, M.J.M. (2004) Determination of the Preferred Electron Acceptor for the Microbial Degradation of Simazine in Subsurface Flow Wetland Microcosms. *9th Int. Conf. Wetland Systems for Water Pollution Control.* Avignon, France. September 26-30. pp. 359-367.

Sparling, D.W. and Lowe, T.P. (1998) Metal Concentration in Aquatic Macrophytes as Influenced by Soil and Acidification. *Wat. Air Soil Poll.* 108:203-221.

Spieles, D.J. and Mitsch, W.J. (2000) The Effects of Season and Hydrologic and Chemical Loading on Nitrate Retention in Constructed Wetlands: A Comparison of Low- and High-Nutrient Riverine Systems. *Ecol. Eng.* 14:77-91.

Stark, L.R. and Williams, F.M. (1995) Assessing the Performance Indices and Design Parameters of Treatment Wetlands for H^+, Fe, and Mn Retention. *Ecol. Eng.* 5:433-444.

Steinberg, S.L. and Coonrod, H.S. (1994) Oxidation of the root zone by aquatic plants growing in gravel-nutrient solution culture. *J. Environ. Qual.* 23: 907-913.

Stott, R. and Tanner, C.C. (2004) Influence of Biofilm on Removal of Surrogate Faecal Microbes in a Constructed Wetland and Maturation Pond. *6th Int. Conf. Waste Stabilization Ponds & 9th Int. Conf. Wetland Systems for Water Pollution Control.* Avignon, France. September 26-30. pp. 209-217.

Stottmeister, W., Wiebner, A., Kuschk, P., Kappelmeyer, U., Kastener, M., Bederski, O., Müller, R.A. and Moormann, H. (2003) Effects of Plants and Microorganisms in Constructed Wetlands for Wastewater Treatment. *Biotechnol. Adv.* 22:93-117.

Sundby, B., Gobeil, C., Silverberg, N. and Mucci, A. (1992) The Phosphorus Cycle in Coastal Marine Sediments. *Limnol. Oceanogr.* 37: 1129-1145

Suthersan, S.S. (2002) *Natural and Enhanced Remediation Systems.* Lewis Publishers, Inc. Washington, D.C.

Suthersan, S.S. (1997) *Remediation Engineering: Design Concepts.* Lewis Publishers, Inc. Washington, D.C.

Tanner, C.C. (1996) Plants for Constructed Wetland Treatment Systems – A Comparison of the Growth and Nutrient Uptake of Eight Emergent Species. *Ecol. Eng.* 7:59-83.

Taylor, J.R., Cardemone, M. and Mitsch, W.J. (1990) Bottomland Hardwood Wetlands, Their Functions and Values. *In Ecological Processes and Cumulative Impacts.* J.G. Gosselink, L.C. Lee, T.A. Muir (Eds). Lewis Publishers Inc. Michigan. p. 13-86.

Vamos, R. and Koves, E. (1972) Role of Light in the Prevention of the Poisoning Action of Hydrogen Sulfide in the Rice Plant. *Ecology.* 53:519-25

vanLoon, G.W. and Duffy, S.J. (2005) *Environmental Chemistry: A Global Perspective.* 2nd Edition. Oxford University Press. New York.

Volesky, B. (1990) *Biosorption of Heavy Metals*. CRC Press, Boca Raton, LA.

Walbridge, M.R. and Struthers, J.P. (1993) Phosphorus Retention in Non-Tidal Palustrine Forested Wetlands of the Mid-Atlantic Region. *Wetlands*.

Walker, D.J. and Hurl, S. (2002) The Reduction of Heavy Metals in a Stormwater Wetland. *Ecol. Eng.* 18:407-414.

Wallace, S., Parkin, G. and Cross, C. (2001) Cold climate wetlands: design and performance. *Wat. Sci. Technol.* 44(11-12):259–265.

Wallace, S. and Kadlec, R.H. (2004) BTEX Degradation in a Cold-Climate Wetland System. *9th Int. Conf. Wetland Systems for Water Pollution Control*. Avignon, France. September 26-30. pp. 335-341.

Wang, N. and Mitsch, W.J. (2000) A Detailed Ecosystem model of phophorous Dynamics in Created Riparian Wetlands. *Ecol. Mod.* 126:101-130.

Weis, J.S. and Weis, P. (2004) Metal Uptake, Transport and Release by Wetland Plants: Implications for Phytoremediation and Restoration. *Environ. Int.* 30:685-700.

Werker, A.G., Dougherty, J.M., McHenry, J.L. and Van Loon, W.A. (2002) Treatment Variability for Wetland Wastewater Treatment Design in Cold Climate. *Ecol. Eng.* 19:1-11.

White, J.S., Bayley, S.E. and Curtis, P.J. (2000) Sediment Storage of Phosphorus in a Northern Prairie Wetland Receiving Municipal and Agro-Industrial Wastewater. *Ecol. Eng.* 14:127–138.

Wildeman, T.R., Duggan, L.A., Updegraff, D.M., and Emerick, J.C. (1993) The role of macrophytes and algae in the removal of metal contaminants in wetland processes. *80th Ann. Meeting Air and Waste Management Association*. Denver, Colorado

Zachritz, W.H., Lundie, L.L., Wang, H. (1996) Benzoic Acid Degradation by Small, Pilot-Scale Artificial Wetlands Filter (AWF) Systems. *Ecol. Eng.* 7:105-116.

Zander, M. (1980) Polycyclic Aromatic and Heteroaromatic Hydrocarbons. *In Handbook of Environmental Chemistry*. Springer-Verlag, New York, NY

CHAPTER 8

Physical Processes and Natural Attenuation

SAY KEE ONG

8.1 Introduction

Physical attenuation processes play an important role in the fate and transport of contaminants in the environment. Physical attenuation processes are "nondestructive" processes which do not change the total mass of the contaminants in the system but instead result in the reduction of the contaminant concentration through dilution and mixing or transfer of the contaminants from one phase to another. This chapter reviews the various naturally-occurring physical processes in groundwater and bodies of water (streams, rivers and lakes) that contribute to the natural attenuation of contaminants. Physical attenuation processes include processes such as diffusion, advection, and hydrodynamic dispersion which affect the contaminant concentration via water movement, and partitioning processes such as sorption and volatilization which affect the contaminant concentration via transfer from water phase to another. Sorption processes are discussed in Chapter 2.

8.2 Diffusion

Diffusion is an important mass-transport mechanism for molecules, ions and particulates in the environment. Molecular diffusion is due to the molecular vibrational, rotational and translation movement of the molecules resulting in random movement (random walk) of the molecules within the fluid. Since it is a spontaneous process with an increase in entropy, molecular diffusion results in the movement of molecules from regions of higher concentrations to regions of lower concentration. In 1855, Fick described mass transfer by diffusion as directly proportional to the concentration gradient. The Fick's first law of diffusion is given by the following equation:

$$J_m = -D_d A \frac{dC}{dx} \qquad \text{(Eq. 8.1)}$$

where J_m is mass flux rate (gm/s), C is the concentration (gm/cm^3), x is defined as distance (cm), A is equal to the cross sectional area (cm^2), and D_d is the diffusion coefficient (cm^2/s).

The diffusion coefficient in Eq. 8.1 is a fundamental property of the chemical and is dependent on the medium in which the chemical is diffusing. Values of molecular diffusion coefficients are found in various reference materials (Lyman et al., 1982 and Thibodeaux, 1979). Examples of diffusion coefficients of various chemicals in air and water are given in Table 8.1. The diffusion of chemicals in water is approximately 4 to 5 orders of magnitude slower than that in the air phase.

Table 8.1 Diffusion coefficients of various chemicals in air and water

Chemical	Air D_d (cm^2/s)	Water D_d (cm^2/s)	References
Ammonia	2.8×10^{-1}	1.76×10^{-5}	Thibodeaux, 1979
Benzene	9.3×10^{-2}	1.1×10^{-5}	ASTM, 1995
Toluene	8.5×10^{-2}	9.4×10^{-6}	ASTM, 1995
Ethylbenzene	7.6×10^{-2}	8.5×10^{-6}	ASTM, 1995
Xylene (mixed)	7.2×10^{-2}	8.5×10^{-6}	ASTM, 1995
Methyl t-butyl ether	7.92×10^{-2}	9.41×10^{-5}	ASTM, 1995
Trichloroethylene	8.18×10^{-2}	1.05×10^{-4}	EPA, 1989
Tetrachloroethene	7.2×10^{-2}	8.2×10^{-6}	EPA, 1989
Cis- trichloroethene	7.36×10^{-2}	1.13×10^{-5}	EPA, 1989
Trans- trichloroethene	7.07×10^{-2}	1.19×10^{-5}	EPA, 1989
Vinyl chloride	1.06×10^{-1}	1.23×10^{-5}	EPA, 1989
Naphthalene	7.20×10^{-2}	9.4×10^{-6}	EPA, 1989
Phenanthrene	3.33×10^{-2}	7.47×10^{-6}	EPA, 1989
Chrysene	2.48×10^{-2}	6.21×10^{-6}	EPA, 1989
Benzo(a)pyrene	5×10^{-2}	5.8×10^{-6}	EPA, 1989

Table 8.1 represents the diffusion coefficients in a homogeneous medium such as air and water. In the subsurface, the presence of a porous medium such as sand creates obstacles for the diffusion of chemicals in soil pore water resulting in the chemicals taking a tortuous path as they diffuse from high to low concentrations. To account for this tortuous path taken, the diffusion coefficient in a homogeneous medium is modified using an empirical coefficient, w, giving an effective diffusion coefficient, D_d^*.

$$D_d^* = w D_d \qquad \text{(Eq. 8.2)}$$

The value of w is found experimentally and generally ranges from 0.01 to 0.5 depending on the porous media (Fetter, 1988). Another example is the diffusion of substrate into a biofilm attached to a surface where the effective diffusion coefficients for the biofilm were found to be about 0.37 to 0.71 the value of the diffusion coefficients for a biofilm modeled by assuming the pores to be made of cylinders (Zhang and Bishop, 1994).

Diffusion is an exceeding slow process for mass transfer. In the transport of contaminants in natural waters such as rivers and lakes, diffusion is not an important process. However, for mass transfer of pollutants at interfaces of air-water, particulate-water, and sediment-water in quiescent conditions, diffusion may be the controlling mechanism. Examples include mass transfer of volatile organics from aqueous to gas phase at the air-water surfaces of lakes and rivers or transfer of oxygen from atmosphere to water and mass transfer of pollutants into the bulk water from pore water of sediments.

8.3   Advection

In most, if not, all natural environmental systems, there is always movement of fluid and fluid mixing. Even lakes or ponds that appear to be quiescent from the surface may undergo mixing internally due to convective currents from temperature gradients or by slight air movement over the surface of the water. Bulk movement of the fluid which results in the transport of pollutants is known as advection. Advection is considered to be the main transport process causing the pollutants to move from an upstream location to a downstream location. As such, knowledge of the movement of the fluid is important in understanding the attenuation and movement of the pollutants.

For chemicals traveling in a river or a stream, the mass transported by advection is the equal to the product of the volumetric flowrate and the mean concentration. This is given by:

$$J = \bar{u}\,A\,C = Q\,C \qquad\qquad (Eq.\ 8.3)$$

where J is the mass discharge rate (kg/s), $\bar{u}$ is the mean current velocity in the stream or river (m/s), C is the concentration (kg/m^3), A is the cross section area of the river or stream (m^2) and Q is the volumetric flowrate (m^3/s). To describe the change in concentration within the flow volume, the one-dimensional advective transport is used:

$$\frac{\partial C}{\partial t} = -\bar{u}\,\frac{\partial C}{\partial x} \qquad\qquad (Eq.\ 8.4)$$

It should be noted that Eq. 8.5 can be expanded to a two- or three-dimensional advective model if traverse and vertical flows are important. Three-dimensional advective models may be important in a lake while a two-dimensional advective model may suffice for flow in a river.   It should be noted that for pollutants that are nonreactive and conservative, advection merely transport the pollutants and have to work in concert with other mechanisms such as dispersion and sorption to attenuate the pollutants in the medium.

For pollutants in the subsurface, an understanding of groundwater flow patterns is important to predict the movement of dissolved contaminants. Groundwater flow is described by Darcy's law (Domenico and Schwartz, 1998):

$$\frac{Q}{A} = q = -K\frac{dh}{dx}$$ (Eq. 8.5)

where Q is the volumetric flow rate (m^3/s), A is equal to the cross-sectional area of flow (m^2) of the porous medium, q is the volumetric flow rate per unit surface area, (specific discharge or Darcy velocity) (m/s), K is defined as the hydraulic conductivity (m/s), and dh/dx is the hydraulic gradient in the direction of groundwater flow.

In Eq. 8.5, the Darcy's velocity is proportional to the hydraulic gradient. The hydraulic conductivity, K, the constant of proportionality is a property of the porous medium. However, flow in the subsurface is limited to the pore space of the medium unlike flow in a river or stream. Instead of the specific discharge or Darcy's velocity, the average pore-water velocity, v, or seepage velocity is used as shown below (Domenico and Schwartz, 1998):

$$v = \frac{q}{n_e} = -\frac{K}{n_e}\frac{dh}{dx}$$ (Eq. 8.6)

where n$_e$ is the effective porosity. The hydraulic conductivity is a function of the medium porosity, particle size and distribution, particle shape, particle arrangement (i.e., packing), and secondary features such as fracturing (US EPA, 1987). Hydraulic conductivity values vary from 3 x 10^{-4} to 3 x 10^{-2} m/s for gravel to as low as 2 x 10^{-7} to 2 x 10^{-4} for fine sand. The hydraulic conductivities for clay are in the range of 1 x 10^{-11} to 4.7 x 10^{-9} m/s. Effective porosity of porous media ranged from 0.25 – 0.55 for fine sand and 0.34 – 0.60 for clay.

For contaminated groundwater with density not significantly different from the ambient groundwater, it is assumed that flow of water and dissolved pollutants move at the same rate (in the absence of other attenuation processes) and in the same direction (Domenico and Schwartz, 1998). Therefore, the average pore-water velocity can be used to estimate the advective component of groundwater flow as in Eq. 8.4 and provide a conservative estimate of the rate of migration of dissolved constituents (US EPA, 1987).

8.4   Hydrodynamic Dispersion

Hydrodynamic dispersion is defined as the spread of a pollutant by any process that results in the movement of the pollutant in a manner different from its average advective velocity (Logan, 1999). Dispersion of a pollutant is produced by the sum of several processes including molecular diffusion, velocity profile, mixing (turbulence), and noncontinuous flow path. Figure 8.1 illustrates the three possible spreading processes. Velocity profiles of fluid flow in pores of the subsurface or close to a fixed surface will move the pollutants at different rates according to the velocities of the fluid even when flow is laminar (Figure 8.1a). For fluid mixed by movement of air or unsteady vortices, the pollutants are redistributed over the distances covered by the fluid eddies (Figure 8.1b). The rate of movement is dependent on the sizes of the eddies and the rate at which the eddies move fluid around in the system. The overall dispersion coefficient, D, can be expressed as follows:

$$D = D_d + \sum_n D_i$$

(Eq. 8.7)

where D_i are dispersion coefficients from various processes such as velocity profiles and turbulence.

8.4.1 *Dispersion in subsurface*

In the subsurface, the noncontinuous nature of the media results in fluid flow around the media, causing the pollutant to move from a fluid streamline into an adjacent fluid streamline. This causes the pollutants to spread out (Figure 8.1c). Several researchers observed, in laboratory-scale column studies with uniform media, that the hydrodynamic dispersion coefficient, D, was relatively constant at low velocities but increased linearly with velocity at higher velocities (Perkins and Johnston, 1963, Fried and Combarnous, 1971). As given in Eq. 8.8, the hydrodynamic dispersion coefficient for flow in the subsurface is assumed to be the sum of the effective diffusion coefficient, D_d, and the mechanical dispersion coefficient, D_m (Palmer and Johnson, 1989):

$$D = D_d + D_m$$

(Eq. 8.8)

Mechanical dispersion is the process where the pollutants are mechanically mixed by the velocity variations at the microscopic level resulting from heterogeneities in the porous medium during advective transport (Domenico and Schwartz, 1998). Mechanical dispersion can occur at several scales, ranging from microscopic scale as in the pores of the porous medium pore to macroscopic scale (intraformational or at well to well scale) and to the megascopic scale (formational).

In homogeneous isotropic media, the mechanical dispersion coefficient, D_m, may be related to the average pore-water velocity by using dispersivity, α, a characteristic property of a medium:

$$D_m = \alpha v^m$$

(Eq. 8.9)

where m is an empirical constant ranging from 1 - 2. Laboratory studies have indicated that m may be assumed to be equal to 1 for granular geologic materials (Freeze and Cherry, 1979). Work done by Klotz and Moser (1974) indicated that the longitudinal dispersivity was generally dependent on the porous medium's grain size and uniformity coefficient. The above relationship (Eq. 8.9) is also applicable for the transverse direction. The longitudinal dispersivity, α_L, in laboratory studies is about 10 to 30 times greater than the transverse dispersivity, α_T (Gillham and Cherry, 1982).

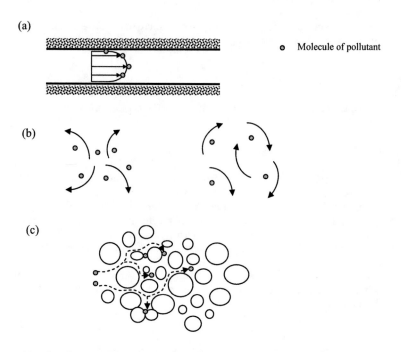

Figure 8.1 Hydrodynamic dispersion due to (a) velocity profile, (b) eddies (turbulence), and (c) noncontinuous nature of porous media

Work done by several researchers indicated that there is an apparent scale dependency in the magnitude of measured values of the dispersion coefficient (Palmer and Johnson, 1989; Gillham and Cherry, 1982; Anderson, 1979). In laboratory-scale experiments, the longitudinal dispersivities are in the range of 0.0001 to 0.01 m while larger values between 10 to 100 m were estimated based on field tracer tests and from model calibration of contaminant plumes (Palmer and Johnson, 1989). Gelhar et al. (1992) evaluated field data from 59 sites and found that longitudinal dispersivities ranged from 0.01 m to 5500 m for testing scales ranging from 0.75 to 100 km.

Field measurements have indicated that longitudinal dispersions are much larger than transverse dispersion (Freeze and Cherry, 1979), which is consistent with the findings at the laboratory scale. Data compiled by Gelhar et al. (1992) showed that field horizontal transverse dispersivities are at least an order of magnitude smaller than longitudinal values. The field horizontal transverse values were found to be 1 or 2 orders of magnitude larger than the vertical transverse dispersivities.

The apparently increasing magnitudes of the longitudinal dispersivity with increasing scale indicate that applying the classic advection-dispersion theory may not be applicable in natural geologic materials (Palmer and Johnson, 1989). With respect to attenuation in the subsurface, hydrodynamic dispersion is the main mixing process for pollutants - impacting pollutant distribution and the availability for degradation.

8.4.2   Dispersion in Rivers and Streams

The longitudinal dispersion coefficient for rivers and streams can be assessed using a dye test. The dispersion coefficient is estimated from the concentration-time data for an upstream station and a downstream station. Using the methods of moments, the dispersion in the reach is given by (Fischer, 1968):

$$E_x = \frac{U^2}{2} \frac{\sigma_{td}^2 - \sigma_{tu}^2}{\bar{t}_d - \bar{t}_u} \qquad \text{(Eq. 8.9)}$$

where σ^2 is the variance of the concentration-time curve, U is the mean longitudinal velocity, $\bar{t}$ is the time of travel to the centroid of the curve, and d, u are subscripts referring to downstream and upstream.

Fischer at al. (1979) provided the following equation as an estimate of the dispersion coefficient in streams which does not account for "dead zones" in the river:

$$E_x = 3.4x10^{-5} \frac{U^2 B^2}{HU^*} \qquad \text{(Eq. 8.10)}$$

where E_x is the dispersion coefficient (mile^2/sec), U is the mean river velocity (ft/s), B is equal to the mean width of the river (ft), H is the mean depth of the river (ft), U^* is equal to the river shear velocity (ft/s) given by $\sqrt{gHS}$, g is the gravitational acceleration, 32 ft^2/s and Sis equal to the river slope (ft/ft). The value 3.4 x 10^{-5} is the proportional constant using the units listed above.

Using a total of 18 streams with flows ranging from 1 to 93 m^3/s (35 to 33,000 ft^3/s) and slopes of 0.00015 to 0.0098 m/m (ft/ft), McQuivey and Keefer (1974) proposed the following equation as an estimate of dispersion coefficient in rivers:

$$E_x = 1.8x10^{-4} \frac{Q}{S_o B} \qquad \text{(Eq. 8.11)}$$

where Q is the steady state base flow (ft^3/s) and S_o is the bed slope (ft/ft).

Values of dispersion coefficients for different systems are presented in Table 8.2. Included in the Table for comparison purposes are the molecular diffusion coefficients in air and water. As seen in Table 8.2, pollutants released in the rivers ate most likely to be dispersed by longitudinal dispersion more than the other mechanisms.

Table 8.2 Ranges of dispersion coefficients of various processes (Schnoor, 1996, Fischer et al., 1979, Thomann and Mueller, 1987)

Systems	Dispersion coefficients (cm^2/s)
Molecular Diffusion in Air	10^{-1}
Molecular Diffusion in Water	$10^{-4} – 10^{-5}$
Rivers – Transverse	$1 - 10^1$
Rivers – Longitudinal	$10^4 - 10^6$
Lakes – Vertical Dispersion	$10^{-2} – 10^1$

8.5   Volatilization

Another important physical process in the fate of organic pollutants is the interphase partitioning process of volatilization. Volatilization is the transfer of a chemical from the aqueous phase to the gas phase. Volatilization is likely to be an important mechanism for fresh spills of volatile organic compounds such as petroleum products in lakes, rivers and soil surfaces and in the subsurface. However, the importance of this mechanism is likely to decrease with time as the volatile fraction present decreased with time. In the case of petroleum products consisting of a mixture of many organic compounds, the more volatile fraction will volatilized leaving behind the less volatile fraction.

In the subsurface, volatile organic compounds partition from the liquid phase into the soil gas/vapor phase. They are then transported by diffusion or by soil vapor movement due to barometric pressure changes in the unsaturated zone or by changes in the water table. They eventually may diffuse into the atmosphere or into the basement of homes (Palmer and Johnson, 1989). The volatilization rate and the amount of mass loss depends on several factors such as the contaminant concentration, Henry's law constant, diffusion coefficient, and the properties of the media such as sorption, temperature, effective porosity, and soil type (ASTM, 1998; Wiedemeier et al., 1996). Of importance in assessing the volatility potential of a compound is its Henry's Law constant and vapor pressure (Kavanaugh and Trussell, 1980)

Henry's law states that "the weight of any gas that will dissolve in a given volume of a liquid, at constant temperature, is directly proportional to the pressure that the gas exerts above the liquid," where the constant of proportionality for each gas-liquid system is referred to as the Henry's law constant or coefficient (Sawyer and McCarty, 1978). For volatilization, Henry's law can be expressed in terms of the mole fraction of the compound i in solution as:

$$P_i = K_{HI} X_i \qquad \text{(Eq. 8.12)}$$

where P_i is the partial pressure of the gas i in air (atm), K_{HI} is the Henry's Law constant (atm/mole fraction), and X_i is the mole fraction of gas i dissolved in solution. Henry's law can be written with the mole fraction expressed as the concentration of compound i dissolved in the liquid in terms of moles/L. In this case, the Henry's constant, K_{HI}, has units of atm·L/mol. Henry's Law constant can have dimensionless units when the partial pressure and the mole fractions are expressed in terms of concentration units (LaGrega et al., 1994):

$$K_H' = \frac{C_g}{C} = \frac{K_{HI}}{RT} \qquad \text{(Eq. 8.13)}$$

where K_H' is dimensionless, R is the universal gas constant (8.25×10^{-5} atm·m^3/mol·K), T is the temperature (K), and the concentration in the gas phase, C_g, and aqueous phase, C, have units of mole/L or mg/L.

Henry's constant, K_{HI}, values for the organic contaminants are presented in Table 8.3. The larger the Henry's constant, the greater is the tendency for the chemical to volatilize. For Henry's Law constant of > 0.01 atm-L/mol, volatilization from all waters is significant (Lyman et al., 1982). Chlorinated compounds such as trichloroethylene and monoaromatic compounds such as benzene and toluene have a greater tendency to volatilize. Polycyclic aromatic compounds especially compounds with three or more rings are less volatile. The Henry's Law constants are a function of environmental conditions such as temperature and the concentration of other dissolved gases and solids. Henry's Law constant is expected to increase with increasing temperature, increase slightly with increasing water salinity, and decrease with increasing concentration of dissolved organic carbon in water. The Henry's Law constant for most volatile hydrocarbons increases about twofold for every 10° C temperature rise (Kavanaugh and Trussell, 1980).

The Henry's Law constant provides an indication of the tendency of the chemical to volatilize. To estimate the rate of volatilization, mass transfer equations are typically used. One approach is to represent the mass transfer process by a first-order expression in which the mass-transfer driving force is proportional to the difference between equilibrium and actual concentrations in solution (Wilkins et al., 1995):

$$\frac{\partial C}{\partial t} = K \left(C - C_e \right) \qquad \text{(Eq. 8.14)}$$

where C is the actual concentration in solution, C_e is the concentration in equilibrium with the gas phase [kg/m^3], and K is the lumped overall mass-transfer coefficient [1/s].

Table 8.3 Vapor pressure and Henry's Law constants for selected organic compounds (T = 20°C)

	Compound	Vapor Pressure[1] (atm)	Henry's Law Constant (atm-L/mole)
Chlorinated Compounds	Tetrachloroethene	24.32[2]	13.3
	Trichloroethene	60.04	7.25
	cis-1,2-dichloroethene	197.6[3]	2.99
	trans-1,2-dichloroethene	197.6[4]	7.32
	1,1-dichloroethene	501.6	20.9
	vinyl chloride	2600[3]	22.0
	1,1,1-trichloroethane	98.8	13.4
	1,1-dichloroethane	182.4	4.35
	carbon tetrachloride	91.2	23.5
	Chloroform	159.6	2.79
	methylene chloride	349.6	1.73
Gasoline Products	Benzene	95.19[3]	5.43[3]
	Toluene	28.4[3]	5.94[3]
	Ethyl-benzene	9.53[3]	8.44[3]
	o-Xylene	6.6[3]	5.1[3]
	m-Xylene	8.3[3]	7.68[3]
	p-Xylene	8.7[3]	7.68[3]
	MTBE	249[3]	5.87[3]
Polycyclic Aromatic Hydrocarbons	Naphthalene	78	0.48
	Acenaphthylene	6.7	0.014
	Anthracene	0.006	0.720
	Fluorene	0.6	0.10
	Phenanthrene	0.12	0.039
	Pyrene	0.004	0.011

[1] From U.S. EPA (1986), Verschueren (1983), Rittmann et al. (1994); Sage et al. (1990), Jarvis et al. (1989), Michalenko et al. (1993), LaGrega et al., 1994
[2] 30°C
[3] 25°C
[4] 14°C

8.6.   Advection-Dispersion-Reaction Equation

A quantitative evaluation of the interactions of the key physical attenuation processes can be described by using a mass balance of the pollutants in the system. A commonly used expression for the mass balance of the dissolved reactive constituents in the aqueous phase of rivers and lakes or in saturated, isotropic porous media is the advection-dispersion-reaction (ADR) equation. The ADR equation of a system describes the effects of the physical processes on concentration changes within the system and provides the time of arrival and concentration of contaminants at the receptor(s) of concern. THE ADR can be expressed as one- to three-dimensional depending on the system and the impact of each of the physical processes. A one-dimensional form of the ADR equation with steady, uniform flow in the x-direction can be written as follows (Freeze and Cherry, 1979):

$$\frac{\partial C}{\partial t} = E_L \frac{\partial^2 C}{\partial x^2} - \bar{u} \frac{\partial C}{\partial x} \pm R \qquad \text{(Eq. 8.15)}$$

where C is the solute concentration, t is the time, E_L is the longitudinal dispersion coefficient, x is the distance in direction of flow, $\bar{u}$ is average water velocity, and R is a source-sink term. The first term on the right hand side of Eq. 8.15 describes the longitudinal hydrodynamic dispersion process while the second term describes advection. The above equation similarly describes pollutant concentration changes in the subsurface. In this case, $\bar{u}$ is average pore water velocity and E_L is substituted for the longitudinal hydrodynamic dispersion in the porous media. The source-sink term represents the physical, chemical and biochemical processes that result in mass changes which includes volatilization, sorption, biodegradation and photolysis reactions.

Although the above equation provides a description of the physical state of the system, it may be simplification of the processes due to various scaling effects. For example, for flow in saturated porous media, Eq. 8.15 can be used to predict solute migration in uniformly packed laboratory columns but its application to field problems is essentially empirical (Gillham and Cherry, 1982), largely due to limitations in the current knowledge of the physical processes occurring within the aquifer. In field applications, the scale-dependent macroscopic dispersion coefficients are not reliable measures of dilution in heterogeneous formations because the rates of dilution and spreading can be quite different (Kitanidis, 1994). Rather the rate of increase of dilution depends only on the local dispersion and the shape of the plume.

8.7   Application of Attenuation Processes

Natural assimilative capacity (or natural attenuation) of the receiving surface water have been used in the past for managing pollutant loads such as carbonaceous pollutants and nutrients from point sources. In recent years, many groundwater and lands in the US have been discovered to be contaminated with petroleum hydrocarbons, chlorinated solvents and other chemicals from leaky tanks, lagoons, and industrial discharges. In addition, many industries produce wastewaters that contained low concentrations of hazardous compounds that must be treated before they are discharged. Remediation of these sites and treatment of diluted wastewater with engineered systems such as pump and treat and chemical oxidation are expensive. Depending on the extent of the contamination of the sites and the risk they pose, natural attenuation processes may be applied to remediate these sites. As for wastewaters, natural processes such as wetlands, phytoprocesses, and sorption can be applied accordingly to treat these wastewaters. Use of natural processes may minimize the cost of clean up of these sites or treatment of the wastewaters.

An application of natural processes that has been adopted widely is the natural attenuation protocol for the remediation of contaminated sites. The natural attenuation protocol is a process that relies on the natural assimilative capacity of a site to reduce or stabilize contaminants to desirable levels (Rao et al., 2004). It is considered to be a viable alternative for site clean up under certain conditions. Natural attenuation for contaminated sites is also referred to as intrinsic or passive remediation, or natural recovery, or natural assimilation. Reduction in contaminant concentrations (and mass) occurs through many site specific natural processes such as dilution, dispersion, sorption, volatilization, biodegradation and bioassimilation, complexation, precipitation and other non specified processes without the intervention of man. The term natural attenuation includes many different processes that reduce or stabilize the contaminants underground, whereas the process referred to as intrinsic bioremediation only the biodegradation of contaminants is the key reaction involved.

For sites that use natural attenuation as a remedial tool, US EPA required that monitoring be a key feature of the process. As such, the process is also known as Monitored Natural Attenuation (MNA). MNA is defined by US EPA as follows (US EPA, 1997):

> The term *monitored natural attenuation, ...* refers to the reliance on natural attenuation processes (within the context of a carefully controlled and monitored site cleanup approach) to achieve site-specific remedial objectives within a time frame that is reasonable compared to that offered by other more active methods. The "natural attenuation processes" that are at work in such a remediation approach include a variety of physical, chemical, or biological processes that, under favorable conditions, act without human intervention to reduce the mass, toxicity, mobility, volume, or concentration of contaminants in soil and groundwater. These in-situ processes include biodegradation, dispersion, dilution, sorption, volatilization and chemical or biological stabilization, transformation, or destruction of contaminants.

To clarify the use the use of MNA at various Superfund, underground storage tank and RCRA sites, a memorandum was issued by US EPA in 1997 (US EPA, 1997). The memorandum indicated that "it was only one component of the total remedy at the site and that great care should be used if it was the sole remedy at a site." MNA was to be used with other remediation objectives such as source control and restoration of contaminated groundwater and was not to be considered as a default or presumptive remedy at any contaminated site, but as appropriate remediation method as long as it was capable of achieving site-specific remediation objectives within the time frame specified. A later directive (US EPA, 1999) further indicated that MNA will be appropriate for sites that have a low potential for contaminant migration. Examples of sites that are amenable to MNA are gasoline-contaminated sites containing benzene, toluene, ethylbenzene and xylene (BTEX) compounds. MNA can reduce the BTEX compounds present by biodegradation and stabilize the spread of the contaminant plume at the site. The heavier components in gasoline may still remain at the site but the risk may be reduced or may continue to be harmful to human health. As such, monitoring of the site becomes an important part of the protocol.

The US EPA (1997) policy directive indicates that three types of "evidence" or site-specific information for the evaluation of the efficacy of monitored natural attenuation as a remedial approach. The three types of "evidence" are:

- Historical groundwater and/or soil chemistry data that demonstrate a clear and meaningful trend of decreasing contaminant mass and/or concentration over time at appropriate monitoring or sampling points.
- Hydrogeologic and geochemical data that can be used to demonstrate indirectly the type(s) of natural attenuation processes active at the site, and the rate at which such processes will reduce contaminant concentrations to required levels.
- Data from field or microcosm studies conducted in or with actual contaminated site media which directly demonstrate the occurrence of a particular natural attenuation process at the site and its ability to degrade the contaminants of concern.

The reasonable time frame mentioned is a site-specific decision. The time frame should not be excessive compared with other remedial options. Some of the factors to be considered in reasonableness in time frame are: current and potential future uses of affected groundwater; public acceptance of the extended time for remediation; reliability of monitoring and institutional controls, and funding availability over the time period to reach the cleanup objectives.

Some of the advantages of using natural attenuation to remediate a contaminated site include (USEPA, 1999):
- an in-situ process that generates lower volume of remediation wastes than active remediation
- reduce the chances of cross-media transfer of contaminants which may occur in ex-situ processes and risk of human exposure to contaminants
- may be cost lower than most active remediation alternatives
- less intrusion and minimal disturbance to the site operations
- used with, or as a follow-up process to other active remedial measures
- can be applied to all or part of a given site depending on the site conditions and remediation goals.

Some of the disadvantages include:
- longer time frame to achieve remediation objectives as compared to active engineered processes
- requires thorough site characterization
- requires extensive and prolonged long-term monitoring
- transformation products may be more toxic and/or more mobile than the original compound
- may require institutional control and the site may not be available for use until contaminant levels are acceptable

- hydrologic and geochemical conditions amenable to natural attenuation may change over time that may adversely impact the remedial effectiveness.

8.8   Case Study: Natural Attenuation of Coal-tar-contaminated Site

MNA was applied for the remediation of a coal tar-contaminated site in Iowa containing polycyclic aromatic hydrocarbons (PAHs) (Rogers et al., 2006). Site investigations have shown free-phase coal-tar source material located under the former gas-holder tanks and shallow source material at ground surface in various areas at the site. Figure 8.2 provides a plan view of the site, showing the original layout of the buildings and extent of the coal-tar-contaminated area. Source materials and contaminated soils were excavated to depths of 2.4 m in 1997 and the excavated area was backfilled with clean sand and capped with a 0.6 m layer of clay and gravel. The geology of the site was characterized by monitoring well boring data, percussion probing direct-push technology (DPT) soil cores, and DPT electrical conductivity probing. Groundwater samples were collected from 5-cm-diameter monitoring wells, direct-push 2.5-cm-diameter monitoring wells, and using one-time direct-push sampling technology. Low flow pumping of the groundwater was used in conjunction with a flow-through cell to collect groundwater samples and on site water quality parameters.

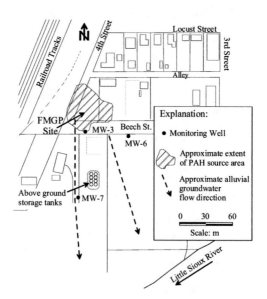

Figure 8.2 Layout of coal-tar-contaminated site (FMGP)

Figure 8.3 shows the changes in dissolved oxygen, ORP, nitrate, nitrite and ammonium along a transect through the groundwater plume. As seen in the transect, oxygen was depleted rapidly in the source area. This was followed by reduction in nitrate and an increase in nitrite and ammonium indicating denitrification within the contaminated plume and area. Other data (not shown here) showed that sulfate-reducing conditions were present as evident by the reduction of sulfate concentration and increase in sulfide downstream of the source area.

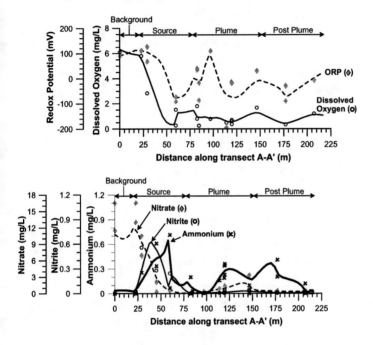

Figure 8.3 Transect of contaminated plume showing changes in dissolved oxygen, ORP, nitrate, nitrite and ammonia.

Using a biogeochemical mass balance which includes stoichiometric relationships describing hydrocarbon mineralization under different terminal electron acceptors (TEA) conditions and changes in TEA concentrations or production of coupled reduced species in the aquifers, the hydrocarbon degradation in the aquifer may be estimated (Rogers at al., 2006). The rate of consumption of TEAs and the rate of degradation of contaminant mass (C_rH_β) in the source or plume region assuming steady-state conditions is balanced by the rate of groundwater recharge of TEAs across flux boundary and can be written:

$$\frac{dM_s}{dt} \approx \frac{\Delta C_\gamma H_\beta}{\Delta t} = \frac{\Delta TEA}{UF_{TEA} \cdot \Delta t} = \frac{\left(\overline{C}_{FB1,TEA} - \overline{C}_{S,TEA}\right)}{UF_{TEA}} \cdot Q \qquad \text{(Eq. 8.16)}$$

where $\dfrac{dM_s}{dt}$ is the rate contaminant mass degradation attributable to reduction of terminal electron acceptor or production of reduced species of interest in the source area, ΔTEA is the mass of TEA consumed, UF_{TEA} is the utilization factor for the TEA of interest, $\overline{C}_{FB1,TEA}$ is the molar TEA concentration on flux boundary, $\overline{C}_{S,TEA}$ is the average molar source region TEA concentration, and Q is the rate of groundwater flow.

The same concept can be applied to the production of reduced species of oxidation-reduction reactions in the source ($\overline{C}_{S,RS}$) and plume ($\overline{C}_{P,RS}$) regions such that:

$$\frac{dM_s}{dt} \approx \frac{\Delta C_\gamma H_\beta}{\Delta t} = \frac{\Delta RS}{UF_{TEA} \cdot \Delta t} = \frac{\left(\overline{C}_{S,RS} - \overline{C}_{FB1,RS}\right)}{UF_{TEA}} \cdot Q \qquad \text{(Eq. 8.17)}$$

$$\frac{dM_p}{dt} \approx \frac{\Delta C_\gamma H_\beta}{\Delta t} = \frac{\Delta RS}{UF_{TEA} \cdot \Delta t} = \frac{\left(\overline{C}_{P,RS} - \overline{C}_{FB2,RS}\right)}{UF_{TEA}} \cdot Q \qquad \text{(Eq. 8.18)}$$

where ΔRS is the mass of reduced species produced.

Another approach is to use a 2-D reactive transport model to estimate the overall aqueous plus solid phase first-order degradation rate coefficient, λ (Stenback et al, 2004) by assuming a constant contaminant source region of width Y, in a homogenous, isotropic medium with groundwater seepage velocity of v_x. The 2-D reactive transport model is expressed as follows: (Stenback et al., 2004, Domenico and Schwartz,1998):

$$C(x,y,t) = \left(\frac{C_o}{4}\right) \exp\left\{\left(\frac{x}{2\alpha_x}\right)\left[1 - \sqrt{1 + \frac{4\lambda\alpha_x}{v_x}}\right]\right\} \operatorname{erfc}\left(\frac{x - v_c t\sqrt{1 + 4\lambda\alpha_x/v_x}}{2\sqrt{\alpha_x v_c t}}\right) \bullet$$
$$\left\{\operatorname{erf}\left(\frac{y + Y/2}{2\sqrt{\alpha_y x}}\right) - \operatorname{erf}\left(\frac{y - Y/2}{2\sqrt{\alpha_y x}}\right)\right\}$$

(Eq. 8.19)

where C_0 is source concentration, x is distance in the direction of groundwater flow, y is the transverse distance from the plume centerline, t is the time since source emplacement, v_c is the retarded contaminant velocity, α_x is the longitudinal dispersivity, and α_y is the transverse dispersivity. For simplicity, a constant aquifer thickness equal to the average aquifer depth is assumed even though the aquifer may be heterogeneous. By superimposing solutions of the 2-D reactive transport model via a least-squares fitting of monitoring well data, the degradation rate coefficients of groundwater contaminants can be estimated (Stenback et al., 2004).

In Eq. 8.19, the retarded contaminant velocity, v_c, for a given compound can be estimated by dividing the seepage velocity by the retardation factor, R:

$$v_c = \frac{v_x}{R} = \frac{v_x}{\left(1 + \frac{B_d K_{oc} f_{oc}}{\phi}\right)}$$ (Eq. 8.20)

where B_d is the soil bulk density, f_{oc} is the fraction of organic carbon in the soil, ϕ is the soil porosity, and K_{oc} is the organic carbon partitioning coefficient of the contaminant.

By multiplying λ by the total dissolved mass within the contaminated aquifer ($\overline{M}_T$), the rate of mass consumption, $\dfrac{dM}{dt}$, at any time (t) can be estimated. This can be estimated by integrating over the volume of contaminated aquifer water such that:

$$\frac{dM}{dt} = -\lambda(\overline{M}_T) = -\lambda\phi \int_x \int_y z(x,y)C(x,y,t)dydx$$ (Eq. 8.21)

where $z(x,y)$ is the thickness of the aquifer. In a discretized form, the rate of mass consumption can be estimated using numerical integration (grid size times concentration) over a grid covering the area of contamination at time (t), based on the best-fit kriged geological surfaces as given by:

$$\frac{dM}{dt} \approx -\lambda\phi A_{elem} \cdot \sum_1^k [z \cdot C(x,y,t)]$$ (Eq. 8.22)

where A_{elem} is the grid size and k is the number of elements within the contaminated region.

Using the biogeochemical mass balance, the mass degraded within the plume was estimated to be approximately 4.5 kg/yr of PAHs. Using a 2-D reactive transport analytical model, the first-order degradation rate coefficients of benzene, naphthalene, and acenaphthene for the dissolved-phase plume were estimated to be 0.0084, 0.0058, and 0.0011 d^{-1}. The total mass of PAH transformed using the degradation rate coefficients was estimated to be approximately 3.6 kg/yr. Based on these two approaches, it appeared that microbially-mediated reactions were responsible for the destruction of PAHs at this site with a significant portion attributed to natural nitrate- and sulfate-reduction biochemical reactions.

8.9   Summary

The two main physical processes that impact the attenuation of the pollutants in a medium are advection and dispersion. Volatilization may be a significant process if the pollutant has a high Henry's Law constant > 0.01 atm-L/mol. Sorption which is described in Chapter 2 can also be significant if sorption sinks such as organic carbon or mineral sorption sites are available. Physical processes play an important role in the fate and transport of contaminants in the environment. These physical processes are "nondestructive" processes but result in the reduction of the contaminant concentration through dilution and mixing or transfer of the contaminants from one phase to another. Although much is known about the physics of the physical processes, estimations of the physical parameters in the field scale are semi-empirical and work is still needed to describe sufficiently the physical processes under field conditions.

Use of natural attenuation for the remediation of contaminated sites and treatment of wastewater requires specific guidelines in order for natural attenuation to be applied properly. Some of the guidelines should include data showing evidence of attenuation by natural processes, estimated short-term and long-term risk to receptors, and estimation of time frame for the implementation of the process so that the technology will not be viewed as a "walk away" technology. Monitoring of a natural attenuation system is important to ensure that the natural processes are behaving as expected in attenuating the contaminants.

References

Anderson, M.P. (1979) Using models to simulate the movement of contaminants through groundwater flow systems. *CRC Crit. Rev. Environ. Control*, 9(2), 97-156.

ASTM (1995) Emergency Standard guide for risk-based corrective action applied at petroleum release sites, ASTM E-1739, Philadelphia, PA

ASTM (1998) Standard guide for remediation of ground water by natural attenuation at petroleum release sites, E 1943-98, West Conshohocken, PA.

Domenico, P.A., and Schwartz, F.W. (1998) Physical and chemical hydrogeology, 2nd ed., John Wiley & Sons, Inc., New York, NY.

Fetter, C. W., Jr., (1988) Applied Hydrogeology, 2nd Edition, Charles E. Merrill and Co., Columbus, Ohio,

Fischer, H.B. (1968) Dispersion predictions in natural streams, *J. Sanit. Eng. Div. Proc. Am. Soc. Civ. Eng.*, 94:927-944.

Fischer, H.B., List, E.J., Koh, C.Y., Imberger, J., and Brooks, N.H. (1979) Mixing in inland and coastal waters, Academic Press, New York, NY.

Freeze, R.A., and Cherry, J.A. (1979) Groundwater, Prentice-Hall, Inc., Englewood Cliffs, NJ.

Fried, J.J., and Combarnous, M.A. (1971) Dispersion in porous media. Advances in Hydroscience, Vol. 7, V.T. Chow, ed., Academic Press, Inc., New York, NY, 169-282.

Gelhar, L.W., Welty, C., and Rehfeldt, K.R. (1992) A critical review of data on field-scale dispersion in aquifers. *Water Resour. Res.* 28(7), 1955-1974.

Gillham, R.W., and Cherry, J.A. (1982) Contaminant migration in saturated unconsolidated geologic deposits. Recent trends in hydrogeology, T.N. Narasimhan, ed., Special Paper 189, The Geological Society of America, Boulder, CO, 31-62.

Jarvis, W. F., Sage, G. W., Basu, D. K., Gray, D. A., Meylan, W., and Crosbie, E. K. (1989) Large production and priority pollutants, in Handbook of Environmental Fate and Exposure Data for Organic Chemicals, P. H. Howard, ed., Lewis Publishers, Inc., Chelsea, MI.

Kavanaugh, M.C., and Trussell, R.R. (1980) Design of aeration towers to strip volatile contaminants from drinking water. *Jour. AWWA*, 72(12), 684-692.

Kitanidis, P.K. (1994) The concept of the dilution index. *Water Resour. Res.*, 30(7), 2011-2026.

Klotz, D., and Moser, H. (1974) Hydrodynamic dispersion as aquifer characteristic: Model experiments with radioactive tracers. Isotope techniques in groundwater hydrology, Vol. 12, International Atomic Energy Agency, Vienna, Austria, 341-354.

LaGrega, M.D., Buckingham, P.L., and Evans, J.C. (1994) Hazardous waste management, McGraw-Hill, Inc., New York, NY.

Logan, B.E. (1999) Environmental Transport Processes, John Wiley & Sons, New York.

Lyman, W.J., Reehl, W.F., Rosenblatt, D.H. (1982) Handbook of Chemical Property Estimation Methods, McGraw-Hill, New York, NY

Michalenko, E. M., Basu, D. K., Sage, G. W., Meylan, W. M., Beauman, J. A., Jarvis, W. F., and Gray, D. A. (1993) Solvents 2. in Handbook of Environmental Fate and Exposure Data for Organic Chemicals, P. H. Howard, ed., Lewis Publishers, Inc., Chelsea, MI.

McQuivey , R.S. and Keefer, T.N. (1974) Simple method for predicting dispersion in streams, *J. Environ. Eng. Div. Proc. Am. Soc. Civ. Eng*, 100(EE4):997-1011.

Palmer C.D., and Johnson, R.L. (1989) Physical processes controlling the transport of contaminants in the aqueous phase, in Transport and fate of contaminants in the subsurface, United States Environmental Protection Agency, Center for Environmental Research Information, Robert S. Kerr Environmental Research Laboratory, EPA/625/4-89/019, Washington, D.C., 29-40.

Perkins, T.K., and Johnston, O.C. (1963) A review of diffusion and dispersion in porous media. *Soc. Pet. Eng. Jour.*, 3(1), 70-84.

Rao, S., Ong, S.K., Seagren, E., Nuno, J., and Banerji, S.K. (2004) Natural Attenuation of Hazardous Wastes, American Society of Civil Engineers, Preston, VA.

Rittmann, B.E., Seagren, E., Wrenn, B.A., Valocchi, A.J., Ray, C., and Raskin, L. (1994) In Situ Bioremediation, 2nd edition, Noyes Publishers, Inc., Park Ridge, NJ.

Rogers, S. W., Ong, S. K., Stenback, G. A., Golchin, J., and Kjartanson, B. H. (2007) Evidence of intrinsic bioremediation of a coal-tar impacted aquifer based on coupled reactive transport and biogeochemical mass balance approaches, Water Environment Federation, 79 (1):13-28.

Sage, G.W., Jarvis, W.F., and Gray, D.A. (1990) Solvents, in Handbook of Environmental Fate and Exposure Data for Organic Chemicals, P. H. Howard, ed., Lewis Publishers, Inc., Chelsea, MI.

Sawyer, C.N., and McCarty, P.L. (1978) Chemistry for environmental engineering, 3rd edition, McGraw-Hill Book Company, New York, NY.

Schnoor, J.L. (1996) Environmental modeling, John Wiley & Sons, Inc., New York, NY.

Stenback, G. A., Ong, S.K., Rogers, S.W., and Kjartanson, B.H. (2004) Impact of transverse and longitudinal dispersion on first-order degradation rate constant estimation, *Journal of Contaminant Hydrology*, 73(1-4):3-14

Thibodeaux, L.J., (1979) Chemodynamics: Environmental Movement of Chemicals in Air, Water and Soil, Wiley, New York, NY.

Thomann, R.V. and J.A. Mueller, (1987) Principles of surface water quality modeling and control, HarperCollins Publishers, Inc., NY, NY.

US EPA (1987) Groundwater. Office of Research and Development, Center for Environmental Research Information, Robert S. Kerr Environmental Research Laboratory, EPA/625/6-87/016, Washington, D.C.

US EPA (1989) Hazardous Waste Treatment, Storage and Disposal Facilities (TSDF), OAQPS, Air Emission Models, EPA/450/3-87/026, Washington, DC.

US EPA (1997) Use of Monitored Natural Attenuation at Superfund, RCRA corrective Action and Underground Storage Tank Sites, Office of Solid Waste and Emergency Response Directive 9200.4-17P, US EPA Washington , D.C.

US EPA (1999) Use of Monitored natural Attenuation at Superfund, RCRA Corrective Action and Underground Storage Tank Sites, Office of Solid Waste and Emergency Response Directive 9200.4-17P, US EPA, Washington, D.C.

Wiedemeier, T.H., Swanson, M.A., Moutoux, D.E., Gordon, E.K., Wilson, J.T., Wilson, B.H., Kampbell, D.H., Hansen, J.E., Haas, P., and Chapelle, F.H. (1996) Technical protocol for evaluating natural attenuation of chlorinated solvents in groundwater. Air Force Center for Environmental Excellence, Technology Transfer Division, Brooks AFB, San Antonio, TX.

Wilkins, M.D., Abriola, L.M., and Pennell, K.D. (1995) An experimental investigation of rate-limited nonaqueous phase liquid volatilization in unsaturated porous media: Steady state mass transfer. *Water Resour. Res.*, 31(9), 2159-2172.

Zhang, T. and Bishop, P., (1994) Evaluation of tortuosity factors and effective diffusivities in biofilms, *Water Research*, 28: 2279-2287.

Errata

Natural Processes and Systems
for Hazardous Waste Treatment

American Society of Civil Engineers
ISBN 978-0-7844-0939-8

Original index is incorrect.
Replacement index enclosed.

Index

Index